# CISM International Centre for Mechanical Sciences

Courses and Lectures

Volume 599

For more than 40 years the book series edited by CISM, "International Centre for Mechanical Sciences: Courses and Lectures", has presented groundbreaking developments in mechanics and computational engineering methods. It covers such fields as solid and fluid mechanics, mechanics of materials, micro- and nanomechanics, biomechanics, and mechatronics. The papers are written by international authorities in the field. The books are at graduate level but may include some introductory material.

More information about this series at http://www.springer.com/series/76

Laura De Lorenzis · Alexander Düster
Editors

# Modeling in Engineering Using Innovative Numerical Methods for Solids and Fluids

Springer

*Editors*
Laura De Lorenzis
Institute of Applied Mechanics
Braunschweig University of Technology
Braunschweig, Germany

Alexander Düster
Numerical Structural Analysis with
Application in Ship Technology (M-10)
Hamburg University of Technology
Hamburg, Germany

ISSN 0254-1971 ISSN 2309-3706 (electronic)
CISM International Centre for Mechanical Sciences
ISBN 978-3-030-37517-1 ISBN 978-3-030-37518-8 (eBook)
https://doi.org/10.1007/978-3-030-37518-8



This Springer imprint is published by the registered company Springer Nature Switzerland AG
The registered company address is: Gewerbestrasse 11, 6330 Cham, Switzerland

# Preface

Due to the ever-increasing computational power of modern computers, numerical simulations are of growing interest in many different fields of science and engineering. The fast-growing computer performance itself, however, is not sufficient to satisfy the increasing requirements for the simulation of complex problems arising in particular in fluid and solid mechanics. To this end, accurate and robust numerical methods are of crucial importance. The development of reliable and efficient discretization methods for solids and fluids supports the understanding of complex physical phenomena and helps to accelerate and improve the development of products and processes in almost all disciplines. Based on numerical simulation, the number of time-consuming and expensive experiments can be significantly reduced, and engineering decisions are supported by computed data which might be very difficult if not impossible to obtain experimentally.

This book stems from the lecture notes of the CISM course: *Modeling in Engineering Using Innovative Numerical Methods for Solids and Fluids* which was held in Udine on October 15–19, 2018. Innovative and promising modeling and simulation approaches are presented, including the basics of the methods as well as advanced topics and complex applications. The contents cover the following topics:

- Particle methods addressing particle-based materials and numerical methods that are based on discrete element formulations.
- Fictitious domain methods, which allow for the efficient discretization of complex problems for which meshing with finite elements is very difficult.
- Phase field models, which have become very popular to model and simulate fracture problems (among other possible applications).
- Computational fluid dynamics based on modern finite volume schemes to efficiently discretize the Navier–Stokes equations.
- Hybridizable discontinuous Galerkin methods, which offer high convergence rates for the simulation of incompressible flow problems.
- Non-intrusive coupling methods for structural models that allow to perform model adaptive simulations based on existing well-developed solvers.

The book is addressed to scientists and engineers from both academia and industry working in the broad field of civil and mechanical engineering or applied physics and mathematics. The intention is to provide a sound introduction to innovative numerical methods for solids and fluids which can be used to model complex problems in engineering.

Braunschweig, Germany — Laura De Lorenzis
Hamburg, Germany — Alexander Düster

# Contents

# Discrete Element Methods: Basics and Applications in Engineering

**Peter Wriggers and B. Avci**

## Introduction

Particle systems are of great importance in many industrial branches like in chemical and food industries as well as in geotechnical engineering problems. Coupling of particles with fluids are related to fluvial erosion, fluidized beds, sedimentation and transport in the blood system. Thus, the numerical simulation of particle systems is of great interest, both from practical and fundamental points of view. Therefore, the understanding, the simulation and analysis of related phenomena is significant, particularly with regard to micro movements, homogenization procedures and coupled moderate or highly concentrated particulate flows. Certainly, such problems require an accurate description of the underlying physics, but the simulation of particulate flow and movement is still a challenging task for a large number of particles.

Popular examples of pure particle methods are Molecular Dynamics (MD), see Alder and Wainwright (1957), Discrete Element Method (DEM), see Cundall and Strack (1979), and Smoothed Particle Hydrodynamics (SPH), see Gingold and Monaghan (1977). In these methods, the positions of the particles and the evolution of their quantities are described by ordinary differential equations and solved in a Lagrangian way. In the framework of DEM and SPH, the particle dynamics are obtained by applying the Newton-Euler equations and the Navier-Stokes equations, respectively. Both of these numerical techniques (DEM and SPH) utilize many common algorithms, such as neighbor search algorithms, distance compu-

P. Wriggers (✉)
Faculty of Mechanical Engineering, Institut for Continuum Mechanics,
Leibniz University Hannover, Hannover, Germany
e-mail: wriggers@ikm.uni-hannover.de

B. Avci
CADFEM GmbH, Hannover, Germany

 
L. De Lorenzis and A. Düster (eds.), *Modeling in Engineering Using Innovative Numerical Methods for Solids and Fluids*, CISM International Centre for Mechanical Sciences 599,
https://doi.org/10.1007/978-3-030-37518-8_1

tations among neighboring particles and force or kernel evaluations. Thus, many subroutines can be implemented, used and maintained in a single framework.

In the recent years, effort has been made to improve the existing methods and to develop new efficient numerical approaches. Among the various methods evolved so far, e.g., see Zhu et al. (2007), Zohdi (2007), Pöschel and Schwager (2005) for an overview, the approaches for the treatment of particle systems through direct numerical simulation models are of high computationally cost. Basically, direct numerical simulations can be performed by different discretization methods.

Concerning the treatment of the particle collision, the methods employed in the DEM can be classified into two main classes, which are characterized by hard particle contact (Alder and Wainwright 1957) and soft particle contact (Cundall and Strack 1979). Methods assigned to the first of the two groups are instantaneous collision models. Here, the particles separate immediately from each other in the event of collision. They undergo no deformation, so they are considered to be rigid. In the other approach the particles are treated as quasi rigid objects such that they are assumed to suffer minute deformations during the collision. The force based methods of the second group can be applied to govern the contact forces implying the particles' strength and eventually also allow to locally break a particle if very strong forces act on it. The difference of the methods for the treatment of particle contact is particularly crucial in highly concentrated systems. Here, a particle will usually collided with more than one partner at the same time. Hence some particles might be as well in a permanent contact situation with neighboring particles like it occurs in a heap of sediments or in case of agglomerated adhesive particles. In these cases, the application of hard contact models may not be suitable. In contrast, force based soft contact models are applicable both for dense and dilute systems. However, the computation of particle interactions is for soft spheres much more expensive compared to the hard particle approach since very small time steps are needed to resolve the contact interaction between the particles in time.

Other applications—like sediment transport or multiscale computation for granular materials—are related to coupling discrete elements to solids and fluids. In case of a direct numerical simulation of 3D particulate flows one has to couple fluids and particles. This can be done in different ways that span the bridge from just tracing of particles to a full interaction of particle and fluid via the tractions. The latter can be based on a complex ALE finite element scheme using a adaptive remesching of the finite elements to follow the particle movement, see e.g. Johnson and Tezduyar (1997), but is—due to the high computational effort—often limited to a small amount of particles. Another approach is the fictitious domain method which can handle much more particles in the flow. In both approaches the numerical tools of DEM and FEM are appropriately coupled by a staggered scheme. To describe the collision between particles, the soft contact approach is applied using repulsive force models. For more details see e.g. Avci and Wriggers (2012).

When coupling finite elements for solids and discrete elements there are two different possibilities. The first is a surface coupling where contact between a finite and a discrete element takes place. This can be handled by standard contact algorithms, see e.g. Wriggers (2006). On the other hand it is possible to couple finite and discrete elements within a volume. Such coupling requires specific treatment of linking the movement of the finite and the discrete elements. On possibility is the Arlequin methodology, see e.g. Dhia and Rateau (2005), which was applied with respect to particle and finite element coupling in Wellmann and Wriggers (2012).

## Governing Equations

The motion of a quasi rigid particle $\mathcal{P}_i$ is described by a position vector $\mathbf{X}_i$ to its center of mass and a rotation $\Psi_i$ at time $t = t_0$. In Fig. 1 the kinematics of the movement of the particle $\mathcal{P}_i$ is depicted for different time instants. Additionally another particle $\mathcal{P}_j$ is depicted that collides with particle $\mathcal{P}_i$.

The trajectory of particle $\mathcal{P}_i$ can be deduced from the Newton-Euler equations. Consequently its translational and angular velocities, $\mathbf{u}_i = \dot{\mathbf{x}}_i$ and $\boldsymbol{\omega}_i = \frac{\mathrm{d}\Psi_i}{\mathrm{d}t}$, have to satisfy

$$M_i \frac{\mathrm{d}^2\mathbf{x}_i}{\mathrm{d}t^2} = \rho_i\, V_i \mathbf{g} + \mathbf{F}_i \tag{1}$$

$$\Theta_i \frac{\mathrm{d}\boldsymbol{\omega}_i}{\mathrm{d}t} + \boldsymbol{\omega}_i \times (\Theta_i \boldsymbol{\omega}_i) = \mathbf{T}_i\,. \tag{2}$$

Therein, $M_i$ is the mass, $\mathbf{x}_i$ the position vector at time t to the center of mass (defined as $\mathcal{M}_i$), $\rho_i$ the mass density, $\mathbf{g}$ the gravity and $V_i$ denotes the volume of the particle $\mathcal{P}_i$. The tensor of inertia is represented by $\Theta_i$. Furthermore, the sum of the contact forces is stated as $\mathbf{F}_i$ The torques that are caused by $\mathbf{F}_i$ with respect to $\mathcal{M}_i$ are associated to the quantities $\mathbf{T}_i$. The traction vector $\mathbf{t}$ on $\partial\boldsymbol{\Omega}_p$ is defined by $\mathbf{t} = \sigma\,\mathbf{n}_f$ where $\mathbf{n}_f$ is the unit outward normal vector and $\mathbf{r}_i$ is the position vector of a point at $\partial\boldsymbol{\Omega}_p$ with respect to $\mathcal{M}_i$.

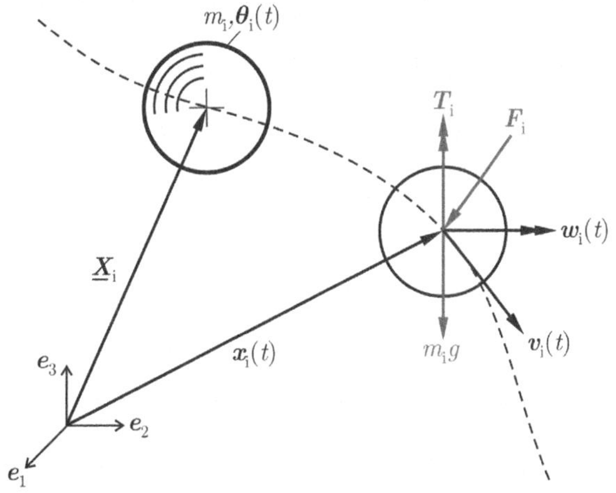


**Fig. 1** Kinematics and applied forces related to a particles

## Constitutive Modeling of the Particle Phase

The particles are modeled as quasi rigid spheres. To describe their collision behavior, a force based approach is used in order to govern the inter-particle forces that are deduced from repulsive models. In the sections below, the relevant concepts of the contact models are stated.

### *Normal Contact Model*

The normal contact force acting between the colliding particles is described by a constitutive viscoelastic model. For adhesive particles being in contact, the attractive van der Waals force is additionally considered in the contact area. For the purpose of governing the elastic contact force $F^n_e$, the Hertzian law (Hertz 1882) constitutes a well-established model. If the particles, to be treated, have also viscous material properties a consistent phenomenological model has to be employed, see e.g., Refs. (Brilliantov et al. 1996, 2007), where the effect of viscosity is considered via an added dissipative force $F^n_d$. Regarding the presence of the attractive van der Waals force in the contact area, the JKR theory, see Johnson et al. (1971), provides a proper treatment of adhesion, even in the case of underwater adhesion, see e.g. Loskofsky et al. (2006). For a detailed description of the JKR model see also Maugis (1992).

If adhesion is considered, the attractive force $F^n_a$ acts against the elastic force $F^n_e$ so that it consequently reduces the particles' compression. Thus, one obtains for the force acting on a particle

$$F^n = F^n_e - F^n_a + F^n_d \,. \tag{3}$$

#### Hertz Law

When using the Hertzian contact law Hertz (1882), the elastic repulsive force is governed by

$$F^n_e = \frac{4}{3}\, E\,\sqrt{R}\,\delta^{3/2}\,, \tag{4}$$

where $\delta = \delta_i + \delta_j$ is the total particle compression which is also called the approach of the particles. The values

$$R = \left(\frac{1}{R_i} + \frac{1}{R_j}\right)^{-1} \quad \text{and} \quad E = \left(\frac{1-\nu_i^2}{E_i} + \frac{1-\nu_j^2}{E_j}\right)^{-1} \tag{5}$$

denote the effective radius and the effective Young's modulus of the contact pair $\mathcal{P}_i$ and $\mathcal{P}_j$, respectively. The Poisson ratio is associated with the particles as $\nu_i$ and $\nu_j$.

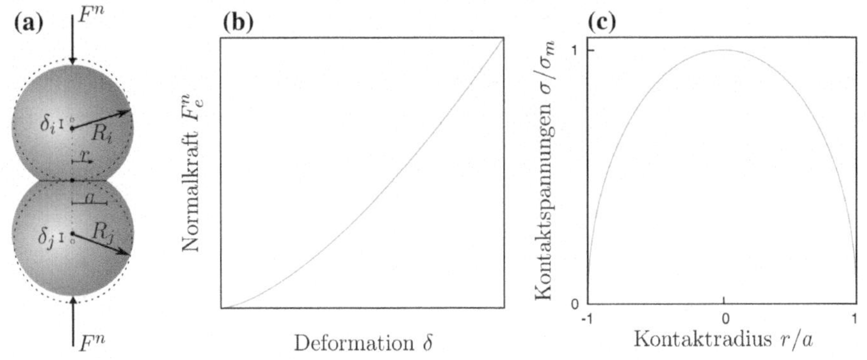


**Fig. 2** Hertz contact law

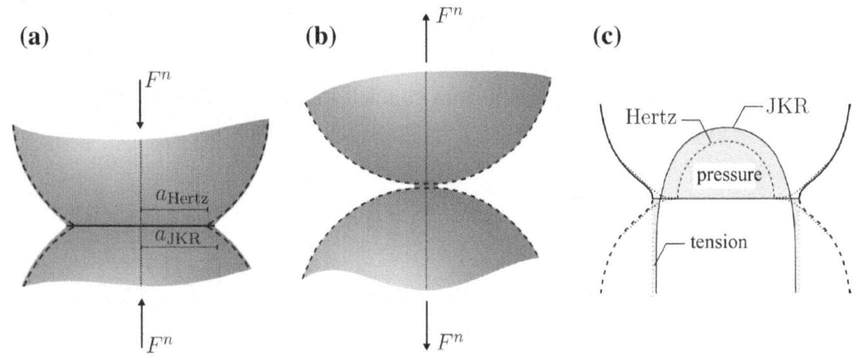


**Fig. 3** Adhesion in the contact interface

As a result of the mutual compression of the particles, a circular area is formed in the contact zone. One can deduct that 1 the radius $a$ of the shaped contact area is related to the total deformation $\delta$ via $a^2 = R\delta$.

Figure 2 depicts the deformation state that is assumed locally within the quasi rigid spheres. It also shows the relation between approach $\delta$ of the two spheres with respect to the force $F^n$. One can easily see the nonlinearity of the Hertz law. Furthermore the distribution of the contact pressure $\sigma/\sigma_m$ depending on the contact radius $r/a$ is depicted in the right part of the figure. Here $\sigma = F^n/(\pi\, a^2)$ and $\sigma_m = \max \sigma$.

## Adhesion Law

According to the JKR model, see Johnson et al. (1971), it is implied that the adhesive force acts only within the contact area, see Fig. 3. Here, the work of adhesion under a liquid or just the free energy changes to separate a unit contact area of $\mathcal{P}_i$ and $\mathcal{P}_j$ in a liquid medium ($l$) is defined as

$$W = \gamma_{il} + \gamma_{jl} - \gamma_{ij}, \tag{6}$$

where $\gamma$ describes the respective interfacial energy, see Loskofsky et al. (2006). Since the adhesive force $F_a^n$ is opposed to the elastic force $F_e^n$, it reduces the elastic deformation $\delta_e$ leading to

$$\delta_e - \delta_a = \frac{a^2}{R} - \sqrt{\frac{2\pi\, Wa}{E}}, \tag{7}$$

where $\delta_e$ is obtained from the Hertzian law, the second term $\delta_a$ is due to adhesion, see Maugis (1992) for details. The corresponding forces follow as

$$F_e^n - F_a^n = \frac{4Ea^3}{3R} - 2\pi a^2 \sqrt{\frac{2WE}{\pi a}}. \tag{8}$$

In case of the absence of external forces, i.e. $F^n = 0$, then $F_a^n \neq 0$ while $F_e^n = 0$ and $F_e^d = 0$. Furthermore, an equilibrium contact area $a_0$ is formed in the contact zone where a mutual compression $\delta_0$ of the particles occurs

$$a_0 = \left(\frac{9\pi\, WR^2}{2E}\right)^{1/3} \quad \text{and} \quad \delta_0 = \left(\frac{3R}{4}\left(\frac{\pi\, W}{E}\right)^2\right)^{1/3}. \tag{9}$$

To pull the particles off each other, one has to apply a traction force under which they suffer minute stretching deformations forming a connecting neck around the contact zone. Once the pulling force has reached a critical level, i.e. $F^n = -F_c^n$, the contact breaks and the particles will separate. The critical force $F_c^n$ yields

$$F_c^n = \frac{3}{2}\pi\, WR \tag{10}$$

and the corresponding critical distance of the particles follows

$$\delta_c = \frac{1}{48^{1/3}} \frac{a_0^2}{R}. \tag{11}$$

Here the pulling distance can be defined as $\delta = -\delta_c$. Figure 4 demonstrates the differences in the force.

## Viscous Effects in the Contact Interface Law

Viscous properties can be modeled using the system depicted in Fig. 5 for two spheres (a) and one sphere (b) being in contact wth a rigid wall. The forces in the damper $d$ are provided below.

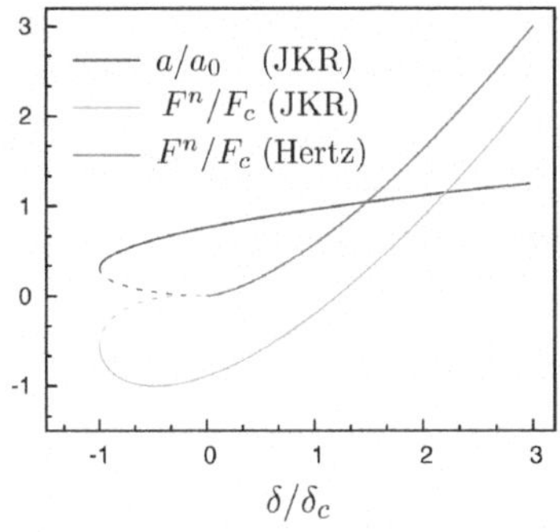


| | $\delta/\delta_c$ | $a/a_0$ | $F^n/F_c$ |
|---|---|---|---|
| Contact | $0$ | $(2/3)^{2/3}$ | $-8/9$ |
| Equilibrium | $(4/3)^{2/3}$ | $1$ | $0$ |
| Lift-Off | $-1$ | $(1/6)^{2/3}$ | $-5/9$ |

**Fig. 4** Forces in the contact interface due to Hertz and the JKR model

**Fig. 5** Viscous effects in the contact interface

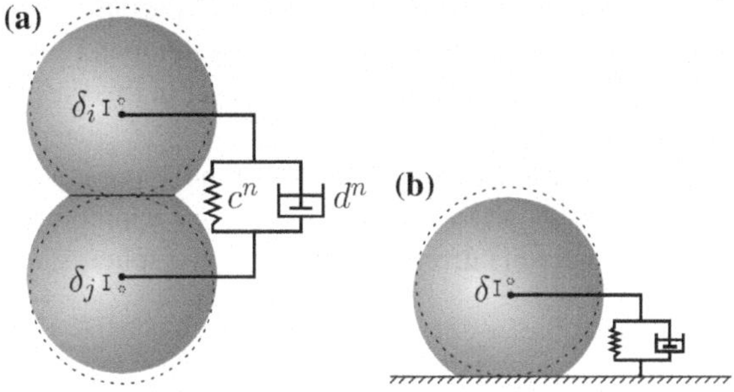


To consider the properties of material viscosity, a dissipative force is adopted according to the work of Brilliantov et al. (1996, 2007) which yields

$$F_d^n = A\dot{a}\frac{\partial}{\partial a}(F_e^n - F_a^n)\,. \tag{12}$$

From this definition, the viscous force follows as

$$F_d^n = \dot{a}A\left(\frac{4Ea^2}{R} - \frac{3}{2}\sqrt{8\pi\,WEa}\right), \tag{13}$$

where the dissipative factor[1] $A$ is related to a constant function of material viscosity. Consequently, considering the above set of equations, one obtains the force-displacement relation for adhesive viscoelastic particles

[1]This factor $A$ can also be used as a fitting parameter within specific simulations—like quasistatic predictions of granular material behaviour—to damp oscillations.

$$
\begin{aligned}
F^n(\delta) =& \frac{4}{3} E \sqrt{R} \delta^{3/2} \\
& - 2R^{3/4}\delta^{3/4}\sqrt{2\pi\, W E} \\
& + A\dot{\delta}\Big( 2E\sqrt{R}\sqrt{\delta} - 3\sqrt{\frac{1}{2} R^{3/2}\pi\, W E \delta^{1/4}} \Big).
\end{aligned} \tag{14}
$$

#### Computation of the Approach

In the present contribution, the constitutive law given in (4) and (14) is treated analogous to the penalty method, e.g., see Wriggers (2006) and Wellmann et al. (2008). In this methodology one computes the force $F_e^n$ between to particles $\mathcal{P}_i$ and $\mathcal{P}_j$ is computed from the approach of two particles by using the interface law (in case of the classical penalty approach this equation is $F_e^n = \varepsilon_P\, \delta$ with the spring stiffness, know as penalty parameter, $\varepsilon_P$). However in this case the penalty spring is nonlinear, see e.g. Eq. (4).

In all constitutive equations above the approach of the spheres or the rate of this approach has to be evaluated. For this one can compute the gap in normal direction $g_n$ from the given current positions of two spheres

$$
g^n = (R_i + R_j) - l > 0, \tag{15}
$$

and define $g^n \equiv \delta$ as the mutual compression or approach of $\mathcal{P}_i$ and $\mathcal{P}_j$. With this kinematic relation the contact forces can be immediately computed by evaluating (4) and (14). In Eq. (15) $l = ||\mathbf{l}||$ is the length of the distance vector between $\mathcal{M}_i$ and $\mathcal{M}_j$ in the current configuration, where $\mathbf{l} = \mathbf{x}_i - \mathbf{x}_j$. The deformation rate $\dot{\delta}$ is computed in Eq. (16) by the projection of the relative velocity $(\mathbf{v}_i - \mathbf{v}_j)$ onto the direction of the normal unit vector $\mathbf{n}$. This yields

$$
\dot{\delta} = \dot{g}^n = -(\mathbf{v}_i - \mathbf{v}_j) \cdot \mathbf{n} \quad \text{where} \quad n = l/l. \tag{16}
$$

We note that the direction of the contact force $F^n$ of the respective particle is opposite to the direction of compression $\delta$. Based on this observation, one can deduct from (14) the contact force $\mathbf{F}^n = F^n \mathbf{n}$ that contributes to the momentum equation (1).

### *Tangential Contact Model*

The constitutive relation of Coulomb's law couples the tangential force $F^t$ to the normal force $F^n$. For this the coefficient of friction has to be introduced as a constitutive parameter. The relation is not smooth since for sliding

$$
F^t = \mu_G F^n \tag{17}
$$

holds while

$$F^t \leq \mu_H F^n \tag{18}$$

is valid for sticking. In this relation the dynamic and the static coefficients of friction have to be introduced. The parameter $\mu_G$ stands for sliding and $\mu_H$ for stick, where $\mu_G \leq \mu_H$. The relations between normal and tangential force are depicted in Fig. 6. Here (a) shows a simplified model of the real behaviour in (b) with $\mu_G = \mu_H$. Another simplified model is depicted in Fig. 6c which differentiates between the coefficient of friction for sticking and sliding.

Within the constitutive treatment for the tangential interface force $F^t$, a tangential spring-dashpot element with an incorporated slider is used in order to model the tangential friction behaviour, see e.g. Cundall and Strack (1979). This model is depicted in Fig. 7 where again the difference between the tangential contact of two particles and of a particle with a rigid wall is made.

Since the tangential part of the interface force $F^t$ is related to two states, sliding and sticking, it can be viewed like an elasto-plastic process where sliding relates to the plastic flow. For these types of problems efficient algorithms were developed in the mid 1980s. The first application to frictional contact can be found in Wriggers (1987) and has been further developed over the years, see e.g. Wriggers et al. (1990), Luding (2004) and Wriggers (2006). The idea is to algorithmically predict first a "trial" stick step followed by a slip check in the second step. Then the regularized penalty formulation for the tangential trial traction takes the form




**Fig. 6** Friction states with stick and sliding

**Fig. 7** Friction states with stick and sliding

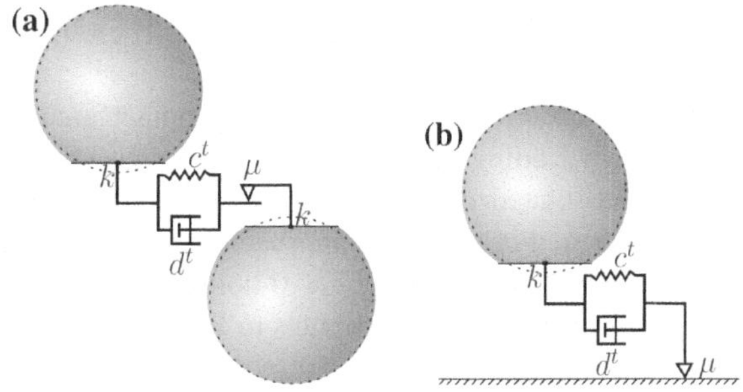

$$\mathbf{F}_o^t = -(c^t \mathbf{g}^t + d^t \mathbf{v}^t). \tag{19}$$

Therein, $\mathbf{g}^t$ is the elongation of the tangential spring, $c^t$ and $d^t$ are the tangential spring stiffness and the tangential dissipation parameter, respectively. The tangential relative velocity at the contact point $\mathcal{C}$ is given by

$$\mathbf{v}^t = \mathbf{v}^s - (\mathbf{v}^s \cdot \mathbf{n})\, \mathbf{n} \tag{20}$$

with the relative velocity at $\mathcal{C}$

$$\mathbf{v}^s = \mathbf{v}_i^{\mathcal{C}} - \mathbf{v}_j^{\mathcal{C}}, \tag{21}$$

where the surface velocities are defined by $\mathbf{v}_i^{\mathcal{C}} = \mathbf{U}_i + \boldsymbol{\omega}_i \times \mathbf{r}_i$ and $\mathbf{v}_j^{\mathcal{C}} = \mathbf{U}_j + \boldsymbol{\omega}_j \times \mathbf{r}_j$. The vectors pointing from $\mathcal{M}_i$ and $\mathcal{M}_j$ to $\mathcal{C}$ are associated with $\mathbf{r}_i = R_i(-\mathbf{n})$ and $\mathbf{r}_j = R_j\mathbf{n}$, respectively. By introducing a trial function $f^{\text{tr}}$, the following relation can be stated for the tangential contact

$$f^{\text{tr}} =: ||\mathbf{F}_o^t|| - \mu_s ||\mathbf{F}^n|| \Rightarrow \begin{cases} \leq 0 \ : & \text{Stick} \\ > 0 \ : & \text{Slip.} \end{cases} \tag{22}$$

If $f^{\text{tr}} \leq 0$ the contact point $\mathcal{C}$ is in the stick region and if $f^{\text{tr}} > 0$ it is in slip region, thus sliding occurs in the contact area. Note that the stick case is valid again if $F_o^t < \mu_d F^n$ holds during the sliding process. If the contact point sticks, the actual tangential spring $\mathbf{g}^t$ is incremented for the succeeding time step by the relation $\Delta\mathbf{g}^t = \mathbf{v}^t \Delta t_{\text{D}}$. Consequently, the new spring length is defined by

$$\bar{\mathbf{g}}^t = \mathbf{g}^t + \Delta\mathbf{g}^t\,. \tag{23}$$

Here, $\Delta t_{\text{D}}$ denotes the time step of the discrete element method. However, if the contact point slides in the current time step, the tangential spring is aligned by

$$\bar{\mathbf{g}}^t = -\frac{1}{c_t}(F^t \mathbf{t} + d^t \mathbf{v}^t) \tag{24}$$

in order to fulfill Coulomb's slip condition. Therein, $\mathbf{t} = \mathbf{F}_o^t / ||\mathbf{F}_o^t||$ is the direction of the trial traction.

In two subsequent time steps, the contact area might be slightly rotated. As proposed in Luding (2004), the tangential spring is continuously projected onto the current rotated contact area at the beginning of each new time step via $\mathbf{g}^t = \bar{\mathbf{g}}^t - (\bar{\mathbf{g}}^t \cdot \mathbf{n})\mathbf{n}$. For the above approach, the tangential contact force is $F^t = ||\mathbf{F}_o^t||$ if $f^{\text{tr}} \leq 0$ and $F^t = \mu_d F^n$ if $f^{\text{tr}} > 0$ holds. By computing $F^t$, one obtains $\mathbf{F}^t = F^t \mathbf{t}$ and $\mathbf{T}^t = \mathbf{r} \times \mathbf{F}^t$ which contributes to $\mathbf{F}$ and $\mathbf{T}$ in Eq. (1) and Eq. (2), respectively.

## *Rolling Resistance Model*

During a rolling motion of two particles over each other, the leading part of the contact area is continuously compressed and the trailing part is decompressed with respect to the rolling direction, see Fig. 8.

In case of an attractive van der Waals force in the contact area, the particles suffer an opposing torque to rolling motion, whereby a rolling resistance is generated. For instance, regarding an agglomerated straight particle chain with a lack of rolling resistance, the chain cannot resist any external impact resulting in a tangential force. As a consequence, the particles will roll smoothly over each other, whereby the chain easily bends and it will finally take a compact shape. For a detailed discussion of the rolling resistance of adhesive particles, see Dominik and Tielens (1995).

For a constitutive treatment of the rolling resistance, a model consisting of a rolling spring-dashpot-slider element is adopted in this contribution, see Iwashita and Oda (1998) and Fig. 9.

At this, the opposing torque is given by

$$\mathbf{M}_o^r = -(c^{\varphi}\phi + d^{\varphi}\dot{\phi}) = -\frac{1}{R}(c^{\phi}\mathbf{g}^r + d^{\varphi}\mathbf{v}^r). \tag{25}$$

Therein, $c^{\varphi}$ is the rolling stiffness, $d^{\varphi}$ relates to the rolling viscosity coefficient, $\phi$ to the particle rotation, $\mathbf{g}^r$ to the rolling distance and $\mathbf{v}^r$ is the rolling velocity which, according to Kuhn and Bagi (2004), can be computed using

$$\mathbf{v}^r = -R\left[(\boldsymbol{\omega}_i - \boldsymbol{\omega}_j) \times \mathbf{n} + \frac{1}{2}\left(\frac{1}{R_j} - \frac{1}{R_i}\right)\mathbf{v}^t\right]. \tag{26}$$

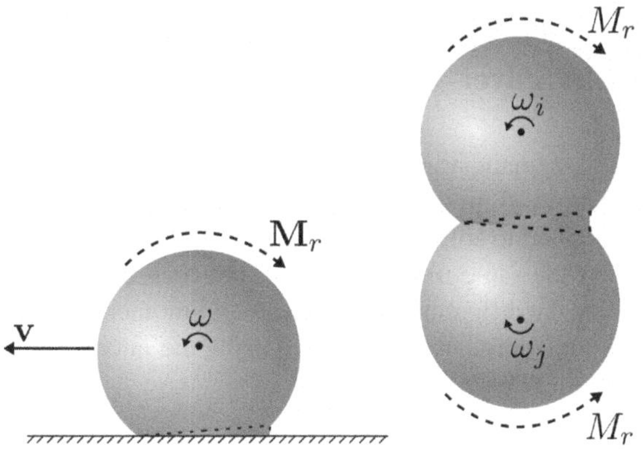


**Fig. 8** Idea of rolling resistance

**Fig. 9** Model of rolling resistance

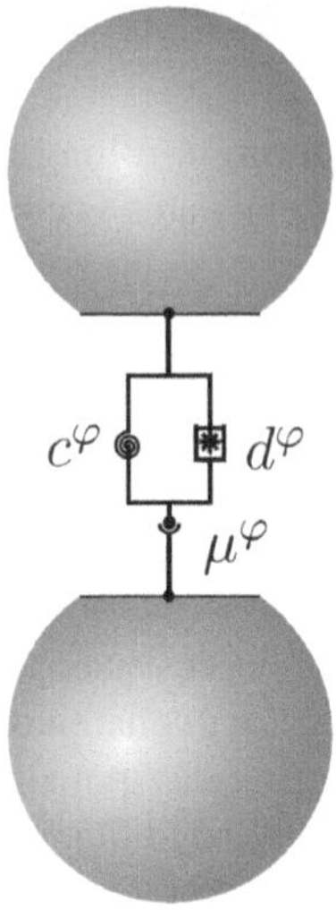


In the context of the JKR theory one can set for $c^{\phi}$ the following expression

$$c^{\varphi} = 4 F_c^n R \left( \frac{a}{a_0} \right)^{3/2} \qquad \text{where} \qquad d^{\varphi} = 0\,, \tag{27}$$

see Dominik and Tielens (1995) for details.

Introducing a trial force $\mathbf{F}_o^r = \mathbf{M}_o^r / R$, the problem of the rolling resistance can algorithmically be treated analogous to the tangential friction model. Here, one can apply as a yield criteria for the slider $M_{\max}^r$ or $\mu^r F^n$, where $\mu^r$ is a rolling friction coefficient, see Iwashita and Oda (1998).

## DEM Solver

The kinematic variables of the particles are governed by the computation of the equations of motion (1) and (2). For this purpose, a time integration scheme has to be applied to solve these equations.

At the same time the contact between the particles has to be considered. This yields the different contact forces and moments between the particles. These forces and moments can be computed using the interaction laws discussed in the previous section.

## *Time Integration*

Since DEM simulations usually need very small time steps $\Delta t$ in order to resolve the constitutive laws between the particles the computational time for a larger system of particles is high. Thus time integration methods need to be considered very carefully. They have on one side to guarantee a certain accuracy and on the other side they need to be efficient and robust to solve the Newton-Euler-equations in (2). An overview with respect to methods that are used in DEM can be found in e.g. Kruggel-Emden et al. (2008).

The integration scheme is composed of three steps:

1. Predicting all the kinematic variables.
2. A force computation step is followed using the predicted variables according to Section "Constitutive Modeling of the Particle". Hence, the evolution of the translational and rotational accelerations can be computed from Eqs. (1) and (2).
3. Applying an error criteria between the predicted and calculated accelerations, the correction of the kinematic variables is ensued.

To illustrate the construction of such time integration scheme we consider a Gear algorithm in more detail. This starts with a predictor computation which is different for translation and rotation.

For the translation one computes

$$\begin{aligned}
\mathbf{x}^P(t+\Delta t) &= \mathbf{X}(t) + \dot{\mathbf{X}}(t)\Delta t + \frac{1}{2}\ddot{\mathbf{X}}(t)\Delta t^2 + \frac{1}{6}\dddot{\mathbf{X}}(t)\Delta t^3 \\
\dot{\mathbf{x}}^P(t+\Delta t) &= \dot{\mathbf{X}}(t) + \ddot{\mathbf{X}}(t)\Delta t + \frac{1}{2}\dddot{\mathbf{X}}(t)\Delta t^2 \\
\ddot{\mathbf{x}}^P(t+\Delta t) &= \ddot{\mathbf{X}}(t) + \dddot{\mathbf{X}}(t)\Delta t \\
\dddot{\mathbf{x}}^P(t+\Delta t) &= \dddot{\mathbf{X}}(t)
\end{aligned} \tag{28}$$

For the rotation the predictor is given by

$$\begin{aligned}
\boldsymbol{\omega}^P(t+\Delta t) &= \boldsymbol{\Omega}(t) + \dot{\boldsymbol{\Omega}}(t)\Delta t + \frac{1}{2}\ddot{\boldsymbol{\Omega}}(t)\Delta t^2 + \frac{1}{6}\dddot{\boldsymbol{\Omega}}(t)\Delta t^3 + \frac{1}{24}\overset{(iv)}{\boldsymbol{\Omega}}(t)\Delta t^4 \\
\dot{\boldsymbol{\omega}}^P(t+\Delta t) &= \dot{\boldsymbol{\Omega}}(t) + \ddot{\boldsymbol{\Omega}}(t)\Delta t + \frac{1}{2}\dddot{\boldsymbol{\Omega}}(t)\Delta t^2 + \frac{1}{6}\overset{(iv)}{\boldsymbol{\Omega}}(t)\Delta t^3 \\
\ddot{\boldsymbol{\omega}}^P(t+\Delta t) &= \ddot{\boldsymbol{\Omega}}(t) + \dddot{\boldsymbol{\Omega}}(t)\Delta t + \frac{1}{2}\overset{(iv)}{\boldsymbol{\Omega}}(t)\Delta t^2 \\
\dddot{\boldsymbol{\omega}}^P(t+\Delta t) &= \dddot{\boldsymbol{\Omega}}(t) + \overset{(iv)}{\boldsymbol{\Omega}}(t)\Delta t \\
\overset{(iv)}{\boldsymbol{\omega}}{}^P(t+\Delta t) &= \overset{(iv)}{\boldsymbol{\Omega}}(t)
\end{aligned} \tag{29}$$

Based on these predictions the force $\mathbf{F}^P(\mathbf{x}^P, \dot{\mathbf{x}}^P)$ and the moment $\mathbf{M}^P(\dot{\mathbf{x}}^P, \boldsymbol{\omega}^P)$ are computed which act both on a particle. Based on these quantities the translational and rotational accelerations can be computed from (2). This yields $\mathbf{a}^* = \mathbf{F}^P/m$ and $\dot{\boldsymbol{\omega}}^* = \mathbf{M}^P/\Theta$. Now the differences $\Delta\ddot{\mathbf{x}} = \mathbf{a}^* - \ddot{\mathbf{x}}^P$ and $\Delta\dot{\boldsymbol{\omega}} = \dot{\boldsymbol{\omega}}^* - \dot{\boldsymbol{\omega}}^P$ can be determined for the corrector step in which the translations and rotations are corrected. For the translations it follows

$$
\begin{aligned}
\mathbf{x}(t+\Delta t) &= \mathbf{x}^P(t+\Delta t) + c_0^t\,\mathbf{s}^t \\
\dot{\mathbf{x}}(t+\Delta t) &= \dot{\mathbf{x}}^P(t+\Delta t) + c_1^t\,\mathbf{s}^t\,\frac{1}{\Delta t} \\
\ddot{\mathbf{x}}(t+\Delta t) &= \ddot{\mathbf{x}}^P(t+\Delta t) + c_2^t\,\mathbf{s}^t\,\frac{2}{\Delta t^2} \\
\dddot{\mathbf{x}}(t+\Delta t) &= \dddot{\mathbf{x}}^P(t+\Delta t) + c_3^t\,\mathbf{s}^t\,\frac{6}{\Delta t^3}
\end{aligned}
\tag{30}
$$

and the rotations are obtained from

$$
\begin{aligned}
\boldsymbol{\omega}(t+\Delta t) &= \boldsymbol{\omega}^P(t+\Delta t) + c_0^r\,\mathbf{s}^r \\
\dot{\boldsymbol{\omega}}(t+\Delta t) &= \dot{\omega}^P(t+\Delta t) + c_1^r\,\mathbf{s}^r\,\frac{1}{\Delta t} \\
\ddot{\boldsymbol{\omega}}(t+\Delta t) &= \ddot{\boldsymbol{\omega}}^P(t+\Delta t) + c_2^r\,\mathbf{s}^r\,\frac{2}{\Delta t^2} \\
\dddot{\boldsymbol{\omega}}(t+\Delta t) &= \dddot{\boldsymbol{\omega}}^P(t+\Delta t) + c_3^r\,\mathbf{s}^r\,\frac{6}{\Delta t^3} \\
\overset{(iv)}{\boldsymbol{\omega}}(t+\Delta t) &= \overset{(iv)}{\boldsymbol{\omega}}{}^P(t+\Delta t) + c_4^r\,\mathbf{s}^r\,\frac{24}{\Delta t^4}
\end{aligned}
\tag{31}
$$

with the constants

$$
\mathbf{s}^t = \frac{\Delta t^2}{2}\Delta\ddot{\mathbf{x}}\,,\; c_0^t = \frac{1}{6}\,,\; c_1^t = \frac{5}{6}\,,\; c_2^t = 1\,,\; c_3^t = \frac{1}{3}
$$

and

$$
\mathbf{s}^r = \Delta t\,\Delta\dot{\boldsymbol{\omega}}\,,\; c_0^r = \frac{251}{720}\,,\; c_1^r = 1\,,\; c_2^r = \frac{11}{12}\,,\; c_3^r = \frac{1}{3}\,,\; c_4^r = \frac{1}{24}
$$

Within this method of Gear the translations are approximated by a Taylor expansion of 3rd order while the rotations have to be modeled by a 4th order Taylor expansion. This choice is necessary to achieve conservation of momentum and moment of momentum and a good accuracy of the solution. A more complete description of the Gear algorithm can be found in Allen and Tildesley (1987), Pöschel and Schwager (2005).

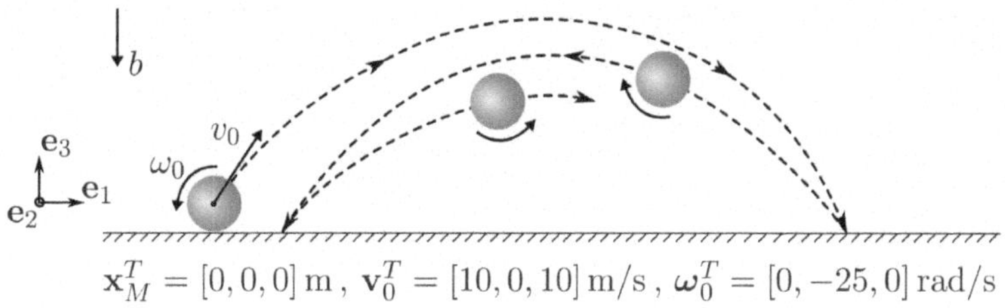


**Fig. 10** Behaviour of a rigid sphere for a given velocity including frictional contact

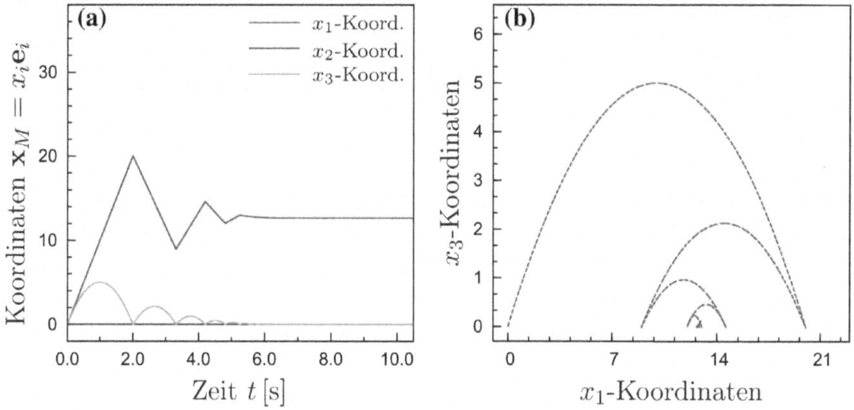


**Fig. 11** **a** Temporal evolvement of the coordinates of the center point of the particle and **b** trajectory of the particle during the motion

## Test Example

A specific example is considered to show that the normal and tangential contact is working correctly. Here only one sphere is considered that comes into contact with a rigid surface. It will be shown that the given formulations and the algorithms are able to reproduce the rebound of the particle, see Fig. 10. The following constants are used to compute the motion of the particle: density $\rho = 2.5\,\mathrm{kg/m^3}$, dissipation factor $A = 6 \times 10^{-4}\,\mathrm{s}$, normal stiffness $c_t = 10^6\,\mathrm{N/m}$, damping $d_t = 50\,\mathrm{Ns/m}$ and friction coefficient for stick and slip $\mu_{(G,H)} = 2.5$. The behaviour can be modeled experimentally using a bouncy ball made of rubber.

As shown in Fig. 10 the particle is subjected to an initial translations and rotational velocity $v_0$ und $\boldsymbol{\omega}_0$ at time $t_0 = 0$ and starts from the rigid plane. The rotational velocity is described as a backward spin.

The temporal evolvement of the numerically computed coordinates of the particle center can be found in Fig. 11a. Figure 11b depicts the trajectory of the particle center and thus shows the behaviour of the particle that jumps back and forth due to the initial velocities and the gravitational force $\rho b$. From the diagrams in Fig. 11 one can observe that the maximal altitude of the particle is $x_{3,\max} = 5.00\,\mathrm{m}$ after that it

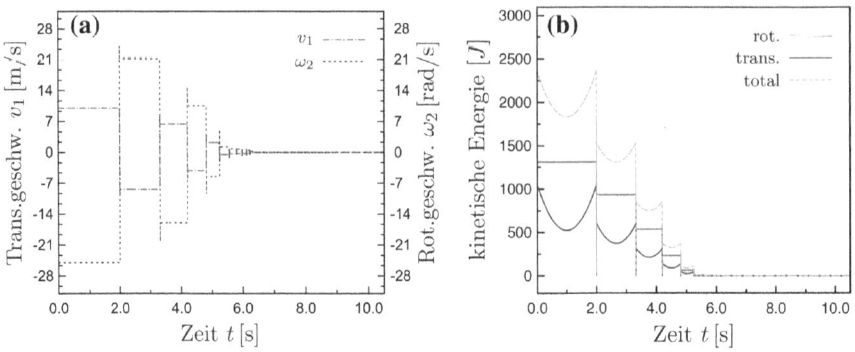


**Fig. 12** **a** Temporal evolvement of the translational and rotational velocities and **b** temporal evolvement of the kinetic energy

starts to drop and contacts the rigid plane after $t_k = 2.00\,\text{s}$ at $x_{1,k} = 20.00\,\text{m}$. This numerical result matches exactly the analytical solutions for the trajectory[2]. When the particle impacts the rigid plane then due to tangential relative displacements a frictional force develops that is opposite to the direction of the relative velocity. This force reduces the tangential translation velocity and contributes to a change in the rotational velocity. In this case the frictional force is large enough to reverse the tangential motion and the rotation, see Fig. 12a. Thus the particle bounces back. However due to the frictional dissipation the kinetic energy is reduced such that the particle does not reach the position at $t_0$. The numerical computed kinetic energy is depicted in Fig. 12b with respect to time. Following the further trajectory of the particle one observes that it bounces back and forth until the kinetic energy is used up which is depicted by the horizontal line in the right part of Fig. 12b .

## *Search Algorithms*

Besides the time integration, the computation of the contact forces is the most CPU time consuming part of a DEM simulation. Thus the contact search and detection governs the efficiency of a DEM code. WIth this in mind, the evaluation of the contact detection Eq. (15) has to be minimized to neighboring particles, since they are the only eligible contact partners. If one simply creates a loop over all $N$ particles in a system, the number of tests amounts to $\sum_{i=1}^{N-1}(N-i) = N(N-1)/2$. Such an algorithm has the complexity of $O(N^2)$ and will result in extremely large computation time, even for small particle systems with less than a few thousand particles. If large particle

[2]Analytical solution for the trajectory:

$$\tilde{x}_{3,\max} = \frac{(v_0 \sin\alpha)^2}{2\,b}\,, \quad \tilde{x}_{1,k} = \frac{v_0^2 \sin(2\,\alpha)}{b}\,, \quad \tilde{t}_k = \frac{2\,v_0 \sin\alpha}{b}\,.$$

**Fig. 13** Particle contact: Verlet distance

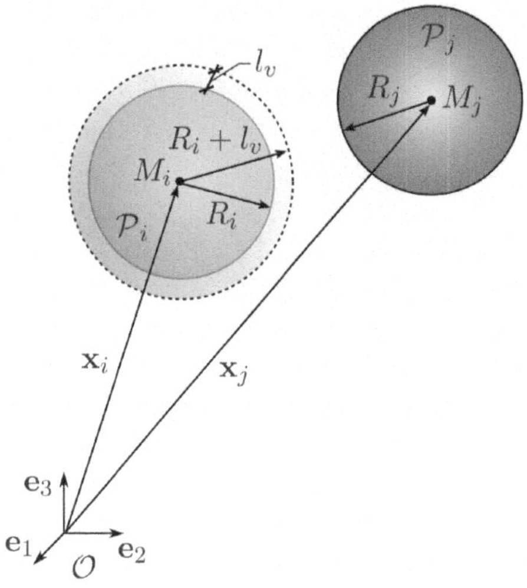


systems with several millions of particles have to be computed then an algorithm of complexity $O(N)$ is needed.

There many different algorithm that reduce the complexity of contact search. Among them are methods like space cell decomposition, combinations with binary tree search and heapsort algorithms for the global search. More advanced algorithms are the NBS algorithm, the alternating digital tree method, space filling curve technique or Double-Ended Spatial Sorting (DESS) algorithm.

Here we will look in more detail at algorithms that stem from the area of molecular dynamics. These are known as Verlet list and linked cells and can be found for the Verlet list in Verlet (1967) and for the link cells in Quentrec and Brot (1973). Here both algorithms will be combined in order to yield a fast contact search algorithm, see also Ref. (Allen and Tildesley 1987) and references therein.

Verlet lists are based on the idea that for each particle $\mathcal{P}_i \, \{i = 1, \ldots, N\}$ one generates a list with neighbouring particles. By this, the local contact check can be reduced to these neighbouring particles and thus is a local operation. The Verlet list can be build, based o the computation of a distance

$$g^n_{\text{Verlet}} = (R_i + l_v) + R_j - (\mathbf{x}_i - \mathbf{x}_j) \cdot \mathbf{n} \begin{cases} \leq 0 \text{ no neighbour} \\ > 0 \text{ neighbour} \begin{cases} j < i \text{ list of } \mathcal{P}_i \\ j > i \text{ list of } \mathcal{P}_j \, . \end{cases} \end{cases}$$

This takes into account the Verlet distance $l_v$ depicted in Fig. 13. The Verlet distance can be defined by

$$l_v = \frac{1}{2} \left( \text{Max} \, [\, R_i \,]_{i=1}^N + \text{Min} \, [\, R_i \,]_{i=1}^N \right) \alpha_v \, .$$

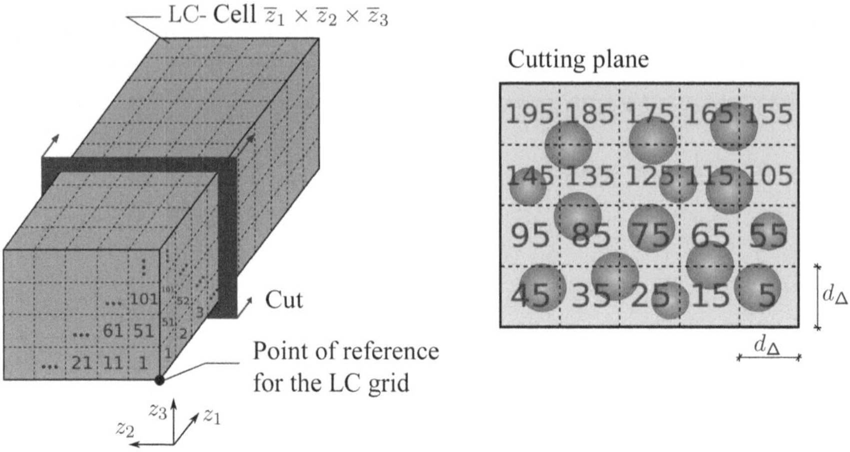


**Fig. 14** Building the Verlet list with the linked cell method

where $\alpha_v$ is a constant that can be chosen. A good choice for this constant is $\alpha_v \approx$ 0.05–0.10.

A list of neighboring particles is maintained for each particle $\mathcal{P}$ holding $l_v > |g^n|$. Once the Verlet list is built, Eq. (15) is only evaluated for neighboring pairs. Certainly, the list has to be updated at some intervals since particles can move to other locations and then have completely new neighbours. Here, a possible rebuild criteria can be defined by

$$\Delta s_{\max} = \mathrm{Max}\left[\,||\mathbf{x}_i - \mathbf{x}_i'||\,\right]_{i=1}^{N} > \beta_v l_v\,.$$

In this equation the position vector $\mathbf{x}_i'$ of a particle $\mathcal{P}_i$ is used which relates to the position vector of the existing Verlet list while $\mathbf{x}_i$ is the position vector of the current time step. The factor $\beta_v$ can be generally chosen in the range 0.5–0.6 which ensures that no complete penetration of a particle by another one occurs. For details see Pöschel and Schwager (2005).

In total the Verlet list method has a complexity of $O(N)$ for the computation of the local contact and thus the contact forces. However building the list is not so efficient. Due to this drawback the linked cell approach is used to construct the list. This amounts to an algorithm that overall approaches a complexity of $O(N)$. The idea of the linked cell algorithm is to divide the computational domain into cubic cells of uniform side lengths that are slightly larger than the maximal particle diameter, see Fig. 14.

After assigning the particles $\mathcal{P}_i$ to the cells with respect to their center of mass $\mathcal{M}_i$, the relevant particles for the construction of the neighbor list are referenced to the 26 surrounding cells and of course to the mid-cell which contains the considered particle $\mathcal{P}_i$.

## *Example: Silo Discharge*

The DEM scheme is validated by means of a laboratory silo discharge experiment. Choi et al. (2004) analyzed the velocity profile in a quasi 2D silo using an image based particle tracking method. For the numerical modeling of this silo discharge, see Wellmann (2011).

The box-shaped silo of size [20 × 2.5 × 90] cm is filled with soda lime glass beads of slight polydispersity ($d = 3 \pm 0.1$ mm) using a distributed filling procedure. In total 190.782 particles are used to model the silo discharge. This leads in three dimensions to 1.144.692 particles. The rectangular orifice of [16 × 25] mm at the bottom center is opened and a steady state flow is allowed to develop before the tracking procedure starts. The tracking covers a rectangular section of [20 × 50] cm above the orifice. The flow is measured at a rate of 125 frames per second for 16.4 s. For the evaluation of the velocity profile the observation window is divided into [48 × 48] mm cells used as averaging domains. For a more detailed description of the data gathering and evaluation see Choi et al. (2004, 2005).

The elastic parameters of soda lime glass are given as $E = 71$ GPa and $\nu = 0.22$. The mass density is $\rho = 2.5$ g/cm3. The friction coefficient between dry soda lime glass beads was measured by Ishibashi et al. (1994) as $\mu = 0.162$. The gravity constant is chosen as $g = 9.81$ m/s$^2$. Since the particles in the vicinity of the orifice move at reasonable velocities the viscoelastic contact law described in Section "Normal Contact Model" is applied. The material constant $A$, see (13), to include viscoelastic effects was chosen as $A = 5.05\ 10^{-8}$ s for the soda lime glass beads. This corresponds to a restitution coefficient of 0.97 at a relative velocity of 1.18 m/s and a particle size of $d = 3.18$ mm. The material properties of the glass silo were assumed to be identical with the glass bead properties.

The stiffness of the particles and their size require a time step of the order of $\Delta t \approx 1\ \mu s$. For a total simulation time of 10 s this results in a number of time steps of the order of $10^7$. Combined with a number of particles of $N \approx 2 \times 10^5$ this amounts to a large computational effort. However, numerical tests show that the system behavior is not altered significantly by reducing the stiffness of the particles to E = 0.9 GPa. This value assures that the overlap corresponding to the maximum contact force at the silo bottom is less than 1% of the particle radius. In order to preserve the dynamic particle behavior the viscoelastic constant is adopted to $A = 2.9\ 10^{-7}$ s. The stiffness reduction enables a time step of $\delta t = 10\ \mu s$ resulting in $3 \times 10^6$ steps for a simulation period of 30 s.

The initial sample, see Fig. 15, is generated using uniformly distributed particles with diameters in the range 2.9–3.1 mm and a solid fraction in the packaging of $\Phi = 0.5$. To account for the solid fraction of a denser package ($\Phi = 0.6$) the box used as package space is enlarged in the z direction to 1.08 m. In order to get realistic fabric properties the sample is settled under the influence of gravity using the DEM scheme and the above material parameters. At the point where the kinetic energy was nearly dissipated the orifice was opened starting the silo discharge simulation for 30 s. During the simulation the output was written at 0.3 s intervals, see Fig. 15. After a

**Fig. 15** Outflow of the particles from the silo at different times

short time a steady state flow develops with a mass flow rate of $Q = 132.6$g/s. This corresponds to a deviation of 6% from the experimental flow rate of Q = 141.1 g/s. Considering the uncertainties regarding the initial distribution of the particles and the container-particle friction this match is satisfactory.

In order to deduce a continuous downward velocity distribution $v(x)$ from the discrete DEM output a coarse graining scheme is applied. For this purpose the box-shaped $[20.5 \times 50]$ cm silo volume above the orifice is divided into a regular grid of $[40 \times 1 \times 80]$ linear hexaeders. An ansatz $v_h$ is defined on the hexaeder mesh and is fittes to the discrete DEM results using a volume weighted least square approach. Finally, the velocity profile is evaluated in the silo mid-plane and averaged over the data points between $t = 5$ s and $t = 20$ s. The resulting profile is compared with the experimental profile from (Choi et al. 2004) in Fig. 16.

There is a good agreement of the maximum velocity at the orifice, of the velocity gradient around the orifice, and of the run of the contour lines. The DEM scheme predicts a realistic shape of the stagnant zones at the lower silo corners. For quantitative comparisons 1D velocity profiles are evaluated at two heights sketched in the right part of Fig. 16.

The results are presented in Fig. 17. While there is a close agreement of the profiles in the vicinity of the orifice at $z = 9.1\,d$ the simulation predicts higher downward velocities at the boundaries at $z = 29.1\,d$. Note that this deviation can result basically from the different evaluation schemes: While the DEM data is averaged over the entire depth of the silo, the experimental image-based particle tracking method is based solely on the trajectories of the particles at the front window.

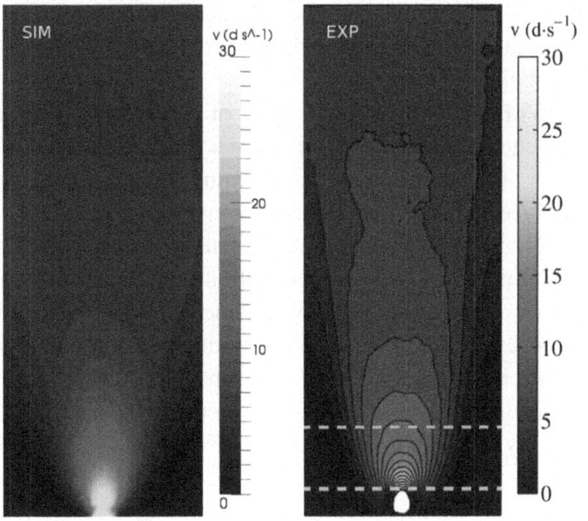


**Fig. 16** Comparison of downward velocity: simulation and experiment

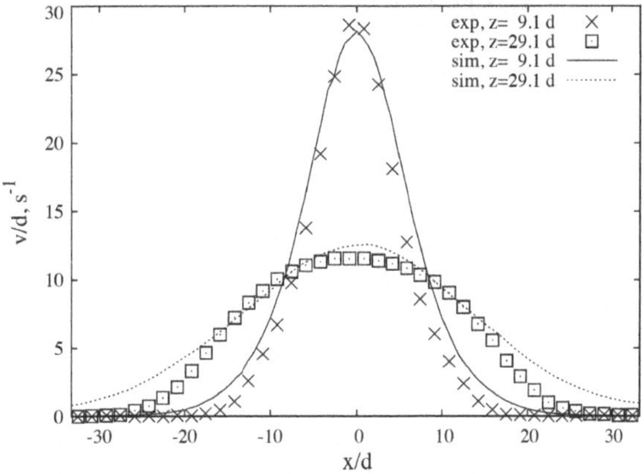


**Fig. 17** Comparison of downward velocity: simulation and experiment

In total it is obvious that the discrete element method can predict flow states in a granular material accurately based on the introduced interface laws between the particles. However due to the large number of particles involved in one discretization it will be necessary to use parallel computing architectures. This is the only possibility to obtain reasonable results within engineering time frames.

## DEM Using Parallel Solvers

Many engineering problems have to be modelled by using a large number of particles. These are e.g. milling applications or outflow problems in silos as discussed in the last section. Since many particles interact which each other high computing power is needed to model a particle flow application in an accurate manner. This leads to computer codes that include algorithms which can make the transition from serial to parallel computing.

Within such models millions of spherical or arbitrary geometry particles have to be generated in order to describe a typical in granular flow problem accurately. The main application areas being simulations in geomechanics, structural, oil and gas engineering and other processes involving granular materials and particulate flows. When using parallel computing chunks of particles have to be distributed to the different processors. This distribution not only affects the computational part but also the set up of the problem. Within the latter task scaling of mesh generators necessarily requires mesh splitting before multiple processors assignment and domain decomposition techniques. During the run time—due to large movements of particles a load balancing procedure has to be executed in order to distribute the computational load equally to all processors.

## *Solver DEMFLOW*

The mesh-free particle code DEMFLOW, developed at our institute over many years, is a parallel and fully open source framework for the simulation of the flow of granular materials, fluid and particulate flows. This particle solver was initially designed for cluster computations and was further developed to run efficiently on massively parallel high performance computer (HPC) architectures allowing large scale computations of engineering problems. The focused application areas of DEMFLOW are particle and multiphase flows and fluid-particle interaction. The DEM-module is completely developed in-house.

A common feature shared by mesh-free particle methods which also include meshless methods like smoothed particle hydrodynamics (SPH), see Gómez-Gesteira et al. (2012a); Gomez-Gesteira et al. (2012b), is that they all represent computationally very intensive simulation techniques. Due to their explicit nature, the corresponding solvers require very small time steps. In classical discrete element methods the restriction is due to the displacement-driven nature of the force computations and in the SPH due to the CFL condition. As a result, a large number of time steps need to be performed in order to simulate a time period of practical interest. Moreover, to discretize real engineering problems in a 3D framework of a particle method, a set of at least several millions particles is required for a sufficiently detailed reconstruction of a complex system. In general, the simulation of systems with more than one million particles can only be carried out in a reasonable way if a massively parallelized strategy is used. This is also necessary in order to handle the aspect of memory requirement, especially in really large scale problems.

A widely used strategy for the parallelization task in computational mechanics is the domain decomposition approach in conjunction with the use of the Message Passing Interface (MPI) library for inter-domain or inter-process communication. In the framework of this strategy, the computational domain is geometrically split into sub-domains according to the number of specified processes (tasks, cores). Each generated sub-domain is then assigned to a core, including the respective particles. Both the communication and the data exchange between the cores are realized by the network interconnect and controlled by the instructions of the MPI library. Communication between sub-domains is necessary since the force computation or kernel evaluation of a particle close to a boundary of its sub-domain requires information about the particles positioned in the next boundary layer of the corresponding adjacent sub-domain.

A crucial point in the domain decomposition strategy, which demands great attention when applied to particle methods, is the aspect of the rescaling load-balance of the processes. As particles can cover large distances by traversing several sub-domains of the computational domain this occurs especially in highly dynamic systems, like in rotary mixing drums (for an illustration, see Fig. 20). Many dynamic systems are often characterized by a heterogeneous particle distribution over the entire domain. As a result, some sub-domains include only a small number of particles and while others have a large number of particles. In the former case, the cores are underused and complete their jobs faster than the cores in the latter case. Thus, they are idle while waiting for the rest of the processes to complete. In general, unequal workload leads to a strong decrease in the efficiency of the calculation with respect to the number of used cores. Consequently, the development of powerful algorithms for a continuous adaptation of the domain decomposition in terms of a homogeneous workload across all processes is inevitable for efficient computations of large scale engineering problems. This critical point represents a great challenge, especially in case of irregular 3D computational domains with a highly dynamic change of the particle distribution.

A second essential aspect that needs to be accounted for in order to design a high performance simulation tool in the framework of particle methods is directly linked to the organization and management of the particle data in the memory. The procedure for the force or kernel evaluation represents the most computationally intensive part in a simulation time step. The corresponding routine is carried out for each particle and over the neighbors of each individual particle. A common approach is to create and maintain an array index list (Verlet list) which stores the neighbors of each particle. The Verlet list is generally built by the use of the Linked-Cell method, see Allen and Tildesley (1987). By using this list, the evaluations are only performed in the respective neighborhood of a discrete element, thus avoiding unnecessary computations and minimizing numerical costs of a simulation, see also Section "Search Algorithm". At the beginning of a simulation, the order of the particle data sequence assembled in the memory corresponds usually to the real particle neighborhood connectivity, which is represented by the indices in the neighbor list. With time, the initial arrangement of the particles will be scattered due to large relative particle motions. As the neighborhood connectivity changes, the particle data in the memory can be considered as dispersed with respect to the accesses to the neighbor list. This leads to a significant amount of cache misses, because the particles are now more or less randomly distributed. Thus, a currently requested data of a neighbor is increasingly less likely to be present in the loaded cache line compared to the beginning of the simulation run. However, in case of a cache miss, fetching the data of a particle from the main memory is several magnitudes slower than retrieving the data from one of the caches (L1-, L2- or L3-cache). Consequently, very efficient particle reordering algorithms and data management strategies are essential to maximize cache hits and to sustain the performance of a code.

There are a number of particle solvers in the literature that use the domain decomposition approach in combination with MPI for parallelization. Some of those advanced tools based on mesh-free particle methods are briefly mentioned below. A sophisticated SPH fluid solver, called SPH-Flow, is presented in Maruzewski et al. (2009). One of the applications of this code was the simulation of the impact of a snooker ball onto the free surface of water. The authors report a parallel efficiency of 58% when solving the problem with 124 million fluid particles on 2048 cores, 72% for 7.5 million particles with 128 cores and 87% for 3.5 million particles with 64 cores. The computations for the performance analysis were carried out on a Blue Gene/L high performance computer. Gadget-2 is another advanced open source particle code reported in the literature. This SPH code extends the work of Springel (2005) and is designed to perform parallel large scale cosmologic simulations. A modified version of Gadget-2 regarding its application to hydrodynamic problems is presented in Ulrich and Rung (2006). In this work, the authors report that the solver shows a super-linear speedup behavior when applying it up to 256 cores. As a scalability test, they considered a fixed-size problem of a sloshing application with 40 million fluid particles. The computations for the performance test were carried out on the HLRN2 high performance computer. Motivated by the libraries that are provided by the well-known frameworks of PETSc and Trilinos for linear algebra, Sbalzarini et al. (2006) have designed and developed a MPI based middleware for particle methods (among others, for DEM, MD and SPH), which consists of highly efficient parallel libraries. The collection of open source libraries associated with this project is known as the Parallel Particle-Mesh (PPM) Library. The overall aim of the PPM middleware is to provide a very flexible framework with which a programmer can develop highly efficient particle-based software for HPC architectures in a short time. For instance, Walther and Sbalzarini (2009) developed a DEM tool in the setting of PPM and demonstrated a parallel efficiency of 40% on 192 cores of a Linux cluster by performing simulations of 3D sand avalanches with up to 122 million particles. The efficiency of a SPH code implemented by adapting the PPM libraries is presented in Sbalzarini et al. (2006) for a vortex ring simulation. The paper shows that this particle solver reaches an efficiency of 91% on 128 cores of a Cray XT4 HPC architecture.

The software DEMFLOW is developed on the basis of PPM libraries. In fact, the optimization of a parallel particle software regarding its load-balancing capabilities can yield a far more efficient tool as optimizing the respective calculation algorithms assigned to the cores. The collection of libraries that the PPM middleware provides in this context are very efficient and complex algorithms that can be included relatively easily in DEMFLOW. The corresponding PPM libraries are highly optimized and tested on HPC architectures with fairly encouraging results, see Sbalzarini et al. (2006) and Walther and Sbalzarini (2009). Those libraries include, among others, a range of adaptive domain decomposition approaches, assignment algorithms for sub-domains onto cores according to their individual performances, dynamic load-balancing and heuristic criteria for particle domain updating. The integration of the parallel PPM libraries in DEMFLOW in terms of load-balancing will significantly contribute to the further development of the particle code regarding its parallel efficiency on HPC architectures.

The best way to minimize the amount of cache misses in particle based computations is to reorder the particle data in the memory according to the particle locality. However, in 3D, it is quite hard to sort scattered particle data for optimal cache use; one has to also consider that the neighborhood of each particle changes with time for relative particle motions. To deal with this problem, Springel (2005) uses (in his astrophysics SPH-software Gadget-2) an approach based on the Peano-Hilbert curve method. By applying this method, one can create a space-filling curve that maps the considered 3D domain onto a 1D curve. In Gadget-2, this curve is employed both to realize the domain decomposition and to reorder the particle data in the memory. In the latter case, the algorithm maps the particle positions onto the Peano-Hilbert curve and sorts the particles based on this mapping. Finally, a particle arrangement in memory which conforms to a large extent to the real neighborhood of the particles is obtained. In Springel (2005), it is reported that computational tests with Peano-Hilbert ordering show a performance gain of the code Gadget-2 by a factor of two when compared with the performances achieved by neglecting data sorting. This approach was implemented in DEMFLOW.

## *Numerical Test: Medium Number of Cores*

Both major factors of a particle solver—adaptive load-balancing and dynamic particle data reordering—have severe impact on the parallel performance of a code.

As the algorithms which should cover these two aspects are not yet developed in a sophisticated manner and implemented into the DEMFLOW solver, the computation domain of the system chosen to study its performance is only slightly affected by a traversing particle flow. THe next example was selected to discuss the scalability of the code on a parallel computer. For that particles were used to roughly model a fluid.

The considered system consists of two relatively light cubes that plunge into a compact particle domain which is fully discretized with fluid particles, as illustrated in Fig. 18. Here, load-balancing and data reordering are of less importance. The performance tests were run on a Linux-Cluster. The cluster is equipped with 16 nodes where most of the nodes contain either two Intel Xeon E5-2642-v2 CPUs ($2 \times 6$ cores) or two Intel Xeon E5-2670 CPUs ($2 \times 8$ cores). The speedups and efficiency of the software are computationally evaluated on the basis of four discretization levels of the system: the original problem consists of 5 million particles, the half-sized model of 2.5 million particles, the double-sized model of 10 million particles and the quadrupled-sized model of 20 million particles. For each discretization level, a

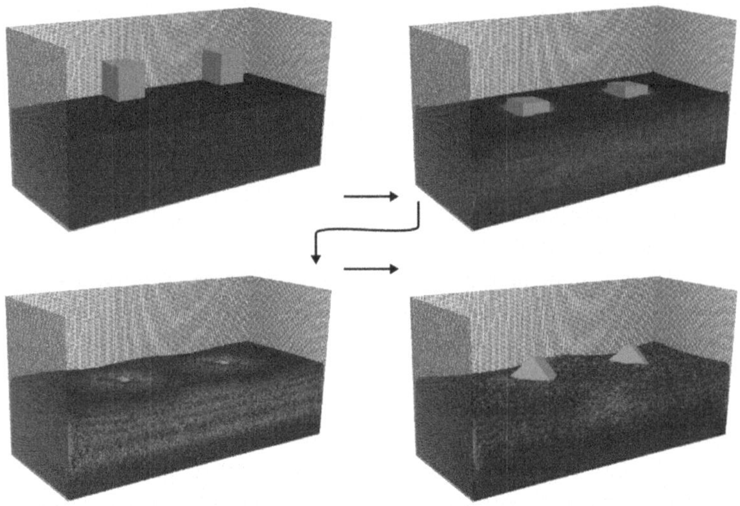

**Fig. 18** Two light solid cubes plunging into water. Simulation results at different configurations in time

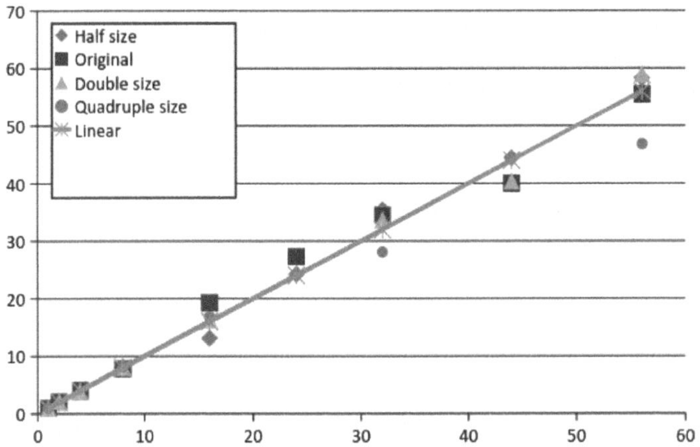


**Fig. 19** Speed-ups of DEMFLOW for different problem sizes

fixed-size problem is examined. The obtained speedups and efficiencies of the test computations are displayed in Fig. 19, respectively. A linear speed-up can be observed for the original, half-sized and double-sized system. For the original problem, the study even shows slight super-linear behavior of the program for a smaller number of cores. Furthermore, a speedup of 47 is obtained for the largest system when 56 cores are used for computation. The latter case results in an efficiency of 84%, whereas the other three cases achieve efficiencies in the range of 90–120%. It can be concluded that the results obtained in this study are successful and promising as they show a very good speedup of the solver up to 56 cores.

## *Numerical Test: Large Number of Cores*

DEMFLOW has been adapted to run on a high performance computer with large core numbers in order to validate its efficiency on HPC architectures. For this task a system which implies a sophisticated strategy both for load-balancing with adaptive domain decomposition and for particle-data reorganization was considered.

It can be shown that the optimization of a parallel particle software regarding its load-balancing capabilities can yield a far more efficient tool as optimizing the respective calculation algorithms assigned to the cores. The collection of libraries that the PPM middleware provides in this context very efficient and complex algorithms.

Hence discrete element methods have become a more and more powerful tool, especially for the treatment of granular materials and in process engineering. The method can nowadays be applied to problems that 10–1000 million particles for an accurate model. These processes are run on high performance parallel computers or GPU systems. The problem of the interchanging contact conditions due to large particle motions presents a challenge for the development of algorithms that scale well for large numbers of processors.

Here we will focus on an application that uses the discrete element method to model the mixing process of different particles in a drum. In this application 25 million of particles are used. The particles are mixed by rigid blades that rotate about the middle axis. The blades move up and down to perform the mixing. Here many small time steps have to be executed to follow the complex motion of the particles in a mixer. Such problems necessitate the use of parallel computing machines with many cores. The problem is depicted in Fig. 20a which also shows the allocation of parts of the particles to a specific core of the parallel computer.

(a) (b)

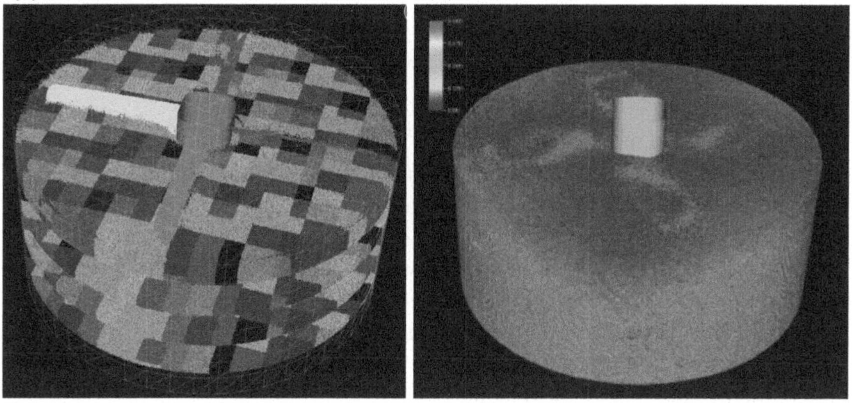

**Fig. 20** **a** Set-up of the particles in the mixer. The color code shows the allocation to a core in the parallel computer. **b** Velocity distribution of the particles during mixing

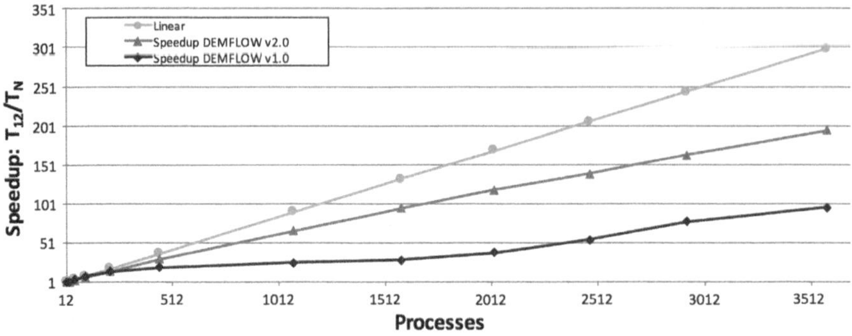


**Fig. 21** Speed-up for different algorithmic treatments

In Fig. 20b the velocity distribution during the mixing process is depicted. One can easily observe the complex movement of the particles that change locations and by that also come into contact with other particles thus leaving the location of the initial allocation and thus have to be moved to another core. This usually requires a reallocation and slows down the computation. In order to get a good performance sophisticated algorithms have to be used that on one hand have to be fast and on the other limit data transfer. By that a speed-up that is linear, even for a large number of processors, can be achieved, see Fig. 21.

The blue curve shows the theoretical limit for a speed-up investigation up to 3500 processors. The red curve shows the performance for one algorithm while the more sophisticated algorithm and related data handling amounts to the green curve that produces an extremely good linear speed-up over all number of processors that reaches 66% of the theoretical limit.

## Conclusion

The discrete element method is a very good tool for the prediction of granular flows and for the determination of macrsocopic material parameters of granular materials. It is a time consuming method since on one side millions of particles have to be used for adequate models and on the other side very small time steps have to be used to resolve the complex interaction laws between particles. This means that discrete element technologies have to be solved on computers with parallel architecture.

It was shown that DEMPACK exhibits a near linear scalability for small scale problems but also for large scale problems with rapidly changing interconnections of the particles due to large movements.

## References

Alder, B. J., & Wainwright, T. E. (1957). Phase transition for a hard sphere system. *Journal of Chemical Physics*, *27*(5), 1208–1209.

Allen, M. P., & Tildesley, D. J. (1987). *Computer simulation of liquids*. New York: Oxford University Press.

Avci, B., & Wriggers, P. (2012). A dem-fem coupling approach for the direct numerical simulation of 3d particulate flows. *Journal of Applied Mechanics*, *79*, 01901.

Brilliantov N. V., Albers N., Spahn F., & Pöschel T. (2007). Collision dynamics of granular particles with adhesion. *Physical Review E*, *76*(5, Part 1).

Brilliantov N. V., Spahn F., Hertzsch J. M., & Pöschel T. (1996). Model for collisions in granular gases. *Physical Review E*, *53*(5, Part B), 5382–5392.

Choi, J., Kudrolli, A., & Bazant, M. Z. (2005). Velocity profile of granular flows inside silos and hoppers. *Journal of Physics: Condensed Matter*, *17*, 2533–2548.

Choi, J., Kudrolli, A., Rosales, R. R., & Bazant, M. Z. (2004). Diffusion and mixing in gravity-driven dense granular flows. *Physical Review Letters*, *92*, 174301.

Cundall, P. A., & Strack, O. D. L. (1979). Discrete numerical model for granular assemblies. *Geotechnique*, *29*(1), 47–65.

Dhia, H. B., & Rateau, G. (2005). The arlequin method as a flexible engineering design tool. *International Journal of Numerical Methods in Engineering*, *62*, 1442–1462.

Dominik, C., & Tielens, A. G. G. M. (1995). Resistance to rolling in the adhesive contact of 2 elastic spheres. *Philosophical Magazine A—Physics of Condensed Matter Structure Defects and Mechanical Properties*, *72*(3), 783–803.

Gingold, R. A., & Monaghan, J. J. (1977). Smoothed particle hydrodynamics: Theory and application to non-spherical stars. *Monthly Notices of the Royal Astronomical Society*, *181*(3), 375–389.

Gómez-Gesteira, M., Crespo, A. J., Rogers, B. D., Dalrymple, R. A., Dominguez, J. M., & Barreiro, A. (2012a). Sphysics-development of a free-surface fluid solver-part 2: Efficiency and test cases. *Computers & Geosciences*, *48*, 300–307.

Gomez-Gesteira, M., Rogers, B. D., Crespo, A. J., Dalrymple, R. A., Narayanaswamy, M., & Dominguez, J. M. (2012b). Sphysics-development of a free-surface fluid solver-part 1: Theory and formulations. *Computers & Geosciences*, *48*, 289–299.

Hertz, H. (1882). Über die Berührung fester elastischer Körper. *Journal für die reine und angewandte Mathematik*, *92*, 156–171.

Ishibashi, I., Perry, C., & Agarwal, T. K. (1994). Experimental determinations of contact friction for spherical glass particles. *Soils and Foundations*, *34*, 79–84.

Iwashita, K., & Oda, M. (1998). Rolling resistance at contacts in simulation of shear band development by dem. *Journal of Engineering Mechanics-ASCE*, *124*(3), 285–292.

Johnson, A. A., & Tezduyar, T. E. (1997). 3d simulation of fluid-particle interactions with the number of particles reaching 100. *Computer Methods in Applied Mechanics and Engineering*, *145*(3–4), 301–321.

Johnson, K. L., Kendall, K., & Roberts, A. D. (1971). Surface energy and contact of elastic solids. *Proceedings of the Royal Society of London Series A-Mathematical and Physical Sciences*, *324*(1558), 301–313.

Kruggel-Emden, H., Sturm, M., Wirtz, S., & Scherer, V. (2008). Selection of an appropriate time integration scheme for the discrete element method (dem). *Computers & Chemical Engineering*, *32*(10), 2263–2279.

Kuhn, M., & Bagi, K. (2004). Alternative definition of particle rolling in a granular assembly. *Journal of Engineering Mechanics-ASCE*, *130*(7), 826–835.

Loskofsky, C., Song, F., & Newby, B. Z. (2006). Underwater adhesion measurements using the JKR technique. *Journal of Adhesion*, *82*(7), 713–730.

Luding, S. (2004). Micro-macro transition for anisotropic, frictional granular packings. *International Journal of Solids and Structures*, *41*(21), 5821–5836.

Maruzewski, P., Le Touze, D., & Oger, G. (2009). Sph high-performance computing simulations of rigid solids impacting the free-surface of water. *Journal of Hydraulic Research*, *47*, 126–134.
Maugis, D. (1992). Adhesion of spheres—The jkr-dmt transition using a dugdale model. *Journal of Colloid and Interface Science*, *150*(1), 243–269.
Pöschel, T., & Schwager, T. (2005). *Computational Granular Dynamics*. Springer.
Quentrec, B., & Brot, C. (1973). New method for searching for neighbors in molecular dynamics computations. *Journal of Computational Physics*, *13*(3), 430–432.
Sbalzarini, I. F., Walther, J. H., Bergdorf, M., Hieber, S. E., Kotsalis, E. M., & Koumoutsakos, P. (2006). PPM—A highly efficient parallel particle-mesh library for the simulation of continuum systems. *Journal of Computational Physics*, *215*, 566–588.
Springel, V. (2005). The cosmological simulation code gadget-2. *Monthly Notices of the Royal Astronomical Society*, *364*, 1105–1134.
Ulrich, C., & Rung, T. (2006). Validation and application of a massively-parallel hydrodynamic SPH simulation code. *Proceedings of NuTTS '09*.
Verlet, L. (1967). Computer experiments on classical fluids. i. Thermodynamical properties of Lennard-Jones molecules. *Physical Review*, *159*(1), 98–103.
Walther, J. H., & Sbalzarini, I. F. (2009). Large-scale parallel discrete element simulations of granular flow. *Engineering Computations*, *26*(6, Sp. Iss. SI), 688–697; *4th International Conference on Discrete Element Methods* (2007). Australia, Brisbane.
Wellmann, C. (2011) *A Two-Scale Model of Granular Materials Using a Coupled Discrete-Finite Element Approach*. Dissertation, B11/1, Institute for Continuum Mechanics, Leibniz University Hannover.
Wellmann, C., Lillie, C., & Wriggers, P. (2008). A contact detection algorithm for superellipsoids based on the common-normal concept. *Engineering Computations*, *25*(5–6), 432–442.
Wellmann, C., & Wriggers, P. (2012). A two-scale model of granular materials. *Computer Methods in Applied Mechanics and Engineering*, *205–208*, 46–58.
Wriggers, P. (1987). On consistent tangent matrices for frictional contact problems. In G. Pande & J. Middleton (Eds.), *Proceedings of NUMETA '87*. M. Nijhoff Publishers, Dordrecht.
Wriggers, P. (2006). *Computational Contact Mechanics* (2nd ed.). Berlin Heidelberg: Springer.
Wriggers, P., Van, T. V., & Stein, E. (1990). Finite-element-formulation of large deformation impact-contact-problems with friction. *Computers and Structures*, *37*, 319–333.
Zhu, H. P., Zhou, Z. Y., Yang, R. Y., & Yu, A. B. (2007). Discrete particle simulation of particulate systems: Theoretical developments. *Chemical Engineering Science*, *62*(13), 3378–3396.
Zohdi, T. I. (2007). *Introduction to the modeling and simulation of particulate flows*. SIAM.

# Adaptive Integration of Cut Finite Elements and Cells for Nonlinear Structural Analysis

**Alexander Düster and Simeon Hubrich**

## Introduction

*Fictitious domain methods* have gained increasing interest since they can be applied to boundary value problems with complicated geometry for which it is difficult to generate body-conforming finite element meshes. The basic idea of fictitious domain methods is not new. To the best of the authors' knowledge, the main idea goes back to Saul'ev (1963a, b) who solved boundary value problems by a fictitious domain method for the first time. Later on, several authors worked on related approaches such as, for example, Neittaanmäki and Tiba (1995), Peskin (2002), Del Pino and Pironneau (2003), Mittal and Iaccarino (2005), Glowinski and Kuznetsov (2007) and Ramière et al. (2007). The so-called CutFEM, which utilizes meshes including finite elements that are cut by the boundary of the domain, was introduced by Burman and Hansbo (2010, 2012). The finite cell method (FCM), proposed by Parvizian et al. (2007) and Düster et al. (2008), is to be seen as a combination of high-order finite elements and the fictitious domain approach. The application of high-order shape functions allows for high convergence rates, provided that the exact solution of the mathematical problem is smooth enough – or that a proper mesh layout is chosen if there are discontinuities or singularities.

A very important issue in all fictitious domain methods is the *numerical integration* of the stiffness and mass matrices as well as load vectors of broken elements/cells. Due to the fact that the applied meshes are not aligned to the geometry of the domain, computing the matrices involves discontinuous integrands which can not be computed efficiently with standard Gauss quadrature rules. To overcome this problem different approaches have been developed.

A rather simple approach to improve the quadrature of discontinuous functions is to split the integration domain on element or cell level into sub-domains on which the integrand is smooth. Then, the standard Gauss quadrature can be applied on each of the so-called sub-cells in a very efficient manner. This composite integration

A. Düster (✉) · S. Hubrich
Numerical Structural Analysis with Application in Ship Technology (M-10), Hamburg University of Technology, Am Schwarzenberg-Campus 4C, 21073 Hamburg, Germany
e-mail: alexander.duester@tuhh.de



L. De Lorenzis and A. Düster (eds.), *Modeling in Engineering Using Innovative Numerical Methods for Solids and Fluids*, CISM International Centre for Mechanical Sciences 599,
https://doi.org/10.1007/978-3-030-37518-8_2

scheme was applied in terms of the FCM, for example, in Düster et al. (2008). The advantage of this approach is that its implementation is straightforward. Unfortunately, if the geometry of the underlying problem is very complex, it might result in a high number of integration points. In cases like this, an adaptive procedure based on spacetrees might be more efficient. Spacetrees, i.e. quadtrees in 2D and octrees in 3D help to organize the spatial refinement for integration purposes by splitting the broken elements/cells into sub-cells which correspond to the leaves of the spacetree. Applications drawing on this approach can be found, for example, in Strouboulis et al. (2000) and Abedian et al. (2013a). This adaptive approach based on spacetrees can still result in a high number of integration points. Further attempts to reduce the number of integration points account for the geometry of the underlying problem by introducing more accurate mapping functions. To this end, Strouboulis et al. (2001) proposed a fast remeshing approach including the blending function method, resulting in curvilinear sub-cells which can represent curved boundaries more accurately. A similar approach was developed in Kudela et al. (2015) and applied within the finite cell method. An alternative approach based on a local meshing strategy was introduced by Loehnert et al. (2011) for the extended finite element method (XFEM). In this contribution, brick elements that are intersected by cracks were subdivided into tetrahedral elements on which the integration was performed. A different approach utilizing implicitly defined geometries suited for high-order accurate integration was suggested by Fries and Omerović (2016). In the proposed scheme, the integration is performed on local meshes with curved elements by applying carefully constructed mapping functions.

Lyness and Jespersen (1975), Lyness and Monegato (1977) used moment fitting equations to derive quadrature rules for triangles and regions exhibiting regular hexagonal symmetry. This approach can also be used to derive quadrature rules for discontinuous integrands defined on quite arbitrary domains. The main idea is to set up an equation system for each broken element/cell, ensuring that solving the system will lead to the desired quadrature rule. In the most general case, the nonlinear system of equations can be used to find the position of the integration points as well as the corresponding weights. If the position of the integration points is prescribed, then the nonlinear system turns into a linear one that can be solved more easily in order to find the desired weights. Mousavi and Sukumar (2010) used the moment fitting method to compute integrals appearing in the XFEM applied to solve two-dimensional problems of fracture mechanics. An extension of the moment fitting for the integration of discontinuous functions on irregular convex polygons and polyhedrons was presented by the same authors, see Mousavi and Sukumar (2011). Müller et al. (2013) extended the moment fitting to derive quadrature rules for elements/cells cut by the boundary of the domain with a geometry that was defined implicitly by means of level set functions. In Joulaian et al. (2016) the moment fitting was extended for an accurate integration of high-order discontinuous integrands appearing in the finite cell method. An attempt to optimize the position of the integration points within the moment fitting was presented in Hubrich et al. (2017). Although this optimization approach resulted in an improved quadrature rule, it was not worth the additional numerical effort. Employing the standard Gauss-Legendre points which might even be located in the fictitious domain turned out to be a good approach. In Hubrich and Düster (2019), a new idea to circumvent the solution of the linear system was suggested together with an adaptive version to improve the robustness of the moment fitting for strongly nonlinear problems of finite strain elastoplasticity.

A completely different approach to accurately integrate elements/cells with discontinuous integrands was proposed by Ventura (2006), Ventura and Benvenuti (2015). The key idea is to replace the discontinuous function by an equivalent polynomial which yields the same result of the integral. This idea was further extended to high-order discontinuous integrands appearing in the finite cell method in Abedian and Düster (2019). Choosing Legendre polynomials and utilizing their orthogonality helps to improve the accuracy and efficiency of the integration approach.

The *outline of this chapter* is as follows: In Section "The Finite Cell Method", we will summarize the finite cell method to set the stage for the discussion of quadrature schemes. In Section "Standard Numerical Integration Schemes", standard numerical integration schemes will be briefly summarized since they are an important ingredient of the advanced numerical integration methods presented in the remaining sections. Section "Adaptive Quadtree/Octree Quadrature Schemes" focuses on spacetrees which can be used for a fully automatic and adaptive integration of broken cells. In Section "Numerical Integration Based on Moment Fitting", we will present the moment fitting method as a promising approach to reduce the number of integration points. In Section "Numerical Examples", several examples including elastoplastic material behavior assuming small as well as finite strains will be investigated. Finally, we summarize this contribution with some concluding comments in Section "Conclusions".

## The Finite Cell Method

The finite cell method proposed by Parvizian et al. (2007), Düster et al. (2008) can be considered as a combination of the fictitious domain approach with high-order elements. In this section, we will briefly summarize the FCM in order to highlight the necessity for reliable and efficient quadrature schemes. A more detailed presentation and overview of the FCM can be found, for example, in Schillinger and Ruess (2015) and Düster et al. (2017).

### *Weak Formulation*

In order to explain the basic idea, let us consider a two-dimensional linear elastostatic problem defined on the physical domain $\Omega$, as depicted in Fig. 1. Dirichlet boundary conditions with prescribed displacements $\bar{\mathbf{u}}$ are defined on $\Gamma_D$, and Neumann boundary conditions accounting for tractions $\bar{\mathbf{t}}$ are applied on $\Gamma_N$. In order to simplify the discretization process, the physical domain $\Omega$ is embedded into a bigger

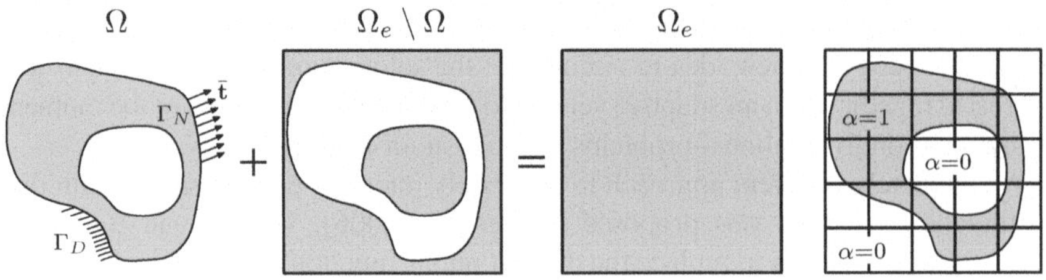


**Fig. 1** The basic idea of the FCM is to embed the physical domain $\Omega$ into a bigger domain $\Omega_e$ which can be easily meshed

domain $\Omega_e$ of a simpler shape, which can be easily meshed into a Cartesian grid consisting of quadrilateral elements. Next, we will consider the weak form of equilibrium which will be discretized to find an approximate solution for the problem. The corresponding bilinear form reads

$$\mathcal{B}_e^{\alpha}(\mathbf{u},\mathbf{v}) = \int_{\Omega_e} [\mathbf{L}\,\mathbf{v}]^T\, \mathbf{C}_e^{\alpha}\, [\mathbf{L}\,\mathbf{u}]\, d\Omega \tag{1}$$

where $\mathbf{u}$ denotes the displacement or trial function, $\mathbf{v}$ corresponds to the virtual displacement or test function, and $\mathbf{L}$ is the standard strain-displacement operator. Furthermore, $\mathbf{C}_e^{\alpha} = \alpha\mathbf{C}$ denotes the elasticity matrix defined on the embedding domain $\Omega_e$, where $\mathbf{C}$ represents the elasticity matrix defined on $\Omega$. The indicator function

$$\begin{aligned} \alpha(\mathbf{x}) &= 1.0 \quad \forall \mathbf{x} \in \Omega \\ 0.0 \leq \ \alpha(\mathbf{x}) &< 1.0 \quad \forall \mathbf{x} \in \Omega_e \setminus \Omega \end{aligned} \tag{2}$$

implicitly represents the geometry of the physical domain $\Omega$. Assuming that $\alpha(\mathbf{x}) = 0$ vanishes for points located in the fictitious domain, i.e. $\mathbf{x} \in \Omega_e \backslash \Omega$, the bilinear form (1) reads

$$\begin{aligned} \mathcal{B}_e^{\alpha}(\mathbf{u},\mathbf{v}) &= \int_{\Omega_e} [\mathbf{L}\,\mathbf{v}]^T\, \alpha\mathbf{C}\, [\mathbf{L}\,\mathbf{u}]\, d\Omega \\ &= \int_{\Omega} [\mathbf{L}\,\mathbf{v}]^T\, \mathbf{C}\, [\mathbf{L}\,\mathbf{u}]\, d\Omega + \int_{\Omega_e \backslash \Omega} [\mathbf{L}\,\mathbf{v}]^T\, \mathbf{0}\, [\mathbf{L}\,\mathbf{u}]\, d\Omega \\ &= \int_{\Omega} [\mathbf{L}\,\mathbf{v}]^T\, \mathbf{C}\, [\mathbf{L}\,\mathbf{u}]\, d\Omega = \mathcal{B}(\mathbf{u},\mathbf{v}). \end{aligned} \tag{3}$$

The load functional is defined as

$$\mathcal{F}_e^{\alpha}(\mathbf{v}) = \int\limits_{\Omega_e} \mathbf{v}^T (\alpha\,\mathbf{f})\, d\Omega + \int\limits_{\Gamma_N} \mathbf{v}^T \bar{\mathbf{t}}\, d\Gamma \quad (4)$$
$$= \int\limits_{\Omega} \mathbf{v}^T \mathbf{f}\, d\Omega + \int\limits_{\Gamma_N} \mathbf{v}^T \bar{\mathbf{t}}\, d\Gamma = \mathcal{F}(\mathbf{v})$$

accounting for volume loads $\mathbf{f}$ and prescribed tractions $\bar{\mathbf{t}}$ on the Neumann boundary of the physical domain $\Omega$. In summary, the weak form of the problem defined on the extended domain $\Omega_e$

$$\mathcal{B}_e^{\alpha}(\mathbf{u}, \mathbf{v}) = \mathcal{F}_e^{\alpha}(\mathbf{v}) \quad (5)$$

recovers the original problem $\mathcal{B}(\mathbf{u}, \mathbf{v}) = \mathcal{F}(\mathbf{v})$, when setting $\alpha(\mathbf{x}) = 0$ for points located in the fictitious domain $\Omega_f = \Omega_e \backslash \Omega$. In order to avoid conditioning problems within the solution process of the resulting linear equation system, $\alpha(\mathbf{x}) = \alpha_f = 10^{-q}\ \forall \mathbf{x} \in \Omega_e \backslash \Omega$ is set to very small (positive) values by choosing $q = 5, 6, \ldots, 15$ accordingly. In this way, the condition number can be improved by a small modification of the original problem. The approach can be interpreted as introducing artificial stiffness to the material of the fictitious domain. Provided that $\alpha_f$ is chosen small enough, very accurate solutions can be achieved, see Dauge et al. (2015).

## *Discretization of the Weak Formulation*

The discretization of the weak form on the extended domain can be carried out in a straightforward manner. The domain $\Omega_e$ is meshed into a Cartesian grid as depicted in Fig. 1. Since the resulting elements might be cut by the boundary of the physical domain, we denote them as finite cells in order to distinguish them from geometry-conforming finite elements. Due to the rectangular shape of the cells, the mapping function from the local to the global coordinate system is simple and results in a constant Jacobi matrix. Cells that are located completely outside of the physical domain can be discarded – which reduces the overall number of degrees of freedom and helps to improve the condition number of the resulting linear equation system. Using such a Cartesian or structured mesh simplifies the meshing process significantly. However, the burden of meshing is shifted towards the integration of the cells cut by the boundary of the physical domain. Since the integrand of the stiffness matrix of cut cells is discontinuous, a standard Gauss quadrature will not perform well. To discretize the displacement (trial) and virtual displacement (test) functions, we can apply different sets of shape functions. To this end, it is possible to use either low-order shape functions based on Lagrange polynomials or a high-order basis applying hierarchic shape functions as proposed by Szabó and Babuška (1991), Szabó et al. (2004) or NURBS as suggested by Hughes et al. (2005), Cottrell et al. (2009). In order to be able to achieve high convergence rates, we apply hierarchic shape functions to discretize the trial and test function

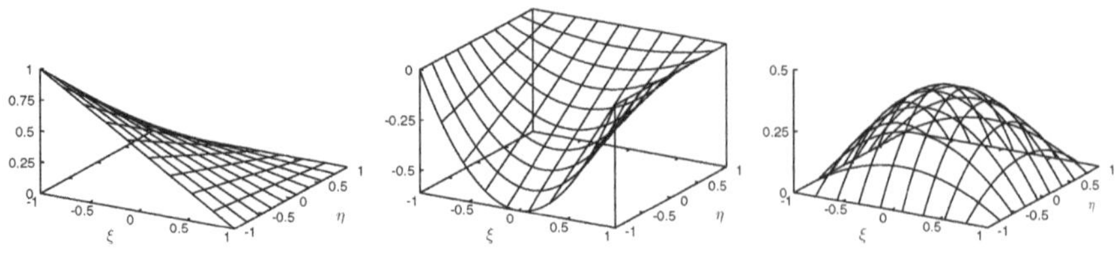


**Fig. 2** Hierarchic shape functions: nodal, edge, and bubble mode

$$\mathbf{u}^c = \mathbf{N}\mathbf{U}^c \tag{6}$$

$$\mathbf{v}^c = \mathbf{N}\mathbf{V}^c \tag{7}$$

for each cell $c$. The two-dimensional hierarchic shape functions for quadrilaterals can be classified into three groups: the nodal modes, the edge modes and the internal or bubble modes, see Szabó and Babuška (1991). Figure 2 shows a representative shape function for each group. The nodal modes are the standard bilinear Lagrange polynomials known from low-order elements. The edge modes are defined for each edge individually. Figure 2 depicts a quadratic mode for edge 1, where $\eta = -1$. The bubble or internal modes are purely local to the element/cell and can therefore be eliminated on element/cell level. Figure 2 shows a quadratic bubble/internal mode. For a more detailed description of the hierarchic shape functions and the related Ansatz spaces, the reader is referred to Szabó and Babuška (1991), Szabó et al. (2004). In Eqs. (6, 7) $\mathbf{N}$ represents the matrix containing the chosen shape functions and $\mathbf{U}^c$, $\mathbf{V}^c$ denote the displacement and virtual displacement vectors, respectively. Inserting (6, 7) into the bilinear form (5) results in

$$\begin{aligned} \mathcal{B}_e^{\alpha}(\mathbf{u}, \mathbf{v}) &= \sum_{c=1}^{n_c} \int_{\Omega^c} \left[\mathbf{L}\,\mathbf{v}^c\right]^T \alpha \mathbf{C} \left[\mathbf{L}\,\mathbf{u}^c\right] d\Omega \\ &= \sum_{c=1}^{n_c} \mathbf{V}^{c,T} \int_{\Omega^c} \mathbf{B}^T \alpha \mathbf{C}\,\mathbf{B}\, d\Omega\, \mathbf{U}^c = \sum_{c=1}^{n_c} \mathbf{V}^{c,T}\, \mathbf{k}^c\, \mathbf{U}^c \end{aligned} \tag{8}$$

where

$$\mathbf{k}^c = \int_{\Omega^c} \mathbf{B}^T \alpha \mathbf{C}\,\mathbf{B}\, d\Omega \tag{9}$$

represents the cell stiffness matrix and $\mathbf{B} = \mathbf{L}\mathbf{N}$ denotes the standard strain-displacement matrix. In a similar way, we can discretize the linear form (4) by inserting the test function (7) resulting in

$$
\begin{aligned}
\mathcal{F}_e^{\alpha}(\mathbf{v}) &= \sum_{c=1}^{n_c} \int_{\Omega^c} \mathbf{v}^{c,T} \alpha \mathbf{f}\, d\Omega + \int_{\Gamma_N^c} \mathbf{v}^{c,T} \bar{\mathbf{t}}\, d\Gamma \\
&= \sum_{c=1}^{n_c} \mathbf{V}^{c,T} \int_{\Omega^c} \mathbf{N}^T \alpha \mathbf{f}\, d\Omega + \mathbf{V}^{c,T} \int_{\Gamma_N^c} \mathbf{N}^T \bar{\mathbf{t}}\, d\Gamma \\
&= \sum_{c=1}^{n_c} \mathbf{V}^{c,T} \mathbf{f}^c
\end{aligned} \tag{10}
$$

where $\mathbf{f}^c$ denotes the load vector of the cell $c$. The second term in Eq. (10) accounts for the traction $\bar{\mathbf{t}}$ defined on $\Gamma_N$. Since the Neumann boundary $\Gamma_N$ will generally not coincide with the cell boundary, the integration has to be performed over a boundary that intersects the corresponding cell. Thus, the second term in (10) is to be understood in such a way that only those cells that are intersected by $\Gamma_N$ have to be considered. Assembling the stiffness matrices and load vectors of all cells

$$
\mathbf{K} = \mathop{\mathsf{A}}_{c=1}^{n_c} \mathbf{k}^c \;,\; \mathbf{F} = \mathop{\mathsf{A}}_{c=1}^{n_c} \mathbf{f}^c \tag{11}
$$

results in the overall linear equation system

$$
\mathbf{K}\mathbf{U} = \mathbf{F} \tag{12}
$$

where $\mathbf{K}$ denotes the global stiffness matrix, and $\mathbf{F}$ is the global load vector. Since the fictitious domain approach is different as compared to a standard finite element method, a few remarks shall be given:

- The geometry of the problem can be defined in various ways – by applying standard *geometric models* such as B-rep models, implicit geometry representations, voxel models, constructive solid geometry representations, or trimmed geometric models, see Düster et al. (2008, 2017). Common to all geometric models is the requirement that for a given point it has to be decided whether it is located in the physical domain. In this way, the indicator function $\alpha$ can be evaluated during the integration of the stiffness matrix and load vector.
- As already indicated in the discussion of the discretization of the load functional, the *treatment of boundary conditions* requires special attention. Neumann boundary conditions can be considered with the procedure presented in Düster et al. (2008). To this end, the surface on which the traction is defined needs to be parametrized such that the integration can be carried out. Dirichlet boundary conditions can be accounted for by applying the Nitsche method, which was applied in the context of FCM in Ruess et al. (2013), Kollmannsberger et al. (2014).
- As the integrand of the stiffness matrix and the load vector involve the indicator function $\alpha$, there is a discontinuity. Thus, a standard Gauss quadrature will not per-

form well and *special integration techniques* are required, which will be addressed in this chapter in more detail.
- In the case of a smooth solution, a pure increase of the polynomial degree of the shape functions will lead to an exponential convergence. In most cases, however, the problem will include discontinuities and singularities. Therefore, a *local enrichment* has to be performed. Several strategies are available. One can take advantage of the partition of unity method (PUM) (Melenk and Babuška 1996) in order to locally enrich the displacement field with tailored shape functions, see, for example, Joulaian and Düster (2013). Alternatively, a more general approach applying an $hp$-type of refinement as proposed by Zander et al. (2016) can be applied.
- The *solution of the linear equation system* (12) might require special attention as well, since – due to the application of the fictitious domain concept – the resulting condition number can be very high. To overcome this problem, one can either apply direct solvers or use specialized preconditioners in combination with iterative solvers, see Heinze et al. (2014), Jomo et al. (2017).

In order to obtain an impression of the efficiency which can be achieved with the FCM, we consider a first numerical example.

## *A Simple Linear Elastostatic Example*

As a first example, we consider a perforated plate under plane stress conditions. Figure 3 shows the geometry and boundary conditions of the plate. The dimensions of the plate are $L = B = 4$ mm and the radius of the hole is $R = 1$ mm. An isotropic, linear elastic material behavior with a Young's modulus of $E = 206{,}900$ MPa and Poisson's ratio of $\nu = 0.29$ is assumed. A traction of 100 MPa acts on the upper edge of the plate, as depicted in Fig. 3. The plate is meshed into $2 \times 2$ finite cells on which high-order hierarchic shape functions suited for a $p$-extension are used. The geometry of the problem is taken into account during the integration of the stiffness matrix of the cells. To this end, an adaptive integration based on a quadtree is applied, which will be explained in more detail in Section "Adaptive Quadtree/Octree Quadrature Schemes". Next, we study the convergence in terms of the relative error in energy norm

$$(e_r)_{E(\Omega_e)} = \sqrt{\frac{|U_{FCM} - U_{EX}|}{U_{EX}}} \cdot 100\,[\%] \tag{13}$$

where $U_{EX}, U_{FCM}$ represent the exact and the approximated strain energy $U = \frac{1}{2}\mathcal{B}(\mathbf{u}, \mathbf{u})$ of the two-dimensional plate. Since the exact solution of the problem is not known, it is replaced by an overkill solution that is accurate enough to compute the error of the FCM approximation. The convergence of the FCM utilizing a $p$-extension with $p = 1, 2, 3, \ldots$ is plotted in Fig. 3 for different levels of accuracy regarding the adaptive integration based on a quadtree. In addition to the FCM,

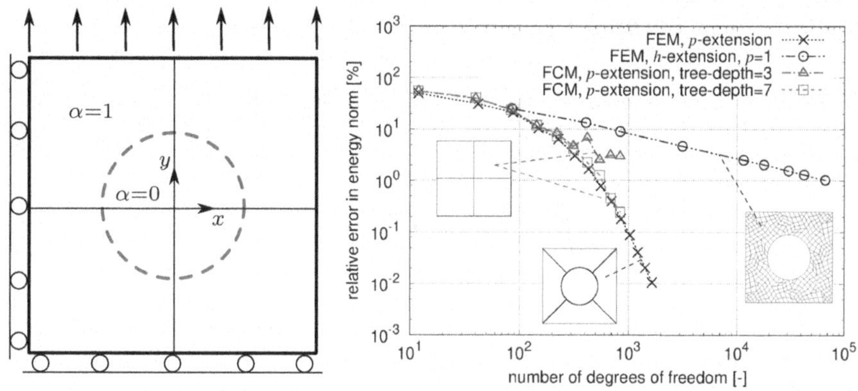


**Fig. 3** Perforated square plate with boundary conditions and convergence of error in energy norm for different methods

results of the FEM computations are given, including an $h$-extension with bilinear quadrilateral elements as well as a $p$-extension on a mesh with 4 quadrilaterals where the geometry is represented exactly by means of the blending function method, see Szabó and Babuška (1991). From the comparison, it is evident that the high-order approaches achieve an exponential convergence, in contrast to the uniform mesh refinement with $p = 1$ which results in an algebraic convergence rate. The FCM closely follows the convergence of the $p$-FEM approach, provided that the integration is carried out accurately enough. A tree-depth level of 3, see Fig. 4, results in an integration error that deteriorates the exponential convergence. However, by choosing a more accurate integration based on a tree-depth level equal to 7, the high convergence rate is maintained, resulting in a very efficient discretization.

In summary, this example demonstrates the appealing properties of the FCM. High convergence rates can be achieved, provided that the integration of the cell matrices is carried out accurately enough. In the remainder of this chapter, we will therefore focus on different quadrature schemes that can be applied in the FCM.

## Standard Numerical Integration Schemes

In this section, we will briefly summarize two standard numerical integration methods, the Newton-Cotes formulas and the Gauss-Legendre quadrature. The main purpose is to recall the underlying ideas of these methods and their respective accuracy. The methods will be presented for one-dimensional integrals.

## *Newton-Cotes Formulas ($p_q = n - 1$)*

The Newton-Cotes formulas are interpolatory quadrature rules to compute the integral

$$\int_a^b f(x)\,dx \approx \sum_{i=1}^{n} w_i\, f(x_i) \tag{14}$$

numerically by a weighted sum of the function $f(x)$ evaluated at a set of $n \geq 2$ equidistant points $\{x_i\}$. The underlying idea is to interpolate $f(x)$ at the $n$ points with a polynomial $p_{n-1}(x)$ of degree $p_q = n - 1$ and to integrate this polynomial

$$\int_a^b f(x)\,dx \approx \int_a^b p_{n-1}(x)\,dx. \tag{15}$$

To this end, the integration domain is sub-divided into intervals of length

$$h = \frac{b-a}{n-1} \tag{16}$$

by utilizing an equidistant distribution of $n$ points

$$x_i = a + (i-1)\,h\,, \quad i = 1, \ldots, n. \tag{17}$$

Using the $n$ points, a polynomial of degree $p_q = n - 1$ is defined, such that

$$p_{n-1}(x_i) = f(x_i)\,, \quad i = 1, \ldots, n \tag{18}$$

holds. In order to set up the interpolation, it is convenient to construct a set of $n$ Lagrange polynomials of degree $p_q = n - 1$

$$l_i^{n-1}(x) = \prod_{\substack{j=1\\ j\neq i}}^{n} \frac{x - x_j}{x_i - x_j}, \quad i = 1, \ldots, n \tag{19}$$

through the $n$ equidistant points $\{x_i\}$. The basic properties of the Lagrange polynomials are $l_i^{n-1}(x_j) = \delta_{ij}$ and the partition of unity $\sum_{i=1}^{n} l_i^{n-1}(x) = 1$. The first property allows to directly represent the interpolation as

$$p_{n-1}(x) = \sum_{i=1}^{n} l_i^{n-1}(x)\, f(x_i). \tag{20}$$

Integrating the interpolatory polynomial yields

$$\int_a^b p_{n-1}(x)dx = \sum_{i=1}^{n} \int_a^b l_i^{n-1}(x)dx\, f(x_i). \tag{21}$$

The antiderivative of the Lagrange polynomials in Eq. (21) can be interpreted as the weights

$$w_i = \int_a^b l_i^{n-1}(x)dx\,, \quad i = 1, \ldots, n \tag{22}$$

required to approximately compute the integral

$$\int_a^b f(x)\,dx \approx \int_a^b p_{n-1}(x)\,dx = \sum_{i=1}^{n} w_i f(x_i). \tag{23}$$

To give two examples, we consider the Newton-Cotes formulas for two and three points. Choosing $n = 2$ points with $x_1 = a$, $x_2 = b$ and $h = (b - a)$ results in the well-known *trapezoidal rule*

$$\int_{x_1}^{x_2} f(x)\,dx \approx \frac{h}{2}(f(x_1) + f(x_2)), \tag{24}$$

which allows to integrate linear ($p_{\mathrm{q}} = n - 1 = 1$) polynomials exactly.

Applying $n = 3$ points with $x_1 = a$, $x_2 = \frac{a+b}{2}$, $x_3 = b$ and $h = \frac{b-a}{2}$ yields the *Simpson rule*

$$\int_{x_1}^{x_3} f(x)\,dx \approx \frac{h}{3}(f(x_1) + 4f(x_2) + f(x_3)). \tag{25}$$

It is interesting to note that the Simpson rule does not only allow to integrate polynomials of degree $p_{\mathrm{q}} = 2$ exactly, but also polynomials of degree $p_{\mathrm{q}} = 3$. In this case, the points $\{x_i\}$ and weights $\{w_i\}$ of the Simpson rule thus coincide with the Gauss-Lobatto rule, which is known to be exact up to polynomials of order $p_{\mathrm{q}} = 2n - 3 = 3$. The integration points and weights of the Gauss-Lobatto rule are listed in Table 4 in Section "Appendix". Setting $x_1 = a = -1$ and $x_3 = b = 1$ results in $x_2 = 0$ and $h = 1$, and, consequently, the weights are $w_1 = \frac{1}{3}$, $w_2 = \frac{4}{3}$, $w_3 = \frac{1}{3}$. Comparing these points and weights for $n = 3$ with Table 4 reveals that in this case the Newton-Cotes formula coincides with the Gauss-Lobatto quadrature which confirms the order of $p_\mathrm{q} = 2n - 3 = 3$. The fact that an additional order of accuracy is gained explains the popularity of the Simpson rule.

### *Gauss-Legendre Quadrature ($p_\mathrm{q} = 2n - 1$)*

The most frequently used quadrature rule in finite element methods is probably the Gauss-Legendre quadrature. It allows to integrate with $n$ quadrature points polynomials

$$\int_{-1}^{1} p_{2n-1}(\xi)\, dx = \sum_{i=1}^{n} w_i\, p_{2n-1}(\xi_i) \tag{26}$$

of order $p_\mathrm{q} = 2n - 1$ exactly. To find the corresponding Gauss-Legendre points $\{\xi_i\}$ and weights $\{w_i\}$, a polynomial

$$p_{2n-1}(\xi) = a_0 + a_1\xi + a_2\xi^2 + \cdots + a_{2n-1}\xi^{2n-1} \tag{27}$$

of order $p_\mathrm{q} = 2n - 1$ is integrated

$$\int_{-1}^{1} p_{2n-1}(\xi)\, d\xi = a_0 \int_{-1}^{1} d\xi + a_1 \int_{-1}^{1} \xi d\xi + \cdots + a_{2n-1} \int_{-1}^{1} \xi^{2n-1} d\xi \tag{28}$$

term by term. For each term in (28), we can apply the quadrature rule – resulting in:

$$\int_{-1}^{1} d\xi = \sum_{i=1}^{n} w_i = w_1 + w_2 + \cdots + w_n \tag{29}$$

$$\int_{-1}^{1} \xi\, d\xi = \sum_{i=1}^{n} w_i\xi_i = w_1\xi_1 + w_2\xi_2 + \cdots + w_n\xi_n \tag{30}$$

$$\vdots \tag{31}$$

$$\int_{-1}^{1} \xi^{2n-1}\, d\xi = \sum_{i=1}^{n} w_i\xi_i^{2n-1} = w_1\xi_1^{2n-1} + w_2\xi_2^{2n-1} + \cdots + w_n\xi_n^{2n-1} \tag{32}$$

The left-hand side, i.e. the integral of the monomials, can be computed analytically, so that we obtain the nonlinear system

$$0 = \sum_{i=1}^{n} w_i\, \xi_i^{\beta},\ \beta = 1, 3, 5, \ldots, 2n-1 \tag{33}$$

$$\frac{2}{\beta+1} = \sum_{i=1}^{n} w_i\, \xi_i^{\beta},\ \beta = 0, 2, 4, \ldots, 2n-2 \tag{34}$$

of $2n$ equations for $n$ unknown abscissas $\{\xi_i\}$ and $n$ unknown weights $\{w_i\}$. The nonlinear system can be solved by applying, for example, the Newton-Raphson method. However, since the Gauss-Legendre points $\{\xi_i, i = 1, \ldots, n\}$ are known to be the roots of the Legendre polynomial of order $n$ (see, for example, Schwarz 2004), the nonlinear system turns into a linear one. Algorithms and implementations to compute the Gauss-Legendre points and weights are available, for example, in Press et al. (2002). In Section "Appendix", Table 5 lists the Gauss-Legendre points and weights for $n = 1, \ldots, 10$.

## Adaptive Quadtree/Octree Quadrature Schemes

In this section, we will discuss the adaptive numerical integration of discontinuous integrands by a combination of spacetrees (Samet 1990) and Gaussian quadrature. Spacetrees help to organize the adaptive spatial refinement in order to capture the discontinuity of the integrand more efficiently than a uniform refinement, see Fig. 4. The cells that are cut by the boundary of the domain or by some internal interfaces are hierarchically sub-divided into sub-cells which correspond to the leaves of the spacetree. In two-dimensional situations, we choose the quadtree, whereas an octree can be applied for three-dimensional problems, see Abedian et al. (2013a). Alternatively, an axis-aligned binary spactree, i.e. a k-d tree can be used instead, see, for example, Gnegel (2019).

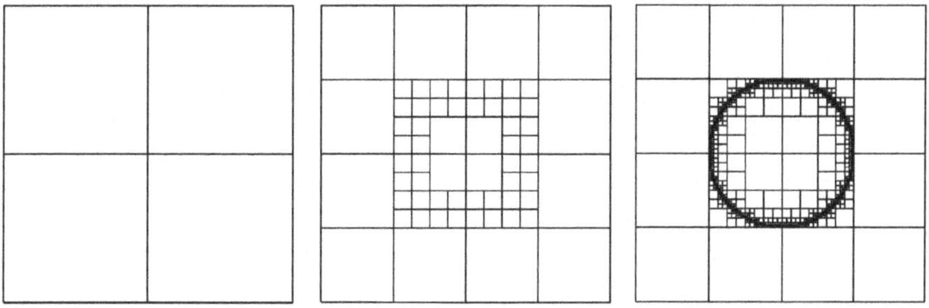

**Fig. 4** Adaptive quadtree refinement of a perforated square plate. Tree-depth level: 0 (left), 3 (middle), and 7 (right)

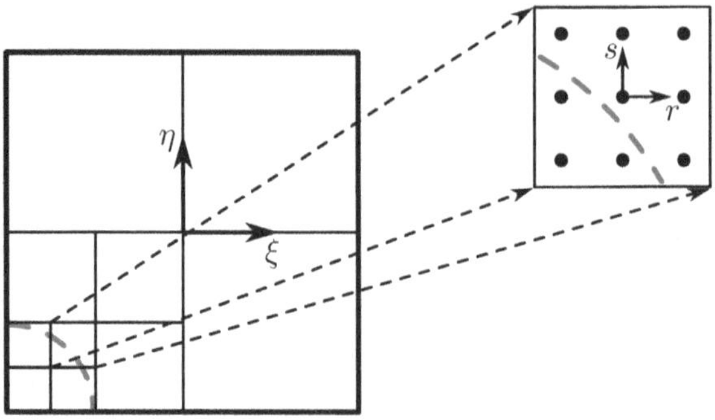


**Fig. 5** Adaptive quadtree integration. On the leaves of the quadtree, i.e. the sub-cells of a cut cell, a Gauss quadrature is performed

The main idea of the adaptive approach based on spacetrees is sketched in Fig. 4 for the two-dimensional problem of the perforated plate presented in Section "A Simple Linear Elastostatic Example". The tree-depth level 0 corresponds to the FCM mesh consisting of 2 × 2 cells which are intersected by the circular hole. Starting from this FCM mesh, each cell that is cut by the circle is recursively refined into four sub-cells. The adaptive refinement process is continued until a pre-defined maximum tree-depth level is reached – which is set to 7 in this example. The adaptive refinement gradually leads to a better representation of the geometry of the circle. On each leaf of the quadtree, which corresponds to a sub-cell, a Gaussian quadrature is applied. The situation is schematically depicted in Fig. 5 where the integration points applied for one sub-cell are indicated.

The integration of the stiffness matrix of cell $c$

$$\mathbf{k}^c = \int_{\Omega^c} \mathbf{B}^T \alpha \mathbf{C}\,\mathbf{B}\, d\Omega = \int_{-1}^{1}\int_{-1}^{1} \mathbf{B}^T(\boldsymbol{\xi})\alpha(\mathbf{x}(\boldsymbol{\xi}))\mathbf{C}\,\mathbf{B}(\boldsymbol{\xi})\, t\, \det\mathbf{J}^c\, d\xi d\eta \tag{35}$$

$$= \sum_{sc=1}^{n_{sc}} \int_{-1}^{1}\int_{-1}^{1} \mathbf{B}^T(\boldsymbol{\xi}(\mathbf{r}))\, \alpha(\mathbf{x}(\boldsymbol{\xi}(\mathbf{r})))\, \mathbf{C}\,\mathbf{B}(\boldsymbol{\xi}(\mathbf{r}))\, t\, \det\mathbf{J}^c\, \det\tilde{\mathbf{J}}^{c,sc}\, drds$$

$$= \sum_{sc=1}^{n_{sc}} \sum_{i=1}^{n_G} \mathbf{B}^T(\boldsymbol{\xi}(\mathbf{r}_i))\, \alpha(\mathbf{x}(\boldsymbol{\xi}(\mathbf{r}_i)))\, \mathbf{C}\,\mathbf{B}(\boldsymbol{\xi}(\mathbf{r}_i))\, t\, \det\mathbf{J}^c\, \det\tilde{\mathbf{J}}^{c,sc} w_i$$

with thickness $t$ is carried out by applying a Gaussian quadrature with $n_G$ integration points on each of the $n_{sc}$ sub-cells. In (35), $\boldsymbol{\xi} = [\xi, \eta]^T$, $\mathbf{r} = [r, s]^T$ represent the local coordinates of the cell $c$ and sub-cell $sc$, respectively. The Jacobi matrix $\mathbf{J}^c$ corresponds to the mapping function $\mathbf{x}(\boldsymbol{\xi})$ that relates the local coordinates $\boldsymbol{\xi} = [\xi, \eta]^T$ of cell $c$ to the global coordinate system $\mathbf{x} = [x, y]^T$. Furthermore, $\tilde{\mathbf{J}}^{c,sc}$ represents the Jacobi matrix of the mapping function $\boldsymbol{\xi}(\mathbf{r})$, relating the local coordinate system of the sub-cell $sc$ to the local coordinate system of the cell $c$. Since the two mapping functions $\mathbf{x}(\boldsymbol{\xi})$ and $\boldsymbol{\xi}(\mathbf{r})$ are related to rectangular cells and sub-cells, the associated Jacobi matrices are constant. The Gaussian quadrature is carried out in the local coor-

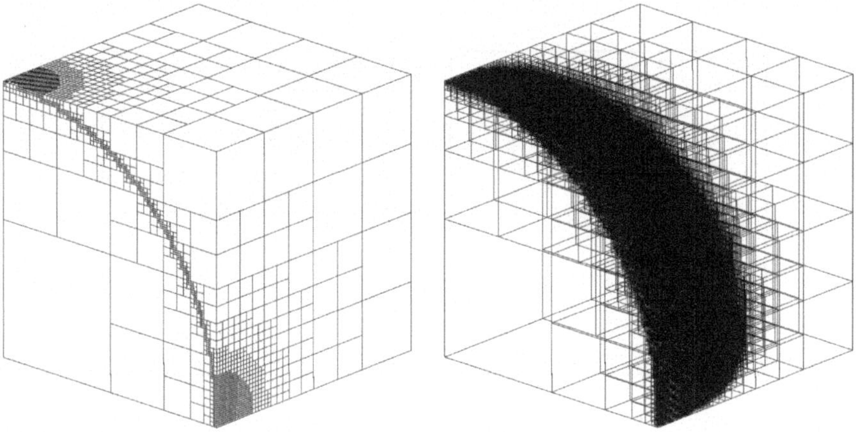

**Fig. 6** Adaptive octree refinement for a cell intersected by a sphere. The adaptive refinement is carried out up to a tree-depth level of 7

dinate system of the sub-cell by applying $\{\mathbf{r}_i = [r_i, s_i]^T, i = 1, \ldots, n_G\}$ integration points with the associated weights $\{w_i, i = 1, \ldots, n_G\}$, which are obtained by the straightforward extension of the 1D quadrature presented in Section "Gauss-Legendre Quadrature ($p_q = 2n - 1$)" to the 2D situation.

The extension of the presented adaptive integration scheme to 3D problems can be implemented by applying an octree to locally refine broken (hexahedral) cells. A first example is given in Fig. 6, where a cell intersected by a sphere is subdivided into sub-cells up to a tree-depth level of 7. The resulting subdivision of the cell into sub-cells can be used to perform an adaptive Gauss integration as explained previously for the two-dimensional case.

## Numerical Integration Based on Moment Fitting

A short literature overview over the moment fitting is presented in Section "Introduction". In this section, we will summarize the moment fitting approach to derive quadrature rules for broken cells.

### *Moment Fitting Equations*

The main idea of the moment fitting approach is to set up an individual quadrature rule for each broken cell. Consider, for example, the situation depicted in Fig. 7a where the task is to integrate a given function over the domain $\Omega_A$, which corresponds to the part of the physical domain that is located within the broken cell under consideration.

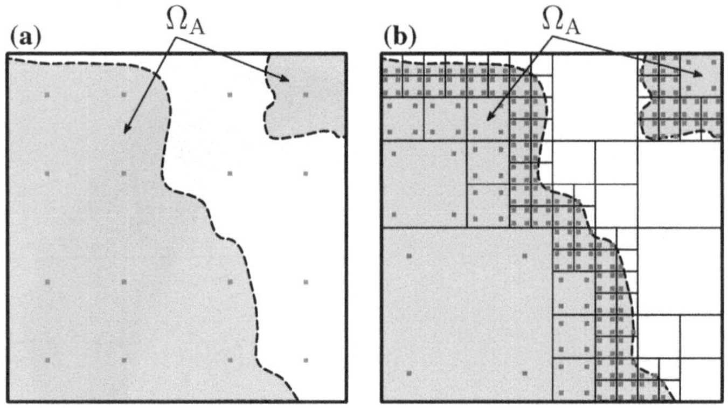


**Fig. 7** **a** Quadrature points of the moment fitting with $p_q = 3$. **b** Quadrature points of the adaptive integration with $k = 4$ and $p_q = 3$

Note that the geometry of the domain over which we want to integrate can be defined by means of the indicator function (2) and can be therefore quite arbitrary. The example under consideration even involves a disconnected domain – which does not cause any problems with the moment fitting approach. To derive a quadrature rule for such a broken cell, an equation system

$$\sum_{i=1}^{n} f_j(\boldsymbol{\xi}_i)\, w_i = \int_{\Omega_A} f_j(\boldsymbol{\xi})\, \mathrm{d}\Omega, \quad j = 1, \ldots, m \tag{36}$$

is set up, where $\boldsymbol{\xi}_i$ and $w_i$ are the positions and the weights of the $n$ integration points, respectively. A set of $m$ linearly independent basis functions $f_j(\boldsymbol{\xi})$ is chosen to define the moment fitting equations, which can be represented in matrix notation as

$$\begin{bmatrix} f_1(\boldsymbol{\xi}_1) & \cdots & f_1(\boldsymbol{\xi}_n) \\ \vdots & \ddots & \vdots \\ f_m(\boldsymbol{\xi}_1) & \cdots & f_m(\boldsymbol{\xi}_n) \end{bmatrix} \begin{Bmatrix} w_1 \\ \vdots \\ w_n \end{Bmatrix} = \begin{Bmatrix} \int_{\Omega_A} f_1(\boldsymbol{\xi})\, \mathrm{d}\Omega \\ \vdots \\ \int_{\Omega_A} f_m(\boldsymbol{\xi})\, \mathrm{d}\Omega \end{Bmatrix} \tag{37}$$

or in symbolic notation as

$$\boldsymbol{A}\boldsymbol{w} = \boldsymbol{b}. \tag{38}$$

In (38), $\boldsymbol{A}$ denotes the coefficient matrix, $\boldsymbol{w}$ the vector of the unknown weights, and $\boldsymbol{b}$ represents the moments which are defined as the integrals of the basis functions $f_j(\boldsymbol{\xi})$. The equation system (38) is nonlinear with respect to the unknown position of the integration points. Therefore, a nonlinear solution approach such as, for example, the Newton-Raphson method has to be used to find an approximate solution. In Hubrich et al. (2017), an optimization algorithm was applied to minimize the norm of the residual of the moment fitting equations. In this way, efficient quadrature rules can be derived. However, the required numerical effort is increased when applying such an optimization approach. Alternatively, we can pre-select the position of the integration points. In Hubrich et al. (2017), it was shown that a suitable approach is to employ standard Gauss-Legendre points, which can also be located in the fictitious domain, see Fig. 7a. Thus, we choose the standard Gauss-Legendre points and define a set of $m$ basis functions, such that in $d$ dimensions we have

$$n = m = \left(p_{\mathrm{q}} + 1\right)^{d}, \tag{39}$$

i.e. a quadrature rule of order $p_{\mathrm{q}}$ with the same number of points and basis functions. Since the integration points are pre-defined, the nonlinear moment fitting equations turn into a linear system with $m$ equations for the $n$ unknown weights. To further accelerate the solution of the linear system, the basis functions are defined as Lagrange polynomials through the Gauss-Legendre points. In order to simplify the notation, we first consider the one-dimensional case, i.e. $d = 1$, and define the basis functions as

$$f_j\left(\xi\right) = l_j\left(\xi\right) \quad \text{with} \quad l_j\left(\xi\right) = \prod_{\substack{k=1\\k\neq j}}^{p_{\mathrm{q}}+1} \frac{\xi - \xi_k^{\mathrm{GL}}}{\xi_j^{\mathrm{GL}} - \xi_k^{\mathrm{GL}}} \tag{40}$$

where $l_j\left(\xi\right)$ denote the one-dimensional Lagrange polynomials and $\xi_j^{\mathrm{GL}}$ as well as $\xi_k^{\mathrm{GL}}$ are the coordinates of the standard one-dimensional Gauss-Legendre points. Recalling that $n = m = \left(p_{\mathrm{q}} + 1\right)^{d}$, the $n$ equations for the unknown weights read

$$\sum_{i=1}^{p_{\mathrm{q}}+1} l_j\left(\xi_i^{\mathrm{GL}}\right) w_i = \int_{\Omega_{\mathrm{A}}} l_j\left(\xi\right) \,\mathrm{d}\Omega, \quad j = 1, \ldots, p_{\mathrm{q}} + 1. \tag{41}$$

Thanks to the fact that Lagrange polynomials through the Gauss-Legendre points are selected as basis functions, the Kronecker delta property holds and, therefore, the coefficient matrix

$$A_{ji} = l_j(\xi_i^{\mathrm{GL}}) = \delta_{ji} \tag{42}$$

results in the identity. This idea was first proposed in Hubrich and Düster (2018, 2019) and allows to reduce the computational overhead of the moment fitting method significantly, since the effort of the solution of the equation system is avoided completely. Furthermore, in Düster and Allix (2019) it was demonstrated that the basis functions can be also chosen to include arbitrary types of discontinuities in an efficient manner preserving the diagonal structure of the coefficient matrix $\boldsymbol{A}$. To determine the $\left(p_{\mathrm{q}} + 1\right)$ one-dimensional moment fitting weights

$$w_i = \int_{\Omega_{\mathrm{A}}} l_i\left(\xi\right) \,\mathrm{d}\Omega\,, \quad i = 1, \ldots, p_{\mathrm{q}} + 1 \tag{43}$$

denoted also as

$$\begin{Bmatrix} w_1 \\ \vdots \\ w_{p_q+1} \end{Bmatrix} = \begin{Bmatrix} \int_{\Omega_A} l_1(\xi)\,\mathrm{d}\xi \\ \vdots \\ \int_{\Omega_A} l_{p_q+1}(\xi)\,\mathrm{d}\xi \end{Bmatrix}, \tag{44}$$

requires that the basis functions have to be integrated over the domain $\Omega_A$.

It is interesting to note that in the case where the cell of interest is *not* broken, i.e. ($\Omega_A = [-1,\ 1]$), the resulting weights coincide with those of the standard Gauss-Legendre quadrature. Furthermore, it can be observed that Eq. (43) is very similar to the definition of the weights (22) of the Newton-Cotes formula. In both approaches, Lagrange polynomials have to be integrated to find the weights of the related quadrature rule. The proposed moment fitting method can therefore be interpreted as a generalization of the Newton-Cotes formula, where Lagrange polynomials defined on Gauss-Legendre points are applied to derive a quadrature rule for broken cells. The utilization of Gauss-Legendre abscissae instead of an equidistant distribution of points allows to increase the accuracy of the moment fitting method. Considering the case where the cell is not broken, the moment fitting method yields the accuracy of the Gauss-Legendre quadrature where with $n$ points a polynomial of degree $p_q = 2n - 1$ can be integrated exactly. Since we use the standard Gauss-Legendre quadrature for non-broken cells anyway, this observation contributes to the understanding of the moment fitting rather than suggesting a new quadrature rule for non-broken cells. In the more interesting case of broken cells, the accuracy of the moment fitting approach reduces to that of the Newton-Cotes formula, i.e. $p_q = n - 1$. In summary, by using Lagrange polynomials through pre-defined quadrature points as basis functions, we can avoid having to solve a system to find the corresponding weights. This diagonalization of the coefficient matrix $\boldsymbol{A}$ in (38) does not only reduce the numerical effort significantly, but it also helps to achieve the desired accuracy of the weights, which can deteriorate when using a set of integration points and basis functions that do not yield a diagonal coefficient matrix. In Hubrich et al. (2017), the influence of the distribution of quadrature points on the condition number of $\boldsymbol{A}$ was investigated in detail. It turned out that selecting Gauss-Legendre points – which might be even located in the fictitious domain – improves the condition number of $\boldsymbol{A}$ and thereby also the quality of the weights.

An extension of the moment fitting to the 2D and 3D case is straightforward. To this end, the basis functions $f_j(\boldsymbol{\xi})$ are defined as the tensor product of the $(p_q + 1)$ one-dimensional Lagrange basis functions. Likewise, we define the quadrature points as the tensor product of the $(p_q + 1)$ one-dimensional Gauss-Legendre points. In the three-dimensional case ($d = 3$), the set of basis functions therefore reads

$$\mathcal{F} = \left\{ l_r(\xi) l_s(\eta) l_t(\zeta),\ r, s, t = 1, \ldots, p_q + 1 \right\} \tag{45}$$

and the quadrature points are defined as

$$\mathcal{X} = \left\{ \left[\xi_r^{\mathrm{GL}},\ \eta_s^{\mathrm{GL}},\ \zeta_t^{\mathrm{GL}}\right],\ r, s, t = 1, \ldots, p_q + 1 \right\}. \tag{46}$$

Again, thanks to the selection of basis functions defined as Lagrange polynomials through the integration points, the weights

$$w_i = \int_{\Omega_\mathrm{A}} L_i(\boldsymbol{\xi})\,\mathrm{d}\Omega\,, \quad i = 1, \ldots, (p_\mathrm{q}+1)^3 \tag{47}$$

can be computed without the necessity of solving a system. In (47), $L_i(\boldsymbol{\xi})$ denotes the product of the one-dimensional Lagrange polynomials (45), where the three indices $r, s, t$ have been replaced by index $i$ to simplify the notation. In order to evaluate the integral (47), there are several strategies available, see, for example, Joulaian et al. (2016). One possibility is to convert the volume integral into a surface integral. Alternatively, the volume integral can be computed by means of an adaptive integration based on spacetrees, as explained in Section "Adaptive Quadtree/Octree Quadrature Schemes". This approach is also sketched in Fig. 7b, where a quadtree with a tree-depth level $k = 4$ is used in combination with a Gaussian quadrature of order $p_\mathrm{q} = 3$ to compute the right-hand side of the moment fitting equation system. Since the integration only involves an evaluation of scalar-valued functions, the adaptive integration based on spacetrees is not expensive, and it is to be seen as an attractive approach since it can be performed in a fully automatic fashion.

### *Adaptive Moment Fitting*

The variant of the moment fitting presented in the preceding section is quite efficient and performs very well for linear problems of structural mechanics, see Hubrich et al. (2017). However, there are situations, especially in nonlinear problems, where the moment fitting method yields a less stable integration scheme. Consider, for example, problems of elastoplasticity, where the moment fitting method turned out to be less robust as compared to the adaptive Gauss integration based on spacetrees. In situations like this, it was observed that the Newton-Raphson method, applied to solve the algebraic set of nonlinear equations resulting from the discretization of the linearized weak form, failed more often when the moment fitting method was used, see Hubrich and Düster (2019). Elastoplastic material models introduce an additional difficulty which is attributed to the evolving elastoplastic front representing a material interface intersecting the cells. Such a material interface introduces a discontinuity that is more difficult to handle than the one related to a boundary of the domain intersecting a cell. Material interfaces require a special treatment of the trial and test function of the spatial discretization as well as of the integration based on moment fitting. The robustness of the overall nonlinear solution process based on the finite cell method becomes even more fragile when considering problems of finite strains and cells that are filled with a small fraction of material only, see Schillinger et al. (2012), Taghipour et al. (2018), Hubrich and Düster (2019). Then, the deformation of broken cells can become quite large, leading to a failure of the nonlinear computation.

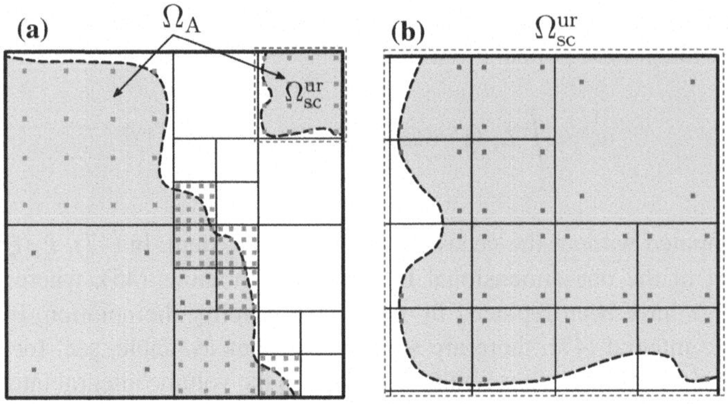


**Fig. 8** **a** Quadrature points of the adaptive moment fitting with $p_q = 3$ and $k_a = 1, 2, 3$. **b** Quadrature points of the adaptive quadtree with $p_q = 3$ – used for the computation of the moment fitting weights of sub-cell $\Omega_{sc}^{ur}$

A simple approach to remedy the aforementioned difficulties is to combine the moment fitting method with the adaptive Gauss quadrature based on spacetrees. The underlying idea of the *adaptive moment fitting* is sketched in Fig. 8 for the same integration problem as discussed in Fig. 7a. The cell under consideration is subdivided by means of a quadtree (or an octree in 3D) up to a certain tree-depth level. We distinguish two different tree-depth levels, termed $k_a$ and $k$. Whereas $k_a$ denotes the tree-depth level used for the adaptive moment fitting, $k$ represents the level of refinement used for the integration of the moments, i.e. the computation of the right-hand side of the moment fitting equation system. Both levels can be chosen independently, see Fig. 8. The decision whether a broken cell or sub-cell has to be (further) subdivided is based on the volume fraction of its integration domain. Based on numerical experiments, it turned out to be of advantage to set different tolerances for the volume fractions for the different tree-depth levels. For $k_a = 0$, we set the tolerance to 0.85 – and to 0.7 as a tolerance for the volume fraction for $k_a = 1, 2$, which means that cells/sub-cells with a volume fraction lower than the defined tolerances are further refined up to a maximum level of $k_a = 3$. This procedure will result in situations where broken cells/sub-cells are integrated either with a standard Gauss quadrature or by applying the moment fitting approach. In Fig. 8a, we can find sub-cells of different refinement levels ($k_a \leq 3$) that are either integrated with a standard Gauss-Legendre quadrature (green points) or by means of the moment fitting approach (red points). Non-broken sub-cells are integrated with a standard Gauss-Legendre quadrature, whereas broken sub-cells are treated with the moment fitting method. For both schemes, the number of integration points is adjusted to achieve an integration order of $p_q = 3$. To compute the moments accurately, an adaptive Gauss quadrature is applied with a tree-depth level of $k = 4$ and $p_q = 3$ on each sub-cell. Considering, for example, the sub-cell denoted as $\Omega_{sc}^{ur}$ in Fig. 8a, its corresponding refinement for the computation of the moments is depicted in Fig. 8b. Since the sub-cell corresponds to tree-depth level 2, two additional refinements are required to achieve a tree-depth level 4. A Gauss-Legendre quadrature is applied on each of the sub-cells to compute the moments.

If we compare Fig. 8 with Fig. 7, we can clearly see that the adaptive moment fitting approach requires more integration points than the moment fitting method, but less than the adaptive Gauss quadrature. As to be demonstrated by some numerical examples in Section "Numerical Examples", the additional integration points help to increase the robustness of the numerical integration scheme.

## *Treatment of Integration Points in the Fictitious Domain*

In the moment fitting method and also in its adaptive version, some of the integration points will be located in the fictitious domain, see Fig. 7a and Fig. 8a. Therefore, the question arises which constitutive model to choose for these points. Based on extensive numerical experiments, it turned out to be of advantage to choose the same material model for the points in the fictitious domain as for those located in the physical domain. If we follow this approach, the results of the nonlinear computations are very similar to those of an approach based on an adaptive Gauss integration scheme, see Abedian et al. (2013a).

In order to improve the condition number of the resulting stiffness matrix and to stabilize the FCM, it is of advantage to add a little bit of stiffness to the fictitious domain. This can be accomplished by introducing additional integration points to the cell, from which only those located in the fictitious domain are considered, see, for example, Abedian et al. (2013b, 2014). The situation is depicted in Fig. 9, where the yellow points represent the additional points introduced for stabilization purposes only. For these additional standard Gauss-Legendre points, the original weights are multiplied with a very small number – such as $\alpha = 10^{-q}$, where $q$ is in the range of $5, \ldots, 12$. The material model assumed for these additional points is either a linear

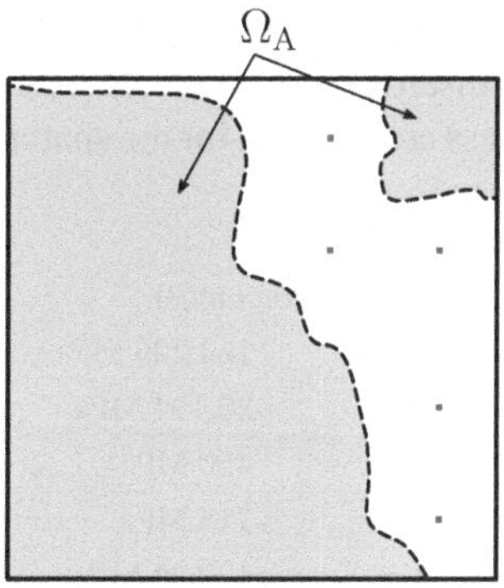


**Fig. 9** Quadrature points used to stabilize the FCM

elastic one or a hyperelastic one, depending on the question whether small strain or finite strain problems are considered. The influence of the additional stiffness on the accuracy of the FCM solution was investigated in Dauge et al. (2015).

## Numerical Examples

The main purpose of this section is to investigate the adaptive integration schemes presented in Sections "Adaptive Quadtree/Octree Quadrature Schemes" and "Numerical Integration Based on Moment Fitting" within the spatial discretization performed with the finite cell method. The adaptive schemes were studied for linear problems of structural mechanics in Abedian et al. (2013a), Joulaian et al. (2016), Hubrich et al. (2017). In this section, we focus on nonlinear problems of structural mechanics with particular attention toward elastoplasticity, including small as well as finite strains. Both material models (for small as well as finite strains) are based on the $J_2$ flow theory of plasticity accounting for nonlinear isotropic hardening

$$K(\bar{\alpha}) = \sigma_0 + h\bar{\alpha} + (\sigma_\infty - \sigma_0)(1 - \exp(-\omega\bar{\alpha})), \tag{48}$$

where $\sigma_0$ represents the initial yield stress, $h$ the linear hardening parameter, $\sigma_\infty$ the saturation stress, and $\omega$ the hardening exponent. The internal variable $\bar{\alpha}$ used to describe the isotropic hardening corresponds to the equivalent plastic strain. The material parameters that are applied within all numerical examples presented in this section are summarized in Table 1.

In the next two subsections, we will study the performance of the adaptive integration schemes. To this end, we will start with the elastoplastic model for small strains, followed by examples regarding finite strain elastoplasticity. The corresponding material models, which are taken from the literature, are described at the beginning of the individual sections. The overall solution process used for the FCM applies a standard incremental/iterative procedure, where the Newton-Raphson method is used in each incremental load step in order to solve the nonlinear set of algebraic equations emerging from the discretization of the linearized weak form. The material models and the solution process are – except for the spatial discretization by the FCM

**Table 1** Material parameters for the elastoplastic model

| | |
|---|---|
| Bulk modulus ($K$) | 164,206 MPa |
| Shear modulus ($\mu$) | 80,194 MPa |
| Initial yield strength ($\sigma_0$) | 450 MPa |
| Saturation stress ($\sigma_\infty$) | 715 MPa |
| Linear hardening parameter ($h$) | 129.24 MPa |
| Hardening exponent ($\omega$) | 16.93 |

and its numerical integration – well established and can be found in textbooks such as Simo and Hughes (1998), Wriggers (2008), de Souza Neto et al. (2008). Therefore, we will keep the related description rather brief and focus on the numerical results achieved with the finite cell method and the different versions of the adaptive integration.

## $J_2$ Flow Theory of Plasticity for Small Strains

In this section, we start with the $J_2$ flow theory of plasticity for small strains. The small strain tensor

$$\boldsymbol{\varepsilon} = \boldsymbol{\varepsilon}^{\mathrm{e}} + \boldsymbol{\varepsilon}^{\mathrm{p}}, \tag{49}$$

is additively decomposed into an elastic $\boldsymbol{\varepsilon}^{\mathrm{e}}$ and a plastic part $\boldsymbol{\varepsilon}^{\mathrm{p}}$. The Cauchy stress tensor

$$\boldsymbol{\sigma} = \boldsymbol{C}\,(\boldsymbol{\varepsilon} - \boldsymbol{\varepsilon}^{\mathrm{p}})\,. \tag{50}$$

is computed by an isotropic linear elastic constitutive model taking only the elastic strains into account. In Eq. (50), $\boldsymbol{C}$ represents the fourth-order elasticity tensor. Admissible stress states are defined by the classical von Mises yield criterion

$$\Phi\,(\boldsymbol{\sigma}, \bar{\alpha}) = \|\mathrm{dev}\,[\boldsymbol{\sigma}]\| - \sqrt{\frac{2}{3}}K(\bar{\alpha}) \leq 0, \tag{51}$$

where $\mathrm{dev}\,[\boldsymbol{\sigma}] = \boldsymbol{\sigma} - \frac{1}{3}\mathrm{tr}\,[\boldsymbol{\sigma}]\,\mathbf{1}$ is the deviator of the Cauchy stress tensor and $K(\bar{\alpha})$ describes the isotropic hardening with Eq. (48). The evolution of the plastic strains

$$\dot{\boldsymbol{\varepsilon}}^{\mathrm{p}} = \dot{\gamma}\frac{\partial\Phi\,(\boldsymbol{\sigma}, \bar{\alpha})}{\partial\boldsymbol{\sigma}}, \tag{52}$$

is modeled by an associative flow rule, where $\gamma \geq 0$ is the non-negative plastic multiplier. A full description of the material model and its numerical integration based on an implicit Euler scheme can be found, for example, in Simo and Hughes (1998), Wriggers (2008). The numerical treatment of small strain elastoplasticity problems with high-order finite elements was presented in Düster and Rank (2002), Düster et al. (2002). First numerical investigations of the finite cell method for small strain elastoplasticity can be found in Abedian et al. (2013b, 2014).

**Porous material**. As a first example, we consider a porous material that was studied with respect to the computation of effective linear elastic material properties in Düster et al. (2012). Here, we want to investigate the FCM and the integration schemes for the same geometry but for small strain elastoplasticity. The porous material of dimension $10 \times 10 \times 10\,\mathrm{mm}^3$ includes randomly distributed ellipsoidal pores and is discretized with $8 \times 8 \times 8$ finite cells, as depicted in Fig. 10a. Symmetry boundary conditions

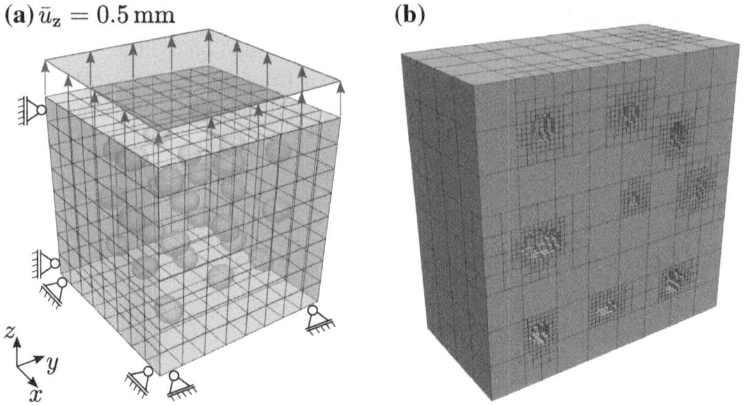


Fig. 10 Porous material. **a** Geometry, boundary conditions, and FCM discretization. **b** Octree mesh

Fig. 11 Porous material. Total number of integration points for adaptive octree and moment fitting

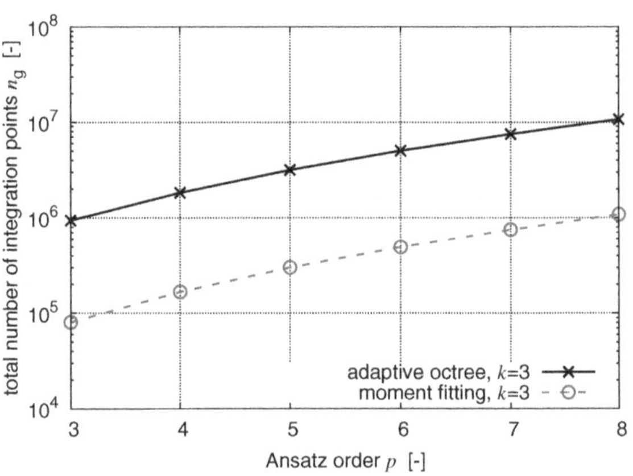


are applied at three faces of the cube and at the top a displacement of $\bar{u}_z = 0.5$ mm is applied in $z$-direction.

In a first step we study the number of integration points required by the different quadrature schemes. To this end, the polynomial degree of the shape functions is uniformly increased from $p = 3, \ldots, 8$, and the number of integration points is plotted in Fig. 11. The adaptive octree scheme – depicted in Fig. 10b for a cut through the porous domain – uses a tree-depth level of $k = 3$. The same refinement level is applied for the moment fitting method to compute the corresponding moments. Please note that the octree is in this case only used to compute the moments. The integration with the moment fitting method is, however, carried out on cell level (i.e. without further refinement, i.e. $k_a = 0$). From Fig. 11, we can see that the moment fitting method helps to reduce the number of integration points by approximately one order of magnitude as compared to the adaptive octree.

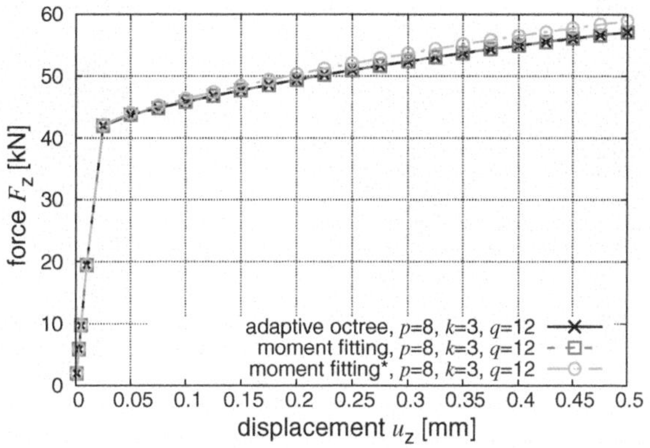


**Fig. 12** Porous material. Load-displacement curves

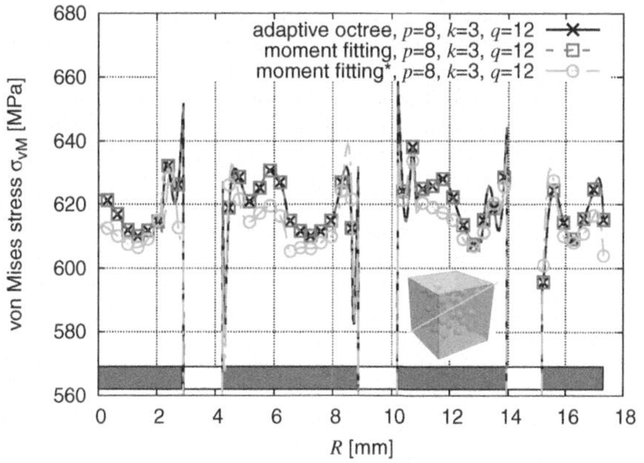


**Fig. 13** Porous material. The von Mises stress $\sigma_{\mathrm{vM}}$ along a diagonal cutline for $\bar{u}_{\mathrm{z}} = 0.5\,\mathrm{mm}$

Next, we study the results in terms of the load-displacement curve and the von Mises stress $\sigma_{\mathrm{vM}}$ along a diagonal cutline through the porous material, see Figs. 12 and 13. The results are obtained with $p = 8$, $k = 3$, and a stabilization parameter $\alpha = 10^{-q}$ where $q = 12$. We compare two different versions of the moment fitting. The first version of the moment fitting (without asterisk) uses the same material model and parameters in the entire domain. The version marked with an asterisk (moment fitting*) uses a linear elastic material in the fictitious domain by setting the yield stress of the related integration points to infinity. Taking the adaptive octree integration scheme as a reference solution, the comparison reveals that choosing the same material model and parameters in the entire domain yields to more accurate results.

Finally, Fig. 14 depicts contour plots of the von Mises stress $\sigma_{\mathrm{vM}}$ and the equivalent plastic strain $\bar{\alpha}$. From this it can be seen that the applied load leads to a stress state resulting in a plastic region that is spread out over almost the entire domain.

**Cube with a cylindrical hole**. While the moment fitting performed very well in the previous example, a second example serves to demonstrate possible problems

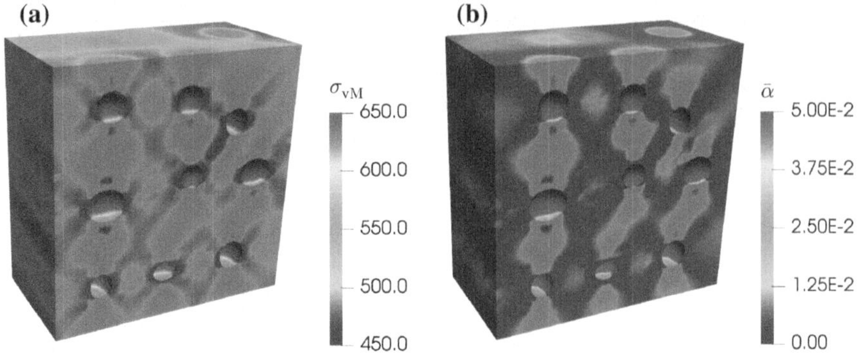


**Fig. 14** Contour plots of the porous material for $\bar{u}_z = 0.5$ mm and $p = 8$. **a** The von Mises stress $\sigma_{\mathrm{vM}}$. **b** Equivalent plastic strain $\bar{\alpha}$

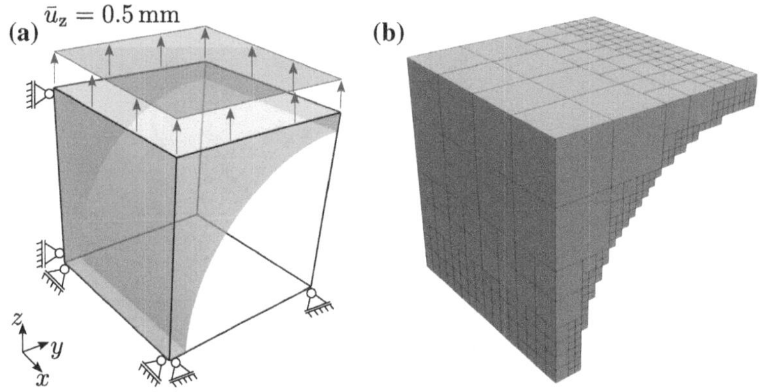


**Fig. 15** Cube with a cylindrical hole. **a** Geometry, boundary conditions, and FCM discretization. **b** Octree mesh

that can appear when applying the moment fitting without adaptive refinement. The example under consideration is a cube with a cylindrical hole discretized by one cell only, see Fig. 15a. The geometry of the cylinder is implicitly defined by the level set function

$$\phi(\boldsymbol{x}) = (y - y_c)^2 + (z - z_c)^2 - r^2 \tag{53}$$

where the center coordinates $y_c$ and $z_c$ of the cylinder and its radius are given as

$$y_c = 10\,\mathrm{mm}\,, \quad z_c = 0\,\mathrm{mm}\,, \quad \text{and} \ r = 9\,\mathrm{mm}. \tag{54}$$

We apply symmetry boundary conditions at three faces of the cube, and a displacement of up to $\bar{u}_z = 0.5$ mm is prescribed at the top.

Figure 16 shows a comparison of the three different integration schemes in terms of the load-displacement curve computed with a single finite cell utilizing $p = 8$ as

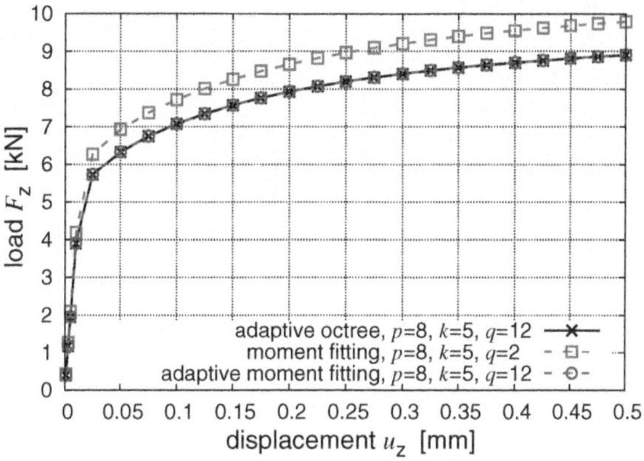


**Fig. 16** Cube with a cylindrical hole. Load-displacement curves

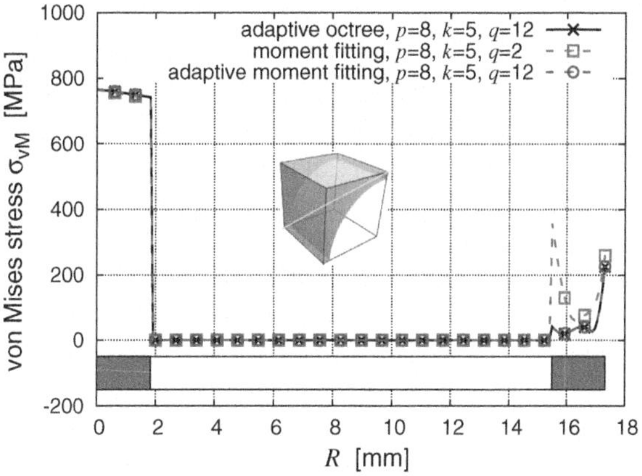


**Fig. 17** Cube with a cylindrical hole. The von Mises stress $\sigma_{vM}$ along a diagonal cutline for $\bar{u}_z = 0.5\,\text{mm}$

the polynomial degree of the shape functions. The tree-depth level of the adaptive octree, the moment fitting, and the adaptive moment fitting are set to $k = 5$. The octree mesh using 5 refinements is given in Fig. 15b. For the adaptive moment fitting, we furthermore choose $k_a = 3$, i.e. a maximum tree-depth level for the adaptive refinement of the cell. As can be seen in Fig. 16, the moment fitting method requires, with $q = 2$, a significantly higher stabilization parameter $\alpha = 10^{-q}$ as compared to the other two integration methods, where $q = 12$ is chosen. The rather high stabilization parameter is required for the moment fitting method in order to achieve a convergence of the global Newton-Raphson method applied to solve the nonlinear set of algebraic equations obtained from the discretization of the linearized weak form of equilibrium. As a consequence of this rather high value of $\alpha$, the stiffness of the fictitious domain is significantly overestimated – resulting in a stiffer behavior, i.e. an increase of the load-displacement curve.

In Fig. 17, the von Mises stress is plotted along the diagonal cutline through the cube. The green bar indicates the physical domain whereas the white bar represents the fictitious domain. The effect of the high stabilization value $\alpha$ can be observed also

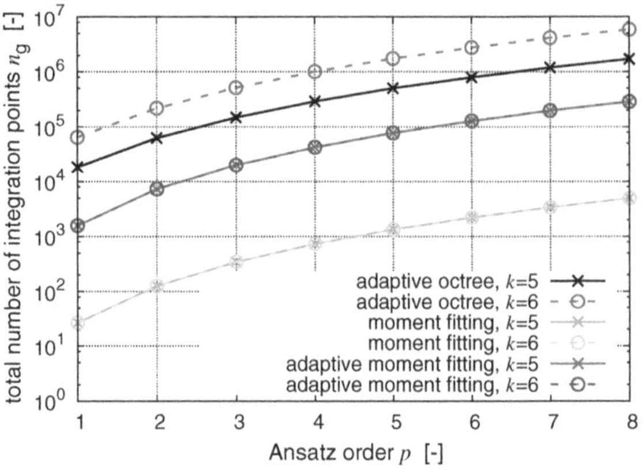


**Fig. 18** Cube with a cylindrical hole. Total number of integration points for different integration schemes

in terms of the von Mises stress where the moment fitting yields to a clear deviation from the results obtained with the other two integration schemes. This effect can be attributed to the extra stiffness introduced in the fictitious domain, resulting in oscillations of the stresses at the boundary of the cylindrical hole.

Finally, in Fig. 18, we compare the required number of quadrature points for the three different integration schemes. The number of integration points is plotted against the polynomial degree of the shape functions. Different tree-depth levels, i.e. $k = 5, 6$ are chosen. Clearly, the number of integration points of the adaptive octree scheme is increased when increasing $k$. However, the number of integration points of the two versions of the moment fitting method does not increase when $k$ is raised. This is due to the fact that $k$ corresponds to the refinement level used to compute the moments. In the adaptive moment fitting, the cell is refined up to a tree-depth level of $k_a = 3$. Therefore, the adaptive moment fitting results in a higher number of integration points as compared to the moment fitting method (where $k_a = 0$). However, the usage of the adaptive refinement in the moment fitting method increases its stability and still results in significantly less integration points as compared to the adaptive octree scheme. Therefore, we believe that the adaptive moment fitting approach presents a good compromise between efficiency and robustness.

**Thick plate with a hole under 3D conditions**. Next, we consider a benchmark that was used by different research groups to compare adaptive finite element schemes developed for elastoplastic problems, see Stein (2002). The geometry and boundary conditions of the three-dimensional plate with a hole are depicted in Fig. 19. According to the definition of the benchmark, we set $a = 100$ mm, the thickness to $t = 10$ mm, and the radius $r = 10$ mm. The traction $\bar{t}_y = \lambda\, 100$ MPa is increased monotonously within 61 load steps up to a factor of $\lambda = 4.15$. Due to symmetry, only one eighth of the plate has to be discretized. The FCM grid is composed of 399 cells, as depicted in Fig. 20a.

First, we compare the number of integration points required for the adaptive octree and the adaptive moment fitting for $k = 5$ as $p$ is increased uniformly, see Fig. 20b.

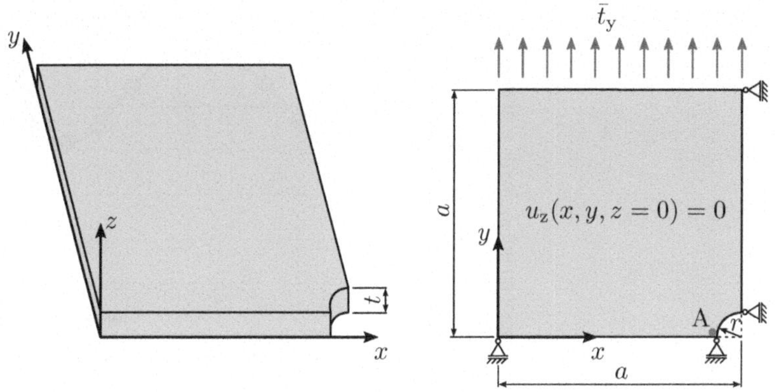


**Fig. 19** Thick-walled plate with a circular hole. Geometry and boundary conditions

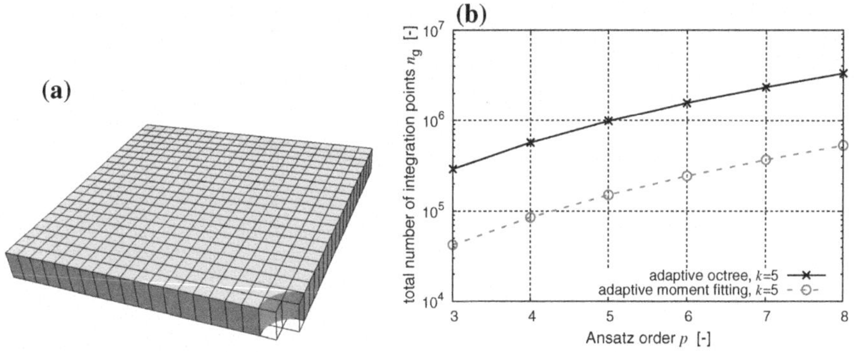


**Fig. 20** 3D perforated plate. **a** FCM discretization. **b** Total number of integration points for adaptive octree and adaptive moment fitting

Again, it can be observed that the number of integration points can be reduced by approximately one order of magnitude when applying the adaptive moment fitting.

Next, we compare the results obtained with the finite cell method with $p = 8$, $k = 5$ and $q = 5$ (i.e. $\alpha = 10^{-5}$) to the reference solution provided by Wieners, see Stein (2002). The results to be compared are the displacement component $u_x$ as well as the stress component $\sigma_{yy}$ at point A, with the coordinates (89.98, 0.01, 0.13) mm. The results obtained with the adaptive octree as well as adaptive moment fitting integration are plotted in Figs. 21 and 22.

From this, it is evident that the results of the finite cell method agree very well the reference solution. Furthermore, it can be observed that the results of the adaptive octree and the adaptive moment fitting method are in good agreement as well. This demonstrates that a factor of about 10 integration points can be saved by the adaptive moment fitting without sacrificing the accuracy of the FCM.

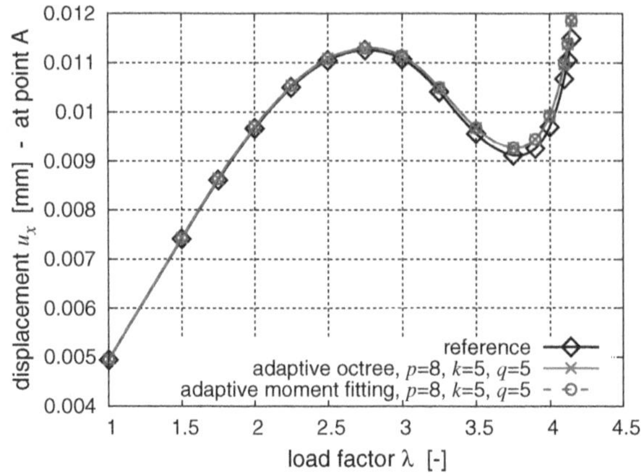


**Fig. 21** 3D perforated plate. Displacement $u_x$ at point A

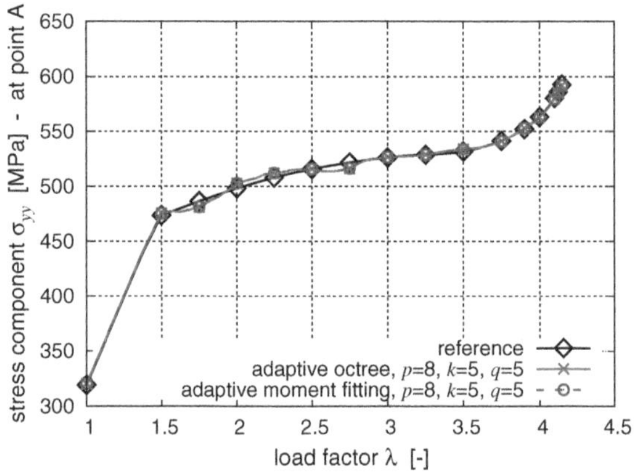


**Fig. 22** 3D perforated plate. Stress component $\sigma_{yy}$ at point A

Finally, we consider the contour plots of the von Mises stress $\sigma_{vM}$ and the equivalent plastic strain $\bar{\alpha}$ in Fig. 23 for a load factor of $\lambda = 4.15$. At this load level, the plastic region clearly occupies a large portion of the plate already.

## *$J_2$ Flow Theory of Plasticity for Large Strains*

As a second nonlinear problem, we consider the $J_2$ flow theory of plasticity for large strains. Here, the deformation gradient

$$\boldsymbol{F} = \frac{\partial \boldsymbol{x}}{\partial \boldsymbol{X}} = \text{Grad}\,[\boldsymbol{X} + \boldsymbol{u}(\boldsymbol{X}, t)] = \mathbf{1} + \text{Grad}\,\boldsymbol{u} \tag{55}$$

is multiplicatively decomposed

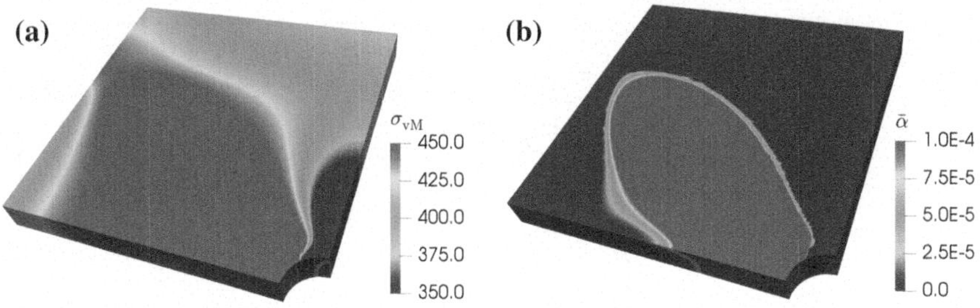


**Fig. 23** Contour plots of the 3D perforated plate for the final load $\lambda = 4.15$. **a** The von Mises stress $\sigma_{vM}$. **b** Equivalent plastic strain $\bar{\alpha}$

$$F = F^e F^p \tag{56}$$

into an elastic $F^e$ and a plastic $F^p$ part. The Kirchhoff stress tensor

$$\tau = D\varepsilon^e \tag{57}$$

is computed based on the specific free energy function of the Hencky material. In Eq. (57), $D$ denotes the fourth-order isotropic elasticity tensor and

$$\varepsilon^e = \frac{1}{2} \ln (B^e) \tag{58}$$

represents the spatial logarithmic elastic strain tensor based on the elastic left Cauchy-Green tensor

$$B^e = F^e (F^e)^T . \tag{59}$$

Admissible stress states are defined by the von Mises yield criterion

$$\Phi (s, \bar{\alpha}) = \sqrt{\frac{3}{2} s \cdot s} - K(\bar{\alpha}) \leq 0 \tag{60}$$

where $s = \text{dev}[\tau] = \tau - \frac{1}{3}\text{tr}[\tau]\,\mathbf{1}$ denotes the deviator of the Kirchhoff stress tensor $\tau$. Again, nonlinear isotropic hardening is modeled by the function $K(\bar{\alpha})$ as defined in Eq. (48). The associative flow rule is defined in the finite strain regime as

$$\dot{F}^p (F^p)^{-1} = \sqrt{\frac{3}{2}} \dot{\gamma} (R^e)^T n R^e \quad \text{with} \quad n = \frac{s}{\|s\|}. \tag{61}$$

where $\gamma \geq 0$ denotes the non-negative plastic multiplier and $R^e$ represents the elastic rotation tensor. The constitutive equations are integrated with a backward Euler scheme. An exponential mapping technique is applied for the flow rule to maintain its volume-preserving feature. A detailed description of the material model and its

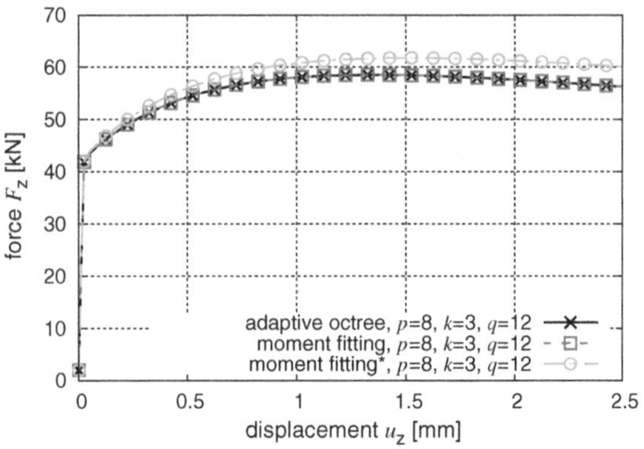


**Fig. 24** Porous material. Load-displacement curves

integration can be found in de Souza Neto et al. (2008). First results of the finite cell method for this kind of material model are summarized in Taghipour et al. (2018).

**Porous material**. Again, we consider the porous material that was investigated in Section "$J_2$ Flow Theory of Plasticity for Small Strains" in the context of small strain elastoplasticity. Here, we apply the finite strain elastoplasticity model as described in Section "$J_2$ Flow Theory of Plasticity for Large Strains". The geometry and discretization of the porous material are the same as depicted in Fig. 10a. However, the magnitude of the applied displacement is increased to $\bar{u}_z = 2.5\,\text{mm}$, resulting in an elongation of 25%. Again, we apply the same integration methods with the same settings as in the small strain case. Therefore, the savings concerning the number of integration points are again about one order of magnitude when applying the moment fitting method instead of the adaptive octree.

Figure 24 shows the load-displacement curves for the different integration schemes. Again, we consider two different versions of the moment fitting. For the integration points in the fictitious domain, the version denoted as moment fitting* assumes an infinite yield stress – resulting in a hyperelastic material behavior. In contrast, the version denoted as moment fitting (without asterisk) assumes the same material (parameters) in the entire domain. Comparing the load-displacement curves to those of the adaptive octree, it is evident that assuming the same material parameters for the moment fitting points in the fictitious part as in the physical domain yields more accurate results. This observation is also confirmed when investigating the von Mises stress $\sigma_{\text{vM}}$ evaluated along a diagonal cutline through the porous material, see Fig. 25.

The contour plots of the von Mises stress $\sigma_{\text{vM}}$ and the equivalent plastic strain $\bar{\alpha}$ are presented in Fig. 26. The black box represents the initial, i.e. undeformed configuration of the porous material, giving an impression of the total deformation.

**Swiss cheese domain**. The last example which serves to investigate the different integration techniques is the so-called 'swiss cheese domain' which is composed of several cheese blocks. According to Burman et al. (2015) the geometric description of one cheese block is given in terms of the following level set function

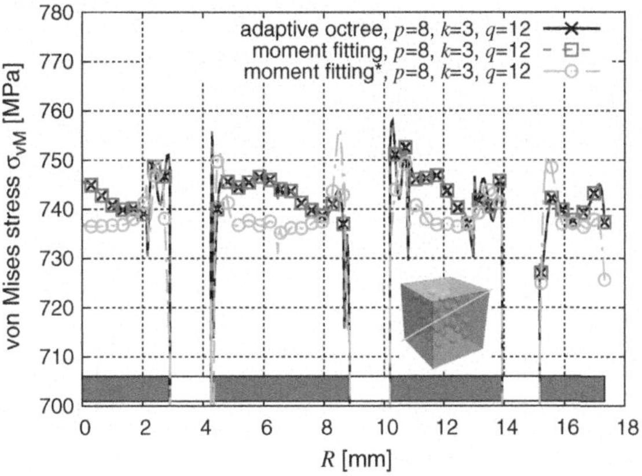


**Fig. 25** Porous material. The von Mises stress $\sigma_{\mathrm{vM}}$ along a diagonal cutline for $\bar{u}_z = 2.5\,\mathrm{mm}$

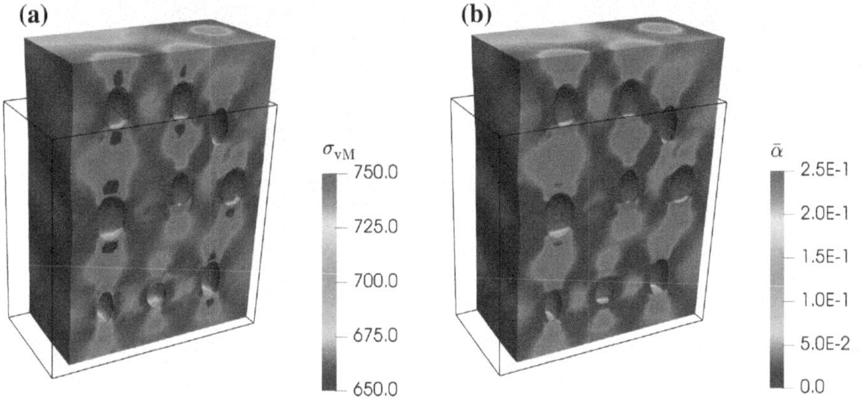


**Fig. 26** Contour plots of the porous material for $\bar{u}_z = 2.5\,\mathrm{mm}$ and $p = 8$. **a** The von Mises stress $\sigma_{\mathrm{vM}}$. **b** Equivalent plastic strain $\bar{\alpha}$

$$\begin{aligned}\phi(\boldsymbol{x}) = &\left[(x-x_c)^2 + (y-y_c)^2 - R^2\right]^2 + \left[(y-y_c)^2 + (z-z_c)^2 - R^2\right]^2 \\ &+ \left[(z-z_c)^2 - r^2\right]^2 + \left[(x-x_c)^2 + (z-z_c)^2 - R^2\right]^2 \\ &+ \left[(x-x_c)^2 - r^2\right]^2 + \left[(y-y_c)^2 - r^2)\right]^2 - d. \end{aligned} \tag{62}$$

We use eight of these cheese blocks, where every block of size $3 \times 3 \times 3\,\mathrm{mm}^3$ has a slightly different geometry, see Table 2 for the corresponding parameters. Figure 27a shows the geometry, together with the boundary conditions and the FCM discretization. The FCM grid consists of 5,376 cells, out of which 4,464 are cut by the geometry. Figure 27b shows the octree using 3 levels of refinement for the resolution of the geometry.

**Table 2** Geometry parameters of the swiss cheese domain

| Cheese block id | $x_c$ (mm) | $y_c$ (mm) | $z_c$ (mm) | $R$ (mm) | $r$ (mm) | $d$ (mm$^4$) |
|---|---|---|---|---|---|---|
| 1 | 1.5 | 1.5 | 1.5 | 1.5 | 1.125 | 5.3 |
| 2 | 4.5 | 1.5 | 1.5 | 1.5 | 1.125 | 4.9 |
| 3 | 1.5 | 4.5 | 1.5 | 1.5 | 1.125 | 5.1 |
| 4 | 4.5 | 4.5 | 1.5 | 1.5 | 1.125 | 4.7 |
| 5 | 1.5 | 1.5 | 4.5 | 1.5 | 1.125 | 5.2 |
| 6 | 4.5 | 1.5 | 4.5 | 1.5 | 1.125 | 4.8 |
| 7 | 1.5 | 4.5 | 4.5 | 1.5 | 1.125 | 5.0 |
| 8 | 4.5 | 4.5 | 4.5 | 1.5 | 1.125 | 4.6 |

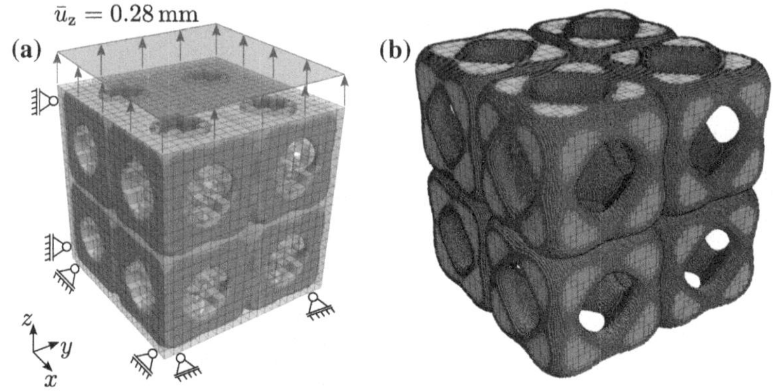


**Fig. 27** Swiss cheese domain. **a** Geometry, boundary conditions, and FCM discretization. **b** Octree mesh

Figure 28 shows the load-displacement curves computed with the FCM and the adaptive octree as well as the adaptive moment fitting integration scheme. The polynomial degree of the shape functions of the finite cells is $p = 4$, the tree-depth level is $k = 3$, and the stabilization parameter is $\alpha = 10^{-q}$, with $q = 5$. The moment fitting method (without refinement, i.e. $k_a = 0$) failed, i.e. the overall Newton-Raphson method did not converge with this integration scheme. However, the adaptive moment fitting with $k_a = 3$ performs very well, as is evident from a comparison with the adaptive octree. Interestingly, 1,440 of the 4,464 broken cells do not even require a refinement in the adaptive moment fitting, i.e. $k_a = 0$.

The number of integration points used for the adaptive octree $n_g^{\text{AOT}}$ and the adaptive moment fitting method $n_g^{\text{AMF}}$ are listed in Table 3 for different polynomial degrees $p$ of the shape functions of the finite cells. From this, we can see that the adaptive moment fitting requires approximately 2.5 times less integration points. The ratio will increase even further if we elevate the tree-depth level $k$ of the octree to resolve the geometry. Recall that the tree-depth level for the computation of the weights in

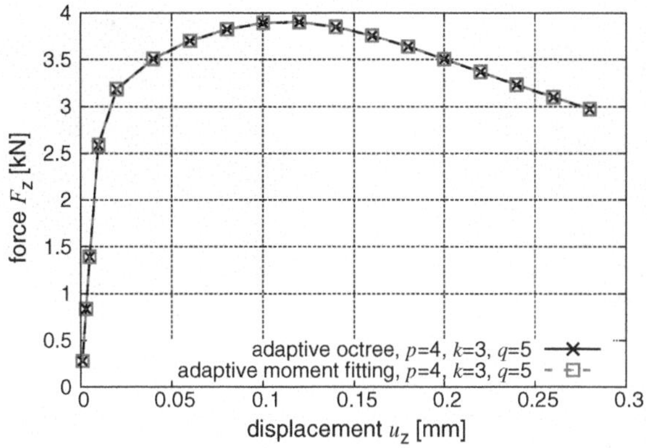


**Fig. 28** Swiss cheese domain. Load-displacement curves

**Table 3** Total number of integration points

| $p$ | $n_g^{AOT}$ | $n_g^{AMF}$ | Ratio |
|---|---|---|---|
| 2 | 10,432,284 | 4,123,260 | ≈2.5 |
| 3 | 24,720,696 | 10,391,292 | ≈2.4 |
| 4 | 48,285,260 | 21,079,600 | ≈2.3 |

the adaptive moment fitting corresponds to $k$, whereas the refinement level of the cells is limited to $k_a$. Therefore, the number of integration points for the computation of the stiffness matrix will remain constant for the adaptive moment fitting method as $k$ is elevated. However, increasing $k$ will help to improve the resolution of the geometry.

Finally, Fig. 29 shows the contour plots of the von Mises stress $\sigma_{\text{vM}}$ and the equivalent plastic strain $\bar{\alpha}$ for the last load step, in which a displacement of $\bar{u}_z = 0.28$ mm is applied. From this, it is evident that necking occurs in the thin parts of the upper cheese blocks. The equivalent plastic strain in this regions is about 50%.

## Conclusions

We presented different integration schemes that can be applied to cut finite elements and cut finite cells appearing in fictitious domain methods. The adaptive octree integration combines a spacetree with a Gaussian quadrature on the leaves of the spacetree. This approach is very robust but introduces a high number of integration points. The moment fitting method, in which an individual quadrature rule is set up for every broken element/cell, reduces the number of integration points significantly. The computational overhead of the moment fitting can be also significantly reduced by circumventing the necessity to solve a (nonlinear) set of equations. This can

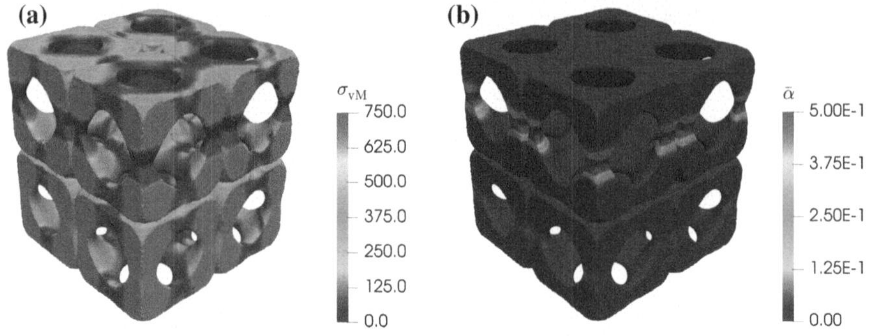


**Fig. 29** Contour plots of the Swiss cheese domain for $\bar{u}_z = 0.28$mm and $p = 4$. **a** The von Mises stress $\sigma_{vM}$. **b** Equivalent plastic strain $\bar{\alpha}$

be accomplished by pre-selecting the position of the integration points and using Lagrange polynomials defined on this set of points.

The different integration methods were tested in the framework of the finite cell method applied to solve elastoplastic problems including small as well as finite strains. In some nonlinear elastoplastic problems, the moment fitting method unfortunately turned out to be less stable than the adaptive octree, leading to a failure of the nonlinear solution process. To circumvent this problem, it is possible to employ the adaptive moment fitting method, which borrows the idea of the spacetree to locally refine broken elements/cells for integration purposes. This, of course, leads to an increased number of integration points – but still results in significantly less points than the adaptive octree approach, especially in cases where a high resolution of the integration domain is required. In this way, the stability of the finite cell method can be retained also for strongly nonlinear problems. Future work will apply the adaptive moment fitting method also to other nonlinear problems to test its stability.

**Acknowledgements** The authors gratefully acknowledge support by the Deutsche Forschungsgemeinschaft in the Priority Program 1748 "Reliable simulation techniques in solid mechanics. Development of non-standard discretization methods, mechanical and mathematical analysis" under the project DU 405/8-2.

# Appendix

Table 4 lists the abscissas and weight factors of the Gauss-Lobatto quadrature. Applying $n$ quadrature points allows to exactly integrate polynomials of degree $p_{\mathrm{q}} = 2n - 3$. Table 5 lists the abscissas and weight factors of the Gauss-Legendre quadrature. Applying $n$ quadrature points allows to exactly integrate polynomials of degree $p_{\mathrm{q}} = 2n - 1$.

**Table 4** Abscissas and weight factors of the Gauss-Lobatto quadrature

| $n$ | Abscissas $\xi_i$ | Weight factors $w_i$ |
|---|---|---|
| 2 | −1.0000000000000000 | 1.0000000000000000 |
| | 1.0000000000000000 | 1.0000000000000000 |
| 3 | −1.0000000000000000 | 0.3333333333333333 |
| | 0.0000000000000000 | 1.3333333333333333 |
| | 1.0000000000000000 | 0.3333333333333333 |
| 4 | −1.0000000000000000 | 0.1666666666666667 |
| | −0.4472135954999580 | 0.8333333333333333 |
| | 0.4472135954999580 | 0.8333333333333333 |
| | 1.0000000000000000 | 0.1666666666666667 |
| 5 | −1.0000000000000000 | 0.1000000000000000 |
| | −0.6546536707079771 | 0.5444444444444444 |
| | 0.0000000000000000 | 0.7111111111111111 |
| | 0.6546536707079770 | 0.5444444444444444 |
| | 1.0000000000000000 | 0.1000000000000000 |
| 6 | −1.0000000000000000 | 0.0666666666666667 |
| | −0.7650553239294646 | 0.3784749562978469 |
| | −0.2852315164806450 | 0.5548583770354865 |
| | 0.2852315164806451 | 0.5548583770354862 |
| | 0.7650553239294645 | 0.3784749562978471 |
| | 1.0000000000000000 | 0.0666666666666667 |

(continued)

**Table 4** (continued)

| $n$ | Abscissas $\xi_i$ | Weight factors $w_i$ |
|---|---|---|
| 7 | −1.0000000000000000 | 0.0476190476190476 |
| | −0.83022389627856708 | 0.2768260473615660 |
| | −0.46884879347071418 | 0.4317453812098626 |
| | 0.0000000000000000 | 0.4876190476190476 |
| | 0.46884879347071423 | 0.4317453812098627 |
| | 0.83022389627856685 | 0.2768260473615661 |
| | 1.0000000000000000 | 0.0476190476190476 |
| 8 | −1.0000000000000000 | 0.0357142857142857 |
| | −0.87174014850960668 | 0.2107042271435060 |
| | −0.59170018143314218 | 0.3411226924835044 |
| | −0.20929921790247885 | 0.4124587946587039 |
| | 0.20929921790247885 | 0.4124587946587039 |
| | 0.59170018143314218 | 0.3411226924835044 |
| | 0.87174014850960646 | 0.2107042271435061 |
| | 1.0000000000000000 | 0.0357142857142857 |
| 9 | −1.0000000000000000 | 0.0277777777777778 |
| | −0.89975799541146018 | 0.1654953615608056 |
| | −0.67718627951073762 | 0.2745387125001618 |
| | −0.36311746382617827 | 0.3464285109730465 |
| | 0.0000000000000000 | 0.3715192743764172 |
| | 0.3631174638261781 | 0.3464285109730464 |
| | 0.67718627951073773 | 0.2745387125001617 |
| | 0.89975799541146007 | 0.1654953615608054 |
| | 1.0000000000000000 | 0.0277777777777778 |
| 10 | −1.0000000000000000 | 0.0222222222222222 |
| | −0.91953390816645864 | 0.1333059908510700 |
| | −0.73877386510550491 | 0.2248893420631265 |
| | −0.47792494981044459 | 0.2920426836796839 |
| | −0.16527895766638698 | 0.3275397611838974 |
| | 0.16527895766638692 | 0.3275397611838974 |
| | 0.47792494981044448 | 0.2920426836796837 |
| | 0.73877386510550491 | 0.2248893420631265 |
| | 0.91953390816645864 | 0.1333059908510700 |
| | 1.0000000000000000 | 0.0222222222222222 |

**Table 5** Abscissas and weight factors of the Gauss-Legendre quadrature

| $n$ | Abscissas $\xi_i$ | Weight factors $w_i$ |
|---|---|---|
| 1 | 0.000000000000000e+00 | 2.000000000000000e+00 |
| 2 | 5.773502691896258e−01 | 1.000000000000000e+00 |
| | −5.773502691896258e−01 | 1.000000000000000e+00 |
| 3 | 7.745966692414834e−01 | 5.555555555555556e−01 |
| | 0.000000000000000e+00 | 8.888888888888889e−01 |
| | −7.745966692414834e−01 | 5.555555555555556e−01 |
| 4 | 8.611363115940526e−01 | 3.478548451374539e−01 |
| | 3.399810435848563e−01 | 6.521451548625461e−01 |
| | −3.399810435848563e−01 | 6.521451548625461e−01 |
| | −8.611363115940526e−01 | 3.478548451374539e−01 |
| 5 | 9.061798459386640e−01 | 2.369268850561891e−01 |
| | 5.384693101056831e−01 | 4.786286704993665e−01 |
| | 0.000000000000000e+00 | 5.688888888888889e−01 |
| | −5.384693101056831e−01 | 4.786286704993665e−01 |
| | −9.061798459386640e−01 | 2.369268850561891e−01 |
| 6 | 9.324695142031520e−01 | 1.713244923791703e−01 |
| | 6.612093864662645e−01 | 3.607615730481386e−01 |
| | 2.386191860831969e−01 | 4.679139345726910e−01 |
| | −2.386191860831969e−01 | 4.679139345726910e−01 |
| | −6.612093864662645e−01 | 3.607615730481386e−01 |
| | −9.324695142031520e−01 | 1.713244923791703e−01 |
| 7 | 9.491079123427585e−01 | 1.294849661688697e−01 |
| | 7.415311855993944e−01 | 2.797053914892767e−01 |
| | 4.058451513773972e−01 | 3.818300505051189e−01 |
| | 0.000000000000000e+00 | 4.179591836734694e−01 |
| | −4.058451513773972e−01 | 3.818300505051189e−01 |
| | −7.415311855993944e−01 | 2.797053914892767e−01 |
| | −9.491079123427585e−01 | 1.294849661688697e−01 |
| 8 | 9.602898564975362e−01 | 1.012285362903763e−01 |
| | 7.966664774136267e−01 | 2.223810344533745e−01 |
| | 5.255324099163290e−01 | 3.137066458778873e−01 |
| | 1.834346424956498e−01 | 3.626837833783620e−01 |
| | −1.834346424956498e−01 | 3.626837833783620e−01 |
| | −5.255324099163290e−01 | 3.137066458778873e−01 |
| | −7.966664774136267e−01 | 2.223810344533745e−01 |
| | −9.602898564975362e−01 | 1.012285362903763e−01 |
| 9 | 9.681602395076261e−01 | 8.127438836157441e−02 |
| | 8.360311073266358e−01 | 1.806481606948574e−01 |
| | 6.133714327005904e−01 | 2.606106964029355e−01 |
| | 3.242534234038089e−01 | 3.123470770400028e−01 |
| | 0.000000000000000e+00 | 3.302393550012598e−01 |

(continued)

**Table 5** (continued)

| $n$ | Abscissas $\xi_i$ | Weight factors $w_i$ |
|---|---|---|
| | −3.242534234038089e−01 | 3.123470770400028e−01 |
| | −6.133714327005904e−01 | 2.606106964029355e−01 |
| | −8.360311073266358e−01 | 1.806481606948574e−01 |
| | −9.681602395076261e−01 | 8.127438836157441e−02 |
| 10 | 9.739065285171717e−01 | 6.667134430868814e−02 |
| | 8.650633666889845e−01 | 1.494513491505806e−01 |
| | 6.794095682990244e−01 | 2.190863625159820e−01 |
| | 4.333953941292472e−01 | 2.692667193099964e−01 |
| | 1.488743389816312e−01 | 2.955242247147529e−01 |
| | −1.488743389816312e−01 | 2.955242247147529e−01 |
| | −4.333953941292472e−01 | 2.692667193099964e−01 |
| | −6.794095682990244e−01 | 2.190863625159820e−01 |
| | −8.650633666889845e−01 | 1.494513491505806e−01 |
| | −9.739065285171717e−01 | 6.667134430868814e−02 |

## References


Abedian, A., & Düster, A. (2019). Equivalent Legendre polynomials: Numerical integration of discontinuous functions in the finite element methods. *Computer Methods in Applied Mechanics and Engineering*, *343*, 690–720. https://doi.org/10.1016/j.cma.2018.08.002.

Abedian, A., Parvizian, J., Düster, A., Khademyzadeh, H., & Rank, E. (2013). Performance of different integration schemes in facing discontinuities in the finite cell method. *International Journal of Computational Methods*, *10*(3), 1350002/1–24. https://doi.org/10.1142/S0219876213500023.

Abedian, A., Parvizian, J., Düster, A., & Rank, E. (2013). The finite cell method for the $J_2$ flow theory of plasticity. *Finite Elements in Analysis and Design*, *69*, 37–47.

Abedian, A., Parvizian, J., Düster, A., & Rank, E. (2014). Finite cell method compared to $h$-version finite element method for elasto-plastic problems. *Applied Mathematics and Mechanics*, *35*(10), 1239–1248. https://doi.org/10.1007/s10483-014-1861-9.

Burman, E., & Hansbo, P. (2010). Fictitious domain finite element methods using cut elements: I. A stabilized Lagrange multiplier method. *Computer Methods in Applied Mechanics and Engineering*, *199*(41–44), 2680–2686.

Burman, E., & Hansbo, P. (2012). Fictitious domain finite element methods using cut elements: II. A stabilized Nitsche method. *Applied Numerical Mathematics*, *62*(4), 328–341. https://doi.org/10.1016/j.apnum.2011.01.008.

Burman, E., Claus, S., Hansbo, P., Larson, M. G., & Massing, A. (2015). CutFEM: Discretizing geometry and partial differential equations. *International Journal for Numerical Methods in Engineering*, *104*, 472–501.

Cottrell, J. A., Hughes, T. J. R., & Bazilevs, Y. (2009). *Isogeometric analysis: Towards integration of CAD and FEM*. Hoboken: Wiley. ISBN 978-0-470-74873-2.

Dauge, M., Düster, A., & Rank, E. (2015). Theoretical and numerical investigation of the finite cell method. *Journal of Scientific Computing*, *65*, 1039–1064. https://doi.org/10.1007/s10915-015-9997-3.

de Souza Neto, E. A., Perić, D., & Owen, D. R. J. (2008). *Computational methods for plasticity, theory and applications*. Hoboken: Wiley. ISBN 978-0-470-69452-7.

Del Pino, S., & Pironneau, O. (2003). A fictitious domain based general pde solver. In P. Neittanmaki, Y. Kuznetsov & O. Pironneau (Eds.), *Numerical methods for scientific computing variational problems and applications*, CIMNE, Barcelona, Spain.

Düster, A., & Allix, O. (2019). Selective enrichment of moment fitting and application to cut finite elements and cells. *Computational Mechanics*. https://doi.org/10.1007/s00466-019-01776-2.

Düster, A., & Rank, E. (2002). A p-version finite element approach for two- and three-dimensional problems of the $J_2$ flow theory with non-linear isotropic hardening. *International Journal for Numerical Methods in Engineering*, *53*, 49–63.

Düster, A., Niggl, A., Nübel, V., & Rank, E. (2002). A numerical investigation of high-order finite elements for problems of elasto-plasticity. *Journal of Scientific Computing*, *17*, 429–437.

Düster, A., Parvizian, J., Yang, Z., & Rank, E. (2008). The finite cell method for three-dimensional problems of solid mechanics. *Computer Methods in Applied Mechanics and Engineering*, *197*, 3768–3782.

Düster, A., Sehlhorst, H.-G., & Rank, E. (2012). Numerical homogenization of heterogeneous and cellular materials utilizing the finite cell method. *Computational Mechanics*, *50*, 413–431. https://doi.org/10.1007/s00466-012-0681-2.

Düster, A., Rank, E., & Szabó, B. (2017). The p-Version of the Finite Element and Finite Cell Methods. In E. Stein, R. de Borst, & T. J. R. Hughes (Eds.), *Encyclopedia of computational mechanics*, 2nd edn, vol Part 1. Solids and Structures (Chap. 4, pp. 137–171). Hoboken: Wiley. https://doi.org/10.1002/9781119176817.ecm2003g. ISBN 978-1-119-00379-3.

Fries, T.-P., & Omerović, S. (2016). Higher-order accurate integration of implicit geometries. *International Journal for Numerical Methods in Engineering*, *106*(5), 323–371.

Glowinski, R., & Kuznetsov, Y. (2007). Distributed Lagrange multipliers based on fictitious domain method for second order elliptic problems. *Computer Methods in Applied Mechanics and Engineering*, *196*, 1498–1506.

Gnegel, S. (2019). *The finite cell method for the computation of cellular materials*. Ph.D. thesis, Fachgebiet für Numerische Strukturanalyse mit Anwendungen in der Schiffstechnik (M-10), TU Hamburg.

Heinze, S., Joulaian, M., Egger, H., & Düster, A. (2014). Efficient computation of cellular materials using the finite cell method. *Proceedings in Applied Mathematics and Mechanics*, *14*, 251–252. https://doi.org/10.1002/pamm.201410113.

Hubrich, S., & Düster, A. (2018). Adaptive numerical integration of broken finite cells based on moment fitting applied to finite strain problems. *Proceedings in Applied Mathematics and Mechanics*, *18*, e201800089. https://doi.org/10.1002/pamm.201800089.

Hubrich, S., & Düster, A. (2019). Numerical integration for nonlinear problems of the finite cell method using an adaptive scheme based on moment fitting. *Computers & Mathematics with Applications*, *77*, 1983–1997. https://doi.org/10.1016/j.camwa.2018.11.030.

Hubrich, S., Di Stolfo, P., Kudela, L., Kollmannsberger, S., Rank, E., Schröder, A., et al. (2017). Numerical integration of discontinuous functions: Moment fitting and smart octree. *Computational Mechanics*, *60*, 863–881. https://doi.org/10.1007/s00466-017-1441-0.

Hughes, T. J. R., Cottrell, J. A., & Bazilevs, Y. (2005). Isogeometric analysis: CAD, finite elements, NURBS, exact geometry and mesh refinement. *Computer Methods in Applied Mechanics and Engineering*, *194*, 4135–4195.

Jomo, J. N., Zander, N., Elhaddad, M., Özcan, A., Kollmannsberger, S., Mundani, R.-P., et al. (2017). Parallelization of the multi-level hp-adaptive finite cell method. *Computers & Mathematics with Applications*, *74*, 126–142. https://doi.org/10.1016/j.camwa.2017.01.004.

Joulaian, M., & Düster, A. (2013). Local enrichment of the finite cell method for problems with material interfaces. *Computational Mechanics*, *52*, 741–762. https://doi.org/10.1007/s00466-013-0853-8.

Joulaian, M., Hubrich, S., & Düster, A. (2016). Numerical integration of discontinuities on arbitrary domains based on moment fitting. *Computational Mechanics*, *57*, 979–999. https://doi.org/10.1007/s00466-016-1273-3.

Kollmannsberger, S., Özcan, A., Baiges, J., Ruess, M., Rank, E., & Reali, A. (2014). Parameter-free, weak imposition of Dirichlet boundary conditions and coupling of trimmed and non-conforming patches. *International Journal for Numerical Methods in Engineering*, *101*(9), 1–30. https://doi.org/10.1002/nme.4817.

Kudela, L., Zander, N., Bog, T., Kollmannsberger, S., & Rank, E. (2015). Efficient and accurate numerical quadrature for immersed boundary methods. *Advanced Modeling and Simulation in Engineering Sciences*, *2*(1), 1–22. https://doi.org/10.1186/s40323-015-0031-y. ISSN 2213-7467.

Loehnert, S., Mueller-Hoeppe, D. S., & Wriggers, P. (2011). 3D corrected XFEM approach and extension to finite deformation theory. *International Journal for Numerical Methods in Engineering*, *86*, 431–452.

Lyness, J. N., & Jespersen, D. (1975). Moderate degree symmetric quadrature rules for the triangle. *Journal of the Institute of Mathematics and Its Applications*, *15*, 19–32.

Lyness, J. N., & Monegato, G. (1977). Quadrature rules for regions having regular hexagonal symmetry. *SIAM Journal on Numerical Analysis*, *14*, 283–295.

Melenk, J. M., & Babuška, I. (1996). The partition of unity finite element method: Basic theory and applications. *Computer Methods in Applied Mechanics and Engineering*, *139*, 289–314.

Mittal, R., & Iaccarino, G. (2005). Immersed boundary method. *Annual Review Fluid Mechanics*, *37*, 239–260.

Mousavi, S. E., & Sukumar, N. (2010). Generalized Gaussian quadrature rules for discontinuities and crack singularities in the extended finite element method. *Computer Methods in Applied Mechanics and Engineering*, *199*(49–52), 3237–3249. https://doi.org/10.1016/j.cma.2010.06.031.

Mousavi, S. E., & Sukumar, N. (2011). Numerical integration of polynomials and discontinuous functions on irregular convex polygons and polyhedrons. *Computational Mechanics*, *47*, 535–554.

Müller, B., Kummer, F., & Oberlack, M. (2013). Highly accurate surface and volume integration on implicit domains by means of moment-fitting. *International Journal for Numerical Methods in Engineering*, *96*, 512–528. https://doi.org/10.1002/nme.4569.

Neittaanmäki, P., & Tiba, D. (1995). An embedding of domains approach in free boundary problems and optimal design. *SIAM Journal on Control and Optimization*, *33*(5), 1587–1602.

Parvizian, J., Düster, A., & Rank, E. (2007). Finite cell method - h- and p-extension for embedded domain problems in solid mechanics. *Computational Mechanics*, *41*, 121–133.

Peskin, C. (2002). The immersed boundary method. *Acta Numerica*, *11*, 1–39.

Press, W. H., Teukolsky, S. A., Vetterling, W. T., & Flannery, B. P. (2002). *Numerical recipes in C++. The art of scientific computing* (2nd ed.). Cambridge: Cambridge University Press. ISBN 0-521-75033-4.

Ramière, I., Angot, P., & Belliard, M. (2007). A fictitious domain approach with spread interface for elliptic problems with general boundary conditions. *Computer Methods in Applied Mechanics and Engineering*, *196*, 766–781.

Ruess, M., Schillinger, D., Bazilevs, Y., Varduhn, V., & Rank, E. (2013). Weakly enforced essential boundary conditions for NURBS-embedded and trimmed NURBS geometries on the basis of the finite cell method. *International Journal for Numerical Methods in Engineering*, *95*(10), 811–846. https://doi.org/10.1002/nme.4522.

Samet, H. (1990). *Applications of spatial data structures: Computer graphics, image processing, and GIS*. Boston, MA: Addison-Wesley Longman Publishing Co., Inc.

Saul'ev, V. K. (1963a). A method for automatization of the solution of boundary value problems on high performance computers. *Doklady Akademii Nauk SSSR*, *144* (1962), 497–500 (in Russian). English translation in *Soviet Mathematics Doklady*, *3*, 763–766.

Saul'ev, V. K. (1963). On solution of some boundary value problems on high performance computers by fictitious domain method. *Siberian Mathematical Journal*, *4*, 912–925.

Schillinger, D., & Ruess, M. (2015). The finite cell method: A review in the context of higher-order structural analysis of CAD and image-based geometric models. *Archives of Computational Methods in Engineering*, *22*, 391–455. https://doi.org/10.1007/s11831-014-9115-y.

Schillinger, D., Ruess, M., Zander, N., Bazilevs, Y., Düster, A., & Rank, E. (2012). Small and large deformation analysis with the p- and B-spline versions of the finite cell method. *Computational Mechanics*, *50*, 445–478. https://doi.org/10.1007/s00466-012-0684-z.

Schwarz, H. R. (2004). *Numerische Mathematik* (5th ed). B.G. Teubner. ISBN 978-3519429609.

Simo, J. C., & Hughes, T. J. R. (1998). *Computational inelasticity*. Berlin: Springer.

Stein, E. (Ed.). (2002). *Error-controlled adaptive finite elements in solid mechanics*. Hoboken: Wiley.

Strouboulis, T., Copps, K., & Babuška, I. (2000). The generalized finite element method: An example of its implementation and illustration of its performance. *International Journal for Numerical Methods in Engineering*, *47*, 1401–1417.

Strouboulis, T., Copps, K., & Babuška, I. (2001). The generalized finite element method. *Computer Methods in Applied Mechanics and Engineering*, *190*, 4081–4193.

Szabó, B. A., & Babuška, I. (1991). *Finite element analysis*. Hoboken: Wiley. ISBN 0-471-50273-1.

Szabó, B. A., Düster, A., & Rank, E. (2004). The p-version of the finite element method. In E. Stein, R. de Borst, & T. J. R. Hughes (Eds.), *Encyclopedia of computational mechanics* (Vol 1, Chap. 5, pp. 119–139). Hoboken: Wiley. https://doi.org/10.1002/0470091355.ecm003g. ISBN 0-470-84699-2.

Taghipour, A., Parvizian, J., Heinze, S., & Düster, A. (2018). The finite cell method for nearly incompressible finite strain plasticity problems with complex geometries. *Computers & Mathematics with Applications*, *75*, 3298–3316. https://doi.org/10.1016/j.camwa.2018.01.048.

Ventura, G. (2006). On the elimination of quadrature subcells for discontinuous functions in the eXtended finite-element method. *International Journal for Numerical Methods in Engineering*, *66*, 761–795.

Ventura, G., & Benvenuti, E. (2015). Equivalent polynomials for quadrature in Heaviside function enrichment elements. *International Journal for Numerical Methods in Engineering*, *102*, 688–710.

Wriggers, P. (2008). *Nonlinear finite-element-methods*. Berlin: Springer. ISBN 3-540-71000-0.

Zander, N., Bog, T., Elhaddad, M., Frischmann, F., Kollmannsberger, S., & Rank, E. (2016). The multi-level hp-method for three-dimensional problems: Dynamically changing high-order mesh refinement with arbitrary hanging nodes. *Computer Methods in Applied Mechanics and Engineering*, *310*, 252–277. https://doi.org/10.1016/j.cma.2016.07.007.

# Numerical Implementation of Phase-Field Models of Brittle Fracture

**Laura De Lorenzis and Tymofiy Gerasimov**

## Introduction

The phase-field framework for modeling systems with sharp interfaces consists in incorporating a continuous field variable – the so-called order parameter – which differentiates between multiple physical phases within a given system through a smooth transition. In the context of fracture, such an order parameter (termed the *crack phase-field*) describes the smooth transition between the fully broken and intact material phases, thus approximating the sharp crack discontinuity, as sketched in Fig. 1a. The evolution of this field as a result of the external loading conditions models the fracture process.

The phase-field approach to brittle fracture dates back to the seminal work of Francfort and Marigo (1998) on the *variational formulation of quasi-static brittle fracture* and to the related *regularized formulation* of Bourdin et al. (2000, 2008), Bourdin (2007a, b). The former is the mathematical theory of quasi-static brittle fracture mechanics, which recasts Griffith's energy-based principle Griffith (1921) as the minimisation problem of an energy functional. The latter presents an approximation, in the sense of $\Gamma$-convergence, of the energy functional and is designed to enable the efficient numerical treatment.

The phase-field simulation of fracture processes holds a number of advantages over classical techniques with discrete fracture description, whose numerical implementation requires explicit (in the classical finite element method, FEM) or implicit (within the extended FEM) handling of the discontinuities. The most obvious one is the ability to track automatically a cracking process by the evolution of the smooth crack field on a fixed mesh. The possibility to avoid the tedious task of tracking

L. De Lorenzis (✉) · T. Gerasimov
Institute of Applied Mechanics, Technische Universität Braunschweig,
Braunschweig, Germany
e-mail: l.delorenzis@tu-braunschweig.de



L. De Lorenzis and A. Düster (eds.), *Modeling in Engineering Using Innovative Numerical Methods for Solids and Fluids*, CISM International Centre for Mechanical Sciences 599,
https://doi.org/10.1007/978-3-030-37518-8_3

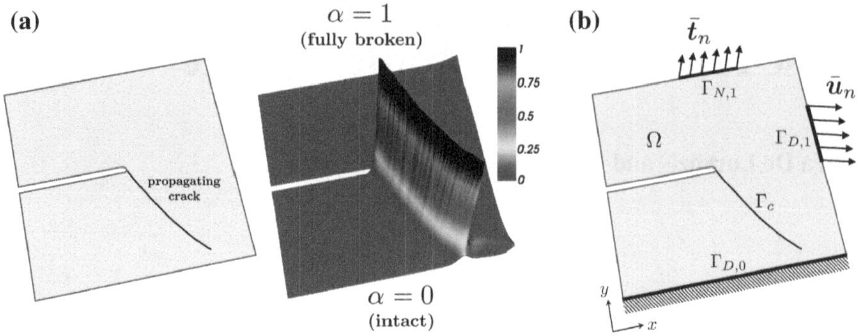


**Fig. 1** **a** Phase-field description of fracture (sketchy) with $\alpha \in C(\Omega, [0, 1])$ as the crack phase-field; **b** mechanical system setup used for representation purposes in Section "Formulation"

complicated crack surfaces in 3d significantly simplifies the finite element implementation. The second advantage is the ability to simulate complicated processes, including crack initiation (also in the absence of a singularity), propagation, coalescence, branching and bifurcation without the need for additional ad-hoc criteria. With the formulation capability to also distinguish between fracture behavior in tension and compression, no supplementary contact problem has to be posed for preventing crack faces interpenetration.

The currently available phase-field formulations of *brittle* fracture encompass static and dynamic models. We mention the papers by Del Piero et al. (2007), Lancioni and Royer-Carfagni (2009), Amor et al. (2009), Freddi and Royer-Carfagni (2009, 2010), Kuhn and Müller (2010), Miehe et al. (2010a, b), Pham et al. (2011), Borden (2012), Borden et al. (2014), Vignollet et al. (2014), Mesgarnejad et al. (2015), Kuhn et al. (2015), Ambati et al. (2015), Marigo et al. (2016), Strobl and Seelig (2016), Weinberg and Hesch (2017), Tanné et al. (2018), Sargado et al. (2018), Gerasimov et al. (2018), Wu and Nguyen (2018), Wu et al. (2019), where various formulations are developed and validated. Recently, the framework has been also extended to ductile (elasto-plastic) fracture Miehe et al. (2015, 2016), Duda et al. (2015), Ambati et al. (2015), Alessi et al. (2015, 2018), Borden et al. (2016), fracture in films Mesgarnejad et al. (2013), León Baldelli et al. (2014), shells Amiria et al. (2014), Ambati and De Lorenzis (2016), Kiendl et al. (2016), Reinoso et al. (2017), fracture under thermal loading Sicsic et al. (2013), Bourdin et al. (2014), Miehe et al. (2015), hydraulic fracture Bourdin et al. (2012), Wheeler et al. (2014), Mikelić et al. (2015a, b, c), Wilson and Landis (2016), fracture in porous media Miehe and Mauthe (2016), Wu and De Lorenzis (2016), Cajuhi et al. (2018), anisotropic fracture Li et al. (2015), Teichtmeister et al. (2017), Zhang et al. (2017), Nguyen et al. (2017), Bleyer and Alessi (2018), Li and Maurini (2019), fracture in laminates Alessi and Freddi (2017), to name a few.

The three important aspects of the formulation in the brittle case, which give rise to algorithmic challenges within the finite element treatment, and which have recently been a subject of intensive studies are the following:

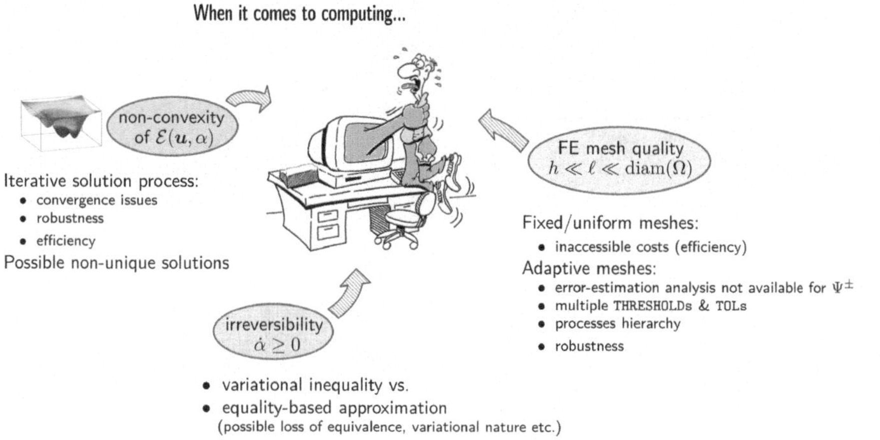


**Fig. 2** Challenging aspects of phase-field computing of brittle fracture with implications

(A) non-convexity of the governing energy functional with respect to the displacement and the phase-field arguments,
(B) irreversibility constraint for the crack phase-field,
(C) smallness of the regularization parameter (i.e., the length scale inherent to the diffusive crack approximation).

Figure 2 sketches the above aspects along with the related implications.

The lack of convexity (A) poses the major difficulty within the so-called *monolithic* treatment of the weak formulation, which aims at solving for both unknowns simultaneously. It is manifested via convergence issues of the direct Newton-Raphson iterative procedure Gerasimov and De Lorenzis (2016), Heister et al. (2015), Wick (2017a, b). Some new results by Gerasimov and De Lorenzis (2016), Heister et al. (2015), and Wick (2017a, b) on modified Newton-Raphson schemes, and by Kopanicakova and Krause (2019) on the trust region method hold a promise for improving the robustness of the monolithic approach. Alternatively (and more commonly), the *staggered* (also termed partitioned, or alternate minimization) solution strategy based on decoupling of the weak formulation into a system and then iterating between the equations is used Bourdin et al. (2000, 2008), Bourdin (2007a, b), Amor et al. (2009), Miehe et al. (2010a, b), Pham et al. (2011), Borden et al. (2014), Mesgarnejad et al. (2015), Ambati et al. (2015). The staggered scheme is intrinsically robust, but typically has a very slow convergence behavior of the iterative solution process, see e.g., Ambati et al. (2015), Gerasimov and De Lorenzis (2016), Farrell and Maurini (2017). Recently, the results by Farrell and Maurini (2017) on accelerated over-relaxed partitioned schemes indicate the possibility for the staggered approach to gain better efficiency. The aforementioned approaches are outlined in Fig. 3.

Due to (B), the formulation is a constrained minimization problem, whose optimality condition is a *variational inequality* Amor et al. (2009), Pham et al. (2011), Mesgarnejad et al. (2015), Marigo et al. (2016), thus requiring special solution algo-

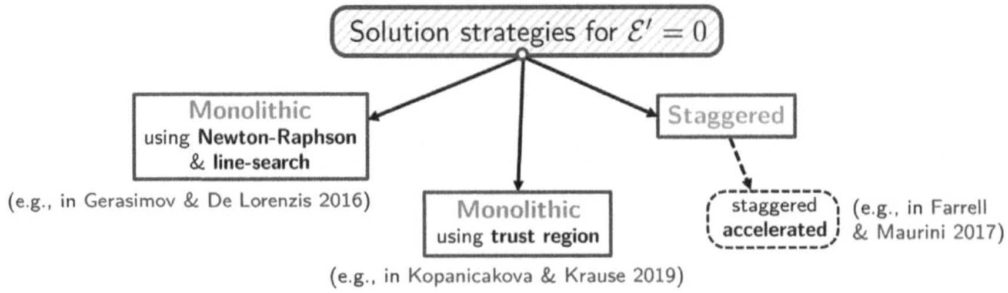


**Fig. 3** Iterative solution approaches for solving the weak formulation $\mathcal{E}' = 0$ with $\mathcal{E}$ denoting the energy functional for the mechanical system under consideration

rithms. Several options of enforcing the constraint that lead to a simpler *equality-based* formulation and which can be classified as relaxed, penalized, and implicit ones are found in the literature, see Gerasimov and De Lorenzis (2019) for an overview. Within the first class Bourdin et al. (2000, 2008), Bourdin (2007a, b), Del Piero et al. (2007), Lancioni and Royer-Carfagni (2009), Gerasimov and De Lorenzis (2016), Burke et al. (2010a, b, 2013), Artina et al. (2014, 2015), the irreversibility of the crack phase-field is enforced only on the so-called 'crack-set' (the points of the domain where the phase-field variable exceeds some threshold value). In this case, irreversibility of only a fully developed crack is modeled, and therefore the technique is termed relaxed. With regard to the second class, we recall the augmented-Lagrangian method in Wheeler et al. (2014), Wick (2017a, b), as well as an advancement of the Lagrange multipliers method using a primal-dual active set strategy Heister et al. (2015). Both approaches allow for both the staggered and the fully monolithic treatment, but may require solving for extra variable(s) Wheeler et al. (2014), Wick (2017a, b), and the necessity of tracing explicitly various (active and inactive) sets in which the corresponding sub-problems are to be solved Heister et al. (2015). Also, a simpler penalization procedure is advocated in Gerasimov and De Lorenzis (2019), with the advantage that the optimally defined penalty parameter guarantees a sufficiently accurate user-prescribed enforcement of the crack phase-field irreversibility. Finally, the implicit enforcement of the constraint using a so-called 'history field' was proposed in Miehe et al. (2010b) and has been adopted in a major amount of works on the topic. This approach, however, yields a problem of non-variational nature, whose equivalence to the original formulation cannot be proven. Also, in this case, only the staggered solution scheme can be employed. The above options of incorporating the constraint are summarized in Fig. 4.

Finally, property (C) calls for extremely fine meshes, at least locally in the crack phase-field transition zone. Modeling a failure process whose final pattern is not known in advance precludes the construction of a suitably pre-refined mesh, thus forcing to compute on fixed uniform meshes (unless adaptivity is introduced). In this case, the computational cost is very high. Already in the seminal paper by Bourdin et al. (2000) and later in Mesgarnejad et al. (2015) parallel computing has been advocated for the staggered solution scheme combined with uniformly fine meshes.

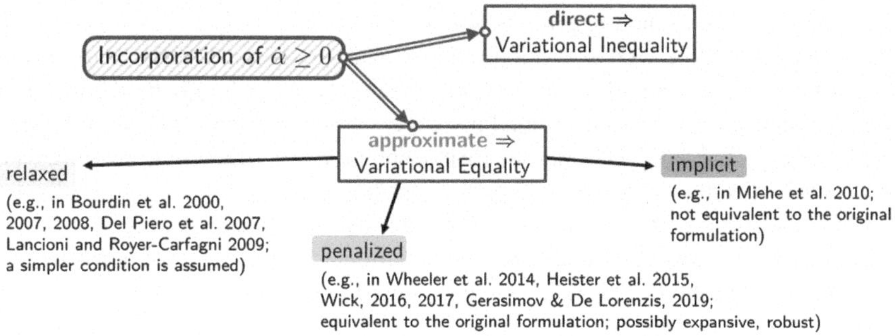


**Fig. 4** Options to incorporate the irreversibility constraint $\dot{\alpha} \geq 0$ for the crack phase-field

Recent findings by Burke et al. (2010a, b, 2013), Artina et al. (2014, 2015) and Wick (2016) on error-controlled adaptive mesh refinement strategies, as well as by Heister et al. (2015), Klinsmann et al. (2015) and Nagaraja et al. (2018) on physics-motivated procedures for mesh adaptivity provide a basis to efficiently tackle (C) as well.

The rest of this chapter is organized as follows. In Section "Formulation", we outline the main concepts of phase-field modeling of brittle fracture including major ingredients of the formulation such as coupling function, tension-compression split of the elastic energy density function, local energy dissipation function etc. The incremental variational problem which models a quasi-static fracture evolution, required for the numerical treatment, is then presented. Section "Treatment of Irreversibility" presents various ideas for incorporating the irreversibility constraint and the resulting weak formulations, whereas Section "Solution Strategies" discusses the available iterative solution strategies for the formulations. Both sections are complemented by illustrative numerical examples. In the simulations, we employ the numerical package FreeFem++ Hecht et al. (2019). Both the displacement field and the crack phase-field are approximated using linear triangles on fixed (non-adaptive) finite element meshes, which are pre-adapted (refined) in the region where crack propagation is expected.

## Formulation

In this section, for a mechanical system undergoing brittle fracture we recall the phase-field formulation that models this process, and outline its main ingredients.

### Governing Energy Functional

Let $\Omega \subset \mathbb{R}^d$, $d = 2, 3$ be an open and bounded domain representing the configuration of a $d$-dimensional linear elastic body, and let $\Gamma_{D,0}$, $\Gamma_{D,1}$ and $\Gamma_{N,1}$ be the (non-overlapping) portions of the boundary $\partial\Omega$ of $\Omega$ on which homogeneous Dirichlet, non-homogeneous Dirichlet and Neumann boundary conditions are prescribed, respectively. The body is assumed to be linearly elastic and isotropic, with the elastic strain energy density function given by $\Psi(\varepsilon) := \frac{1}{2}\varepsilon : \mathbb{C} : \varepsilon = \frac{1}{2}\lambda \mathrm{tr}^2(\varepsilon) + \mu \mathrm{tr}(\varepsilon \cdot \varepsilon)$, where, in turn, $\varepsilon$ is the second-order infinitesimal strain tensor, $\mathbb{C}$ is the fourth-order elasticity tensor, and $\lambda$ and $\mu$ are the Lamé constants. Also, let $G_c$ be the material fracture toughness. A quasi-static loading process with the discrete pseudo-time step parameter $n = 1, 2, \ldots$, such that the displacement $\bar{\boldsymbol{u}}_n$ and traction $\bar{\boldsymbol{t}}_n$ loading data are prescribed on the corresponding parts of the boundary is considered. Finally, let $\Gamma_c \subset \Omega$ be the crack surface that is evolving during the process, see Fig. 1b.

For the mechanical system at hand, the variational approach to brittle fracture in Francfort and Marigo (1998) relies on the energy functional

$$\mathcal{E}(\boldsymbol{u}, \Gamma_c) = \underbrace{\int_{\Omega\backslash\Gamma_c} \Psi(\varepsilon(\boldsymbol{u}))\,\mathrm{d}\boldsymbol{x}}_{E_{\mathrm{el.}}(\boldsymbol{u},\Gamma_c)} + \underbrace{G_c \int_{\Gamma_c} \mathrm{d}s}_{E_S(\Gamma_c)} - \int_{\Gamma_{N,1}} \bar{\boldsymbol{t}}_n \cdot \boldsymbol{u}\,\mathrm{d}s, \tag{1}$$

with $\boldsymbol{u} : \Omega\backslash\Gamma_c \to \mathbb{R}^d$ such that $\boldsymbol{u} = \boldsymbol{0}$ on $\Gamma_{D,0}$ and $\boldsymbol{u} = \bar{\boldsymbol{u}}_n$ on $\Gamma_{D,1}$ as the displacement field, $\Gamma_c$ as the crack set, and the related minimization problem at each $n \geq 1$. In (1), the functionals termed $E_{\mathrm{el.}}$ and $E_S$ represent the elastic energy stored in the body and the fracture surface energy dissipated within the fracture process. The latter rigorously reads $E_S(\Gamma_c) = G_c \mathcal{S}^{d-1}(\Gamma_c)$ with $\mathcal{S}^p$ as the so-called $p$-dimensional Hausdorff measure of the crack set $\Gamma_c$. In simple terms, $\mathcal{S}^1(\Gamma_c)$ and $\mathcal{S}^2(\Gamma_c)$ represent the length and the surface area of $\Gamma_c$ when $d = 2$ and 3, respectively.

The regularization of (1) *á la* Bourdin et al. (2000, 2008), Bourdin (2007a, b), which is the basis for a variety of fracture phase-field formulations, reads as follows:

$$\mathcal{E}(\boldsymbol{u}, \alpha) = \underbrace{\int_{\Omega} g(\alpha)\Psi(\varepsilon(\boldsymbol{u}))\,\mathrm{d}\boldsymbol{x}}_{E_{\mathrm{el.}}(\boldsymbol{u},\alpha)} + \underbrace{\frac{G_c}{c_w}\int_{\Omega}\left(\frac{w(\alpha)}{\ell} + \ell|\nabla\alpha|^2\right)\mathrm{d}\boldsymbol{x}}_{E_S(\alpha)} - \int_{\Gamma_{N,1}} \bar{\boldsymbol{t}}_n \cdot \boldsymbol{u}\,\mathrm{d}s, \tag{2}$$

with $\boldsymbol{u} : \Omega \to \mathbb{R}^d$ and $\alpha : \Omega \to [0, 1]$ standing for the smeared counterparts of the discontinuous displacement and the crack set in (1). The phase-field variable $\alpha$ takes the value 1 on $\Gamma_c$, decays smoothly to 0 in a subset of $\Omega\backslash\Gamma_c$ and then takes the 0-value in the rest of the domain. With this definition, the limits $\alpha = 1$ and $\alpha = 0$ represent the fully broken and the intact (undamaged) material phases, respectively, whereas the intermediate range $\alpha \in (0, 1)$ mimics the transition zone between them. The function $g$ is responsible for the material stiffness degradation. The function $w$ defines the decaying profile of $\alpha$, whereas the parameter $0 < \ell \ll \mathrm{diam}(\Omega)$ controls the size of the localization zone of $\alpha$, in other words, the thickness of the transition zone between the two material states.

**Degradation and local dissipation functions**

The functions $g$ and $w$ are the major ingredients of (2), and their specific choice establishes the rigorous link between (1) and (2) when $\ell \to 0$ via the notion of $\Gamma$-convergence, see e.g., Braides (1998), Chambolle (2004), also giving a meaning to the induced constant $c_w$. Thus, $g$ is a continuous monotonic function that fulfills

**Table 1** Ingredients of formulation (2) and (5)

| $g$ | $w$ | Name |
|---|---|---|
| $(1-\alpha)^2$ | $\alpha$ | AT-1 model |
| | $\alpha^2$ | AT-2 model |

the properties: $g(0) = 1$, $g(1) = 0$, $g'(1) = 0$ and $g'(\alpha) < 0$ for $\alpha \in [0, 1)$, see e.g., Pham et al. (2011) for argumentation and discussion. The quadratic polynomial

$$g(\alpha) := (1-\alpha)^2, \tag{3}$$

is the simplest choice of the kind. The function $w$, also called the local part of the dissipated fracture energy density function Pham et al. (2011), is continuous and monotonic such that $w(0) = 0$, $w(1) = 1$ and $w'(\alpha) \geq 0$ for $\alpha \in [0, 1]$. The constant $c_w := 4\int_0^1 \sqrt{w(t)}\,\mathrm{d}t$ is a normalization constant in the sense of $\Gamma$-convergence. The two suitable candidates for $w$ reading

$$w(\alpha) := \begin{cases} \alpha, \\ \alpha^2, \end{cases} \quad \text{such that} \quad c_w = \begin{cases} \frac{8}{3}, \\ 2, \end{cases} \tag{4}$$

are widely adopted. Formulation (2) combined with the aforementioned choices for $g$ and $w$ are typically termed the AT-1 and AT-2 models, see Table 1. AT stands for *Ambrosio-Tortorelli* and the corresponding type of regularization, see Ambrosio and Tortorelli (1990). The main difference between the two models is that AT-1 leads to the existence of an elastic stage before the onset of fracture, whereas using AT-2 the phase-field starts to evolve as soon as the material is loaded, see e.g., Amor et al. (2009), Pham et al. (2011), Marigo et al. (2016) for a more detailed explanation.

Other representations for $g$ and $w$ are available in the literature, see e.g., Borden (2012), Borden et al. (2016), Kuhn et al. (2015), Sargadoa et al. (2018), Ambati et al. (2015), Alessi et al. (2015), Wilson and Landis (2016), Burke et al. (2010b, 2013).

**Tension-compression split**. Regarding the elastic strain energy function $\Psi$ the following must be noted. Due to the symmetry of $\Psi$ with respect to the variable $\boldsymbol{u}$, formulation (2) does not distinguish between fracture behavior in tension and compression. In the numerical simulations, this is manifested by the mesh interpenetration inside of the compressed fractured zones, as reported e.g., in Bourdin et al. (2000, Sect. 3.3), Del Piero et al. (2007, Sect. 7) and Lancioni and Royer-Carfagni (2009, Sects. 4.1–4.3). In the discrete crack setting, this would be equivalent to compressive interpenetration of the crack faces. One of the proposed remedies for avoiding such a non-physical behavior implies placing $\Psi$ into the context of non-linear (finite) elasticity, as first presented in Del Piero et al. (2007). To remain within the framework of linear elasticity, the alternative is to break the symmetry by introducing an additive split of $\Psi$ into the so-called 'tensile' and 'compressive' parts $\Psi^+$ and $\Psi^-$, respectively, and enabling the degradation of $\Psi^+$ only. This option is advocated in Lancioni and Royer-Carfagni (2009), Amor et al. (2009), Freddi and Royer-Carfagni (2009, 2010), Miehe et al. (2010a, b) and yields the following enhanced representation of (2):

$$\mathcal{E}(\boldsymbol{u},\alpha) = \int_\Omega \left[g(\alpha)\Psi^+(\varepsilon(\boldsymbol{u})) + \Psi^-(\varepsilon(\boldsymbol{u}))\right] \mathrm{d}\boldsymbol{x} + \frac{G_c}{c_w}\int_\Omega \left(\frac{w(\alpha)}{\ell} + \ell|\nabla\alpha|^2\right) \mathrm{d}\boldsymbol{x} - \int_{\Gamma_{N,1}} \bar{\boldsymbol{t}}_n \cdot \boldsymbol{u}\, \mathrm{d}s. \tag{5}$$

The two widely adopted options for tension-compression splits $\Psi^\pm$ are based respectively on the so-called volumetric-deviatoric decomposition of the strain tensor $\varepsilon$, i.e.,

$$\varepsilon = \varepsilon^{\mathrm{dev}} + \varepsilon^{\mathrm{vol}},$$

where $\varepsilon^{\mathrm{vol}} := \frac{1}{d}\mathrm{tr}(\varepsilon)\boldsymbol{I}$, $\varepsilon^{\mathrm{dev}} := \varepsilon - \varepsilon^{\mathrm{vol}}$ and $\boldsymbol{I}$ is the second-order identity tensor, and the spectral decomposition of $\varepsilon$, namely,

$$\varepsilon = \varepsilon_+ + \varepsilon_-,$$

where $\varepsilon_\pm := \sum_{i=1}^{3}\langle\varepsilon_i\rangle_\pm \boldsymbol{n}_i \otimes \boldsymbol{n}_i$ with $\{\varepsilon_i\}_{i=1}^3$ and $\{\boldsymbol{n}_i\}_{i=1}^3$ as the principal strains and principal strain directions, respectively, and $\langle a\rangle_\pm := \frac{1}{2}(a \pm |a|)$. The resulting representations read

$$\begin{cases} \Psi^+(\varepsilon) := \frac{1}{2}K_d\langle\mathrm{tr}(\varepsilon)\rangle_+^2 + \mu\mathrm{tr}(\varepsilon^{\mathrm{dev}}\cdot\varepsilon^{\mathrm{dev}}) \\ \Psi^-(\varepsilon) := \frac{1}{2}K_d\langle\mathrm{tr}(\varepsilon)\rangle_-^2 \end{cases} \tag{6}$$

with $K_d := \lambda + \frac{2\mu}{d}$, and

$$\Psi^\pm(\varepsilon) := \frac{1}{2}\lambda\langle\mathrm{tr}(\varepsilon)\rangle_\pm^2 + \mu\mathrm{tr}(\varepsilon_\pm\cdot\varepsilon_\pm), \tag{7}$$

respectively. Notice that (6) is independently presented in Amor et al. (2009) and Freddi and Royer-Carfagni (2009) (based on an extension of the split in Lancioni and Royer-Carfagni 2009), and (7) is considered in Miehe et al. (2010a, b). An idea similar to the latter one is also developed in Freddi and Royer-Carfagni (2010, Sect. 3.4). We refer the interested reader to the aforementioned publications and to Li (2016), where options for constructing $\Psi^\pm$ and the related implications are explained.

Our final note here is that formulation (5) is what is usually referred to as *phase-field model of brittle fracture* (at least in the engineering literature). Despite the wide employment of the formulation, the Γ-convergence result that relates (5) to the original Francfort-Marigo formulation (1) is, in general, not available. Some particular results have recently been established in Chambolle et al. (2018).

## Quasi-Static Evolution (Incremental Variational Problem)

With $\mathcal{E}$ defined by (5), the state of the system at a given loading step $n \geq 1$ is represented by

$$\arg\min\{\mathcal{E}(\boldsymbol{u}, \alpha) : \boldsymbol{u} \in \boldsymbol{V}_{\bar{\boldsymbol{u}}_n}, \alpha \in \mathcal{D}_{\alpha_{n-1}}\}, \tag{8}$$

where

$$\boldsymbol{V}_{\bar{\boldsymbol{u}}_n} := \{\boldsymbol{u} \in \boldsymbol{H}^1(\Omega) : \boldsymbol{u} = \boldsymbol{0} \text{ on } \Gamma_{D,0},\ \boldsymbol{u} = \bar{\boldsymbol{u}}_n \text{ on } \Gamma_{D,1}\}$$

is the kinematically admissible displacement space with $\boldsymbol{H}^1(\Omega) := [H^1(\Omega)]^d$ and $H^1$ denoting the usual Sobolev space, and

$$\mathcal{D}_{\alpha_{n-1}} := \{\alpha \in H^1(\Omega) : \alpha \geq \alpha_{n-1} \text{ in } \Omega\}$$

is the admissible space for $\alpha$ with $\alpha_{n-1}$ known from the previous step. The condition $\alpha \geq \alpha_{n-1}$ in $\Omega$ is used to enforce the *irreversibility* of the crack phase-field evolution. It is the backward difference quotient form of $\dot{\alpha} \geq 0$ in $\Omega$.

Thus, quasi-static evolution of fracture within the system is then given by the sequence of the solution snapshots $\{(\boldsymbol{u}_n, \alpha_n)\}$, $n \geq 1$.

Due to the $\alpha \geq \alpha_{n-1}$ requirement, the incremental variational problem (8) is a constrained minimization problem and its necessary optimality condition for computing the solution $(\boldsymbol{u}, \alpha) \in \boldsymbol{V}_{\bar{\boldsymbol{u}}_n} \times \mathcal{D}_{\alpha_{n-1}}$ is a variational inequality. Written down in the partitioned form, it reads

$$\boxed{\begin{cases} \mathcal{E}'_{\boldsymbol{u}}(\boldsymbol{u}, \alpha; \boldsymbol{v}) = 0, & \forall\, \boldsymbol{v} \in \boldsymbol{V}_0, \\ \mathcal{E}'_{\alpha}(\boldsymbol{u}, \alpha; \beta - \alpha) \geq 0, & \forall\, \beta \in \mathcal{D}_{\alpha_{n-1}}, \end{cases}} \tag{9}$$

see e.g. Burke et al. (2010b, 2013), Pham et al. (2011), Farrell and Maurini (2017), where $\mathcal{E}'_{\boldsymbol{u}}$ and $\mathcal{E}'_{\alpha}$ are the directional derivatives of the energy functional with respect to $\boldsymbol{u}$ and $\alpha$, respectively,

$$\mathcal{E}'_{\boldsymbol{u}}(\boldsymbol{u}, \alpha; \boldsymbol{v}) := \int_\Omega \left[ g(\alpha)\frac{\partial \Psi^+}{\partial \boldsymbol{\varepsilon}}(\boldsymbol{\varepsilon}(\boldsymbol{u})) + \frac{\partial \Psi^-}{\partial \boldsymbol{\varepsilon}}(\boldsymbol{\varepsilon}(\boldsymbol{u})) \right] : \boldsymbol{\varepsilon}(\boldsymbol{v})\, \mathrm{d}\boldsymbol{x} - \int_{\Gamma_{N,1}} \bar{\boldsymbol{t}}_n \cdot \boldsymbol{v}\, \mathrm{d}s, \tag{10}$$

$$\mathcal{E}'_{\alpha}(\boldsymbol{u}, \alpha; \beta) := \int_\Omega \left[ g'(\alpha)\Psi^+(\boldsymbol{\varepsilon}(\boldsymbol{u}))\beta + \frac{G_c}{c_w}\left(\frac{1}{\ell}w'(\alpha)\beta + 2\ell\nabla\alpha \cdot \nabla\beta\right)\right] \mathrm{d}\boldsymbol{x}. \tag{11}$$

The displacement test space in (9) is defined as $\boldsymbol{V}_0 := \{\boldsymbol{v} \in \boldsymbol{H}^1(\Omega) : \boldsymbol{v} = \boldsymbol{0}$ on $\Gamma_{D,0} \cup \Gamma_{D,1}\}$.

## Treatment of Irreversibility

The variational inequality $\mathcal{E}'_{\alpha} \geq 0$ in (9) that stems from the irreversibility constraint $\alpha \geq \alpha_{n-1}$ requires special solution algorithms, see e.g., Kinderlehrer and Stampacchia (1980), Glowinski et al. (1981) and Burke et al. (2010b, Sect. 5). In the following, we provide details about the three available options of handling $\alpha \geq \alpha_{n-1}$ mentioned in the introduction, which lead to the equality-based formulations and, hence, allow for simpler algorithmic treatment. The equivalence between the corresponding formulations and the reference one in (9) is highlighted and then illustrated by the numerical examples.

### *Relaxed ('Crack-Set' Irreversibility)*

This is a version of irreversibility introduced by Bourdin et al. (2000, 2008), Bourdin (2007a, b) and also adopted e.g. by Burke et al. (2010a), which relies on the notion of a crack set: if at the current loading step $n$ the (crack) set

$$\mathrm{CR}_{n-1} := \{\boldsymbol{x} \in \overline{\Omega} : \ \alpha_{n-1}(\boldsymbol{x}) \geq \mathtt{CRTOL}\}, \tag{12}$$

see Fig. 5 for a sketch where $0 \ll \mathtt{CRTOL} < 1$ is a specified threshold, is non-empty, one explicitly sets $\alpha = 1$ for all $\boldsymbol{x} \in \mathrm{CR}_{n-1}$, and the corresponding analogue to (8) to be solved at step $n$ is

$$\arg\min\{\mathcal{E}(\boldsymbol{u}, d) : \ \boldsymbol{u} \in \boldsymbol{V}_{\bar{\boldsymbol{u}}_n}, \alpha \in \{H^1(\Omega) : \alpha|_{\mathrm{CR}_{n-1}} = 1\}\}. \tag{13}$$

The weak system of equations for $(\boldsymbol{u}, \alpha)$ in this case reads

$$\boxed{\begin{cases} \mathcal{E}'_{\boldsymbol{u}}(\boldsymbol{u}, \alpha; \boldsymbol{v}) = 0, & \forall\, \boldsymbol{v} \in \boldsymbol{V}_0 \\ \mathcal{E}'_{\alpha}(\boldsymbol{u}, \alpha; \beta) = 0, & \forall\, \beta \in \{H^1(\Omega) : \beta|_{\mathrm{CR}_{n-1}} = 0\}, \end{cases}} \tag{14}$$

where $\mathcal{E}'_{\boldsymbol{u}}$ and $\mathcal{E}'_{\alpha}$ are given by (10) and (11), respectively. As noted in Amor et al. (2009), the present option can be viewed as a *relaxed* version of the requirement $\alpha \geq \alpha_{n-1}$ in $\Omega$ since it only enforces irreversibility of a fully developed crack, whereas phase-field patterns with $\alpha(\boldsymbol{x}) < \mathtt{CRTOL}$ for all $\boldsymbol{x} \in \Omega$, which from the mechanical standpoint may be viewed as partially damaged regions, are allowed to heal.

*Remark 3.1* Various algorithmic treatments of the 'crack-set' irreversibility can be found in Del Piero et al. (2007), and in Lancioni and Royer-Carfagni (2009). Its more sophisticated version is considered in Burke et al. (2010b, 2013).

*Remark 3.2* In Artina et al. (2014, 2015) and Gerasimov and De Lorenzis (2016), the Dirichlet condition $\alpha|_{\mathrm{CR}_{n-1}} = 1$ is enforced via penalization by introducing into (5) the functional

$$P(\alpha; \tau) := \frac{1}{2\tau} \int_{\mathrm{CR}_{n-1}} (1 - \alpha)^2 \,\mathrm{d}\boldsymbol{x}, \quad 0 < \tau \ll 1,$$

thus yielding the penalized counterpart of (13) and (14).

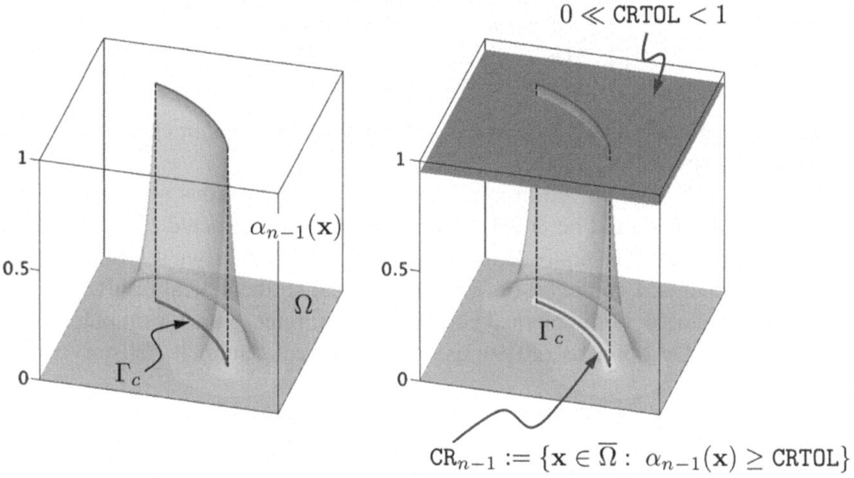


**Fig. 5** Illustration of a fully-developed crack phase-field $\alpha_{n-1}$ (left) and the related crack set $\mathrm{CR}_{n-1}$ (right); $\Gamma_c$ indicates the crack $\alpha_{n-1}$ relates to

The choice of the threshold value CRTOL in (12) is subtle and may have a strong impact on the computational results, as shown in Burke et al. (2010b, 2013). A similar concern applies to the choice of $\tau$ in the corresponding penalized realization.

### *Implicit ('History-Field' Irreversibility)*

In Miehe et al. (2010b), Miehe and co-workers proposed the idea of enforcing the irreversibility constraint $\alpha \geq \alpha_{n-1}$ implicitly, via the notion of a *history*-field. Their major assumption is that the tensile energy $\Psi^+$ can be viewed as the *driving force* of the phase-field evolution and, hence, the maximal $\Psi^+$ accumulated within the loading history, denoted as

$$\mathcal{H}_n(\boldsymbol{x}) := \max_{n\geq 1}\{\mathcal{H}_{n-1}(\boldsymbol{x}), \Psi^+(\varepsilon(\boldsymbol{u}))\}, \quad \mathcal{H}_0 \equiv 0, \tag{15}$$

must guarantee the fulfillment of $\alpha \geq \alpha_{n-1}$. Technically, one substitutes $\mathcal{H}_n$ to $\Psi^+$ in the original $\mathcal{E}'_\alpha$ in (11) such that

$$\widetilde{\mathcal{E}}'_\alpha(\boldsymbol{u}, \alpha; w) := \int_\Omega \left[ g'(\alpha)\mathcal{H}_n\beta + \frac{G_c}{c_w}\left(\frac{1}{\ell}w(\alpha)\beta + 2\ell\nabla\alpha\cdot\nabla\beta\right)\right] \mathrm{d}\boldsymbol{x}, \tag{16}$$

and forms the system for computing the solution $(\boldsymbol{u}, \alpha) \in \boldsymbol{V}_{\bar{\boldsymbol{u}}_n} \times H^1(\Omega)$:

$$\begin{cases} \mathcal{E}'_{\boldsymbol{u}}(\boldsymbol{u},\alpha;\boldsymbol{v}) = 0, & \forall\, \boldsymbol{v} \in \boldsymbol{V}_0 \\ \widetilde{\mathcal{E}}'_{\alpha}(\boldsymbol{u},\alpha;\beta) = 0, & \forall\, \beta \in H^1(\Omega), \end{cases} \tag{17}$$

where $\mathcal{E}'_{\boldsymbol{u}}$ is given by (10). System (17) is composed of equalities and also uses unconstrained spaces for $\alpha$ and $\beta$.

This approach is employed in a large number of publications on the topic due to its simplicity. It should be noted, however, that the constructed $\widetilde{\mathcal{E}}'_{\alpha}$ is no longer of variational nature. The equivalence between (17) and the original formulation in (9) is not evident and, to the best of our knowledge, no theoretical results that can prove it are available. For numerical comparisons we refer to the recent publication Gerasimov and De Lorenzis (2019), as well as to Sect. "Implicit ('History-Field' Irreversibility)".

Interestingly, in the seminal paper of Miehe et al. (2010a), the authors considered the penalization option similar to representation (18) below, yet already in Miehe et al. (2010b) they switched to the notion of $\mathcal{H}_n$ and have been using it in all their following publications.

### *Penalized*

The third alternative of addressing $\alpha \geq \alpha_{n-1}$, which results in the *equality* to be solved instead of the inequality, is via penalization. Several available options are listed below.

**Option** ①: In the most straightforward case, one can add the penalty term

$$P(\alpha;\gamma) := \frac{\gamma}{2}\int_{\Omega} \langle \alpha - \alpha_{n-1}\rangle_{-}^{2}\, \mathrm{d}\boldsymbol{x}, \quad \gamma \gg 1, \tag{18}$$

to the energy functional $\mathcal{E}$ in (5).[1] In (18), $\langle y\rangle_{-} := \min(0, y)$. The corresponding variational problem reads

$$\arg\min\{\mathcal{E}(\boldsymbol{u},\alpha) + P(\alpha;\gamma) : \boldsymbol{u} \in \boldsymbol{V}_{\bar{\boldsymbol{u}}_n}, \alpha \in H^1(\Omega)\}. \tag{19}$$

and the resulting weak system for $(\boldsymbol{u},\alpha)$ is as follows

$$\begin{cases} \mathcal{E}'_{\boldsymbol{u}}(\boldsymbol{u},\alpha;\boldsymbol{v}) = 0, \quad \forall\, \boldsymbol{v} \in \boldsymbol{V}_0, \\ \mathcal{E}'_{\alpha}(\boldsymbol{u},\alpha;\beta) + \gamma\displaystyle\int_{\Omega} \langle \alpha - \alpha_{n-1}\rangle_{-}\beta\, \mathrm{d}\boldsymbol{x} = 0, \quad \forall\, \beta \in H^1(\Omega), \end{cases} \tag{20}$$

[1] A general form of the integrand in (18) is $\langle \alpha - \alpha_{n-1}\rangle_{-}^{p}$ with $p \geq 1$ and $P$ represents a regularization of the indicator function

$$I_{\mathcal{D}_{\alpha_{n-1}}}(\alpha) := \begin{cases} 0, & \text{in } \mathcal{D}_{\alpha_{n-1}}, \\ +\infty, & \text{otherwise}, \end{cases} \tag{C}$$

to be added to $\mathcal{E}$ in (5).

with $\mathcal{E}'_u$ and $\mathcal{E}'_\alpha$ given again by (10) and (11), respectively. Note that the admissible space for $\alpha$ and the test space for $\beta$ are no longer constrained. The obtained penalized un-constrained problem (20) approximates the original constrained problem (9) and thus is *equivalent* to it in the limit of $\gamma \to \infty$. Thus the appropriate choice of $\gamma$ is always viewed as a critical point of the technique also within the numerical experiments. A too small penalty parameter will lead to inaccurate (i.e., insufficient) enforcement of the constraint, a too large one will result in ill-conditioning (i.e., a so-called stability issue).

In Gerasimov and De Lorenzis (2019), an analytical procedure for the 'reasonable' choice of a *lower bound* for $\gamma$ that guarantees a sufficiently accurate enforcement of the crack phase-field irreversibility constraint and, seemingly, does not manifest ill-conditioning is devised. It is shown that the optimal penalty parameter $\gamma$ is a function of two formulation parameters (the fracture toughness $G_c$ and the regularization length scale $\ell$), but is independent of the problem setup (geometry, boundary conditions etc.), the formulation ingredients (degradation function $g$, tension–compression split $\Psi_\pm$ etc.), as well as the discretization (mesh size). The carried out numerical studies validate the findings.

**Option** ②: The augmented Lagrangian approach is adopted by Wheeler et al. (2014) and Wick (2017a, b). It implies adding to (5) the functional

$$P(\alpha; \Xi, \gamma) := \frac{1}{2\gamma} \int_\Omega \langle \Xi + \gamma\alpha \rangle_+^2 \, \mathrm{d}\boldsymbol{x} + \frac{1}{2\gamma} \int_\Omega \langle \Xi + \gamma(\alpha - \alpha_{n-1}) \rangle_-^2 \, \mathrm{d}\boldsymbol{x} - \frac{1}{2\gamma} \int_\Omega \Xi^2 \, \mathrm{d}\boldsymbol{x}, \tag{21}$$

with an unknown $\Xi \in L^2(\Omega)$ and a user-prescribed penalty constant $\gamma > 0$. The aforementioned $P$ is the so-called Moreau-Yosida approximation of the indicator function $I_{\mathcal{D}_{\alpha_{n-1}}}$ in (C), see Wheeler et al. (2014) for details. It is advocated that using (21), the stability issues possibly occurring in case of (18) are avoided. We notice that simplified representations for $P$ in (21) which contain only the first two terms and only the second term are considered in Wheeler et al. (2014) and in Wick (2017a, b), respectively. Furthermore, in either case, a 'fixed-point iteration'-like strategy for calculating $\Xi$ using $P'$ is also employed, see e.g., Wheeler et al. (2014, Algorithm 1) and Wick (2017a, Algorithm 4.1). This is done to prevent the possibly significant increase of the computational effort in the case the field $\Xi$ is to be solved for.

**Option** ③: In Heister et al. (2015), $\alpha \geq \alpha_{n-1}$ is incorporated using a Lagrange multiplier and, following Hintermüller et al. (2002), the obtained formulation is then tackled using a primal-dual active set strategy. As noticed by the authors, this approach can be viewed as a semi-smooth Newton method with the advantage that the fully-monolithic solution scheme would "allow for fast (super-linear) convergence", and that, "in contrast to other (penalization) methods, no adjustment of parameters is required". On the other hand, the necessity of tracing explicitly various (active and inactive) sets in which the corresponding sub-problems are to be solved makes the approach computationally more expensive than a simple penalization approach.

## *Examples*

In this section we compare the solutions of the formulations in (17) and (20) for two benchmark problems. Recall that the former problem is the weak formulation based on the use of a history-field of Miehe and co-workers (2010b), and the latter is the penalized formulation recently proposed in Gerasimov and De Lorenzis (2019). As explained in Section "Treatment of Irreversibility", (17) is not equivalent to the original inequality-based problem (9), whereas (20) is equivalent [more precisely, it is a good approximation of (9)]. For every benchmark problem, formulations (17) and (20) are computed using the staggered scheme, see Section "Solution Strategies" for details.

The considered benchmark problems are the so-called single edge notched (SEN) specimen under shear originally considered in Bourdin et al. (2000), and later adopted with some modifications in many related papers, see, e.g., Miehe et al. (2010b), Borden et al. (2014), Ambati et al. (2015), Gerasimov and De Lorenzis (2016, 2019), and the traction test on a fiber-reinforced matrix Bourdin et al. (2008) (also in Bourdin et al. 2000; Bourdin 2007a; Amor et al. 2009), see Fig. 6. The first example is a crack propagation problem under external displacement-controlled loading, where the pre-existing crack is modeled discretely and the propagating crack is represented by the phase-field evolution. The problem setup is simple, but the failure pattern is not symmetric, being the result of a non-trivial combination of local tension–compression within the specimen during shear. The second example describes a more complicated process, since it encompasses crack *initiation* in the absence of a strong crack-tip singularity, which precedes the propagation stage. For either case, the pure monotonic loading regimes are simulated.

The SEN shear test is computed using both formulations in (17) and (20) in which functional $\mathcal{E}$ in (5) uses both the volumetric-deviatoric and spectral splits (for the sake of compactness, we denote them as `VD`- and `S`-split, respectively) and the models `AT`-1 and `AT`-2 given in Table 1. That is, there are four available combinations of split and model type in this case. The fiber-reinforced matrix test is computed using (17) and (20) and only for the `S`-split and the `AT`-2 model combination.

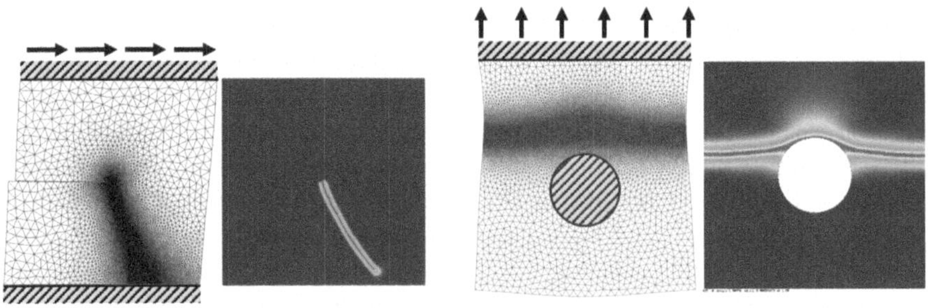

**Fig. 6** Problem setup (sketchy) and the computed failure patterns for the single edge notched (SEN) specimen subject to shear (left) and the traction experiment on a fiber-reinforced matrix (right)

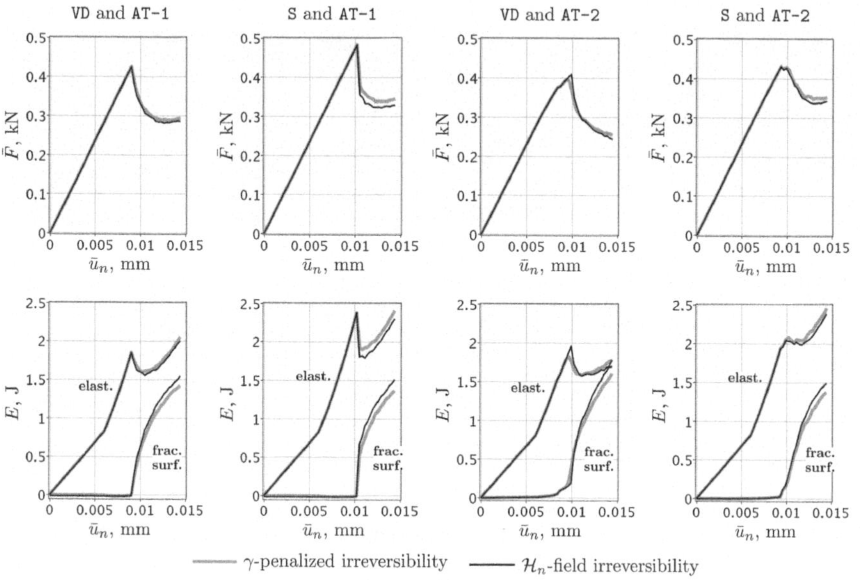


**Fig. 7** Comparison of the load–displacement and energy–displacement curves obtained for the SEN specimen shear test using the formulations in (17) and (20) and various combinations of split and model type

In Figs. 7 and 8, we present and compare, respectively, the load-/energy-displacement curves and the crack phase-field profiles obtained for the SEN shear test using the two irreversibility approaches. Similarly, Figs. 9 and 10 depict and compare the corresponding results for the traction test on a fiber-reinforced matrix.

In all considered cases, both irreversibility approaches yield qualitatively and quantitatively rather similar, but not identical results. Taking the $\gamma$-penalized results as reference, it can be seen that the history field approach yields under-estimation of the elastic energy and over-estimation of the fracture surface energy. We attribute this to the observation drawn from the crack phase-field plots in Figs. 8 and 10 that the support of the phase-field profile in the latter case is visibly thicker.

## Solution Strategies

In this section we discuss two iterative strategies for solving the equation $\mathcal{E}' = \mathcal{E}'_{\boldsymbol{u}} + \mathcal{E}'_{\alpha} = 0$: the so-call staggered and the monolithic ones.[2] Their advantages and disadvantages are already outlined in the introductory part and were also extensively

[2]In the latter case, according to the sketch in Fig. 3, we imply the monolithic procedure based on the Newton-Raphson method with the line-search procedure in Gerasimov and De Lorenzis (2016) to cope with the iterative convergence issues.

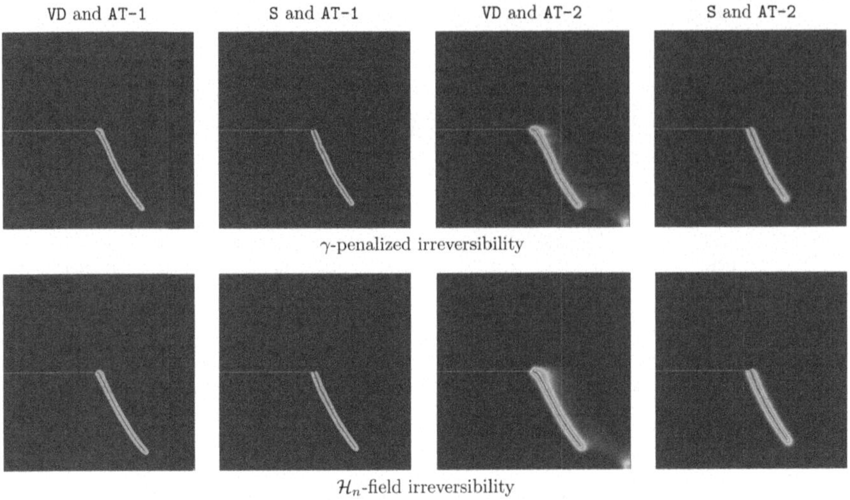


**Fig. 8** Comparison of the crack phase-field profile obtained at the last loading step for the SEN specimen shear test using the formulations in (17) and (20) and various combinations of split and model type

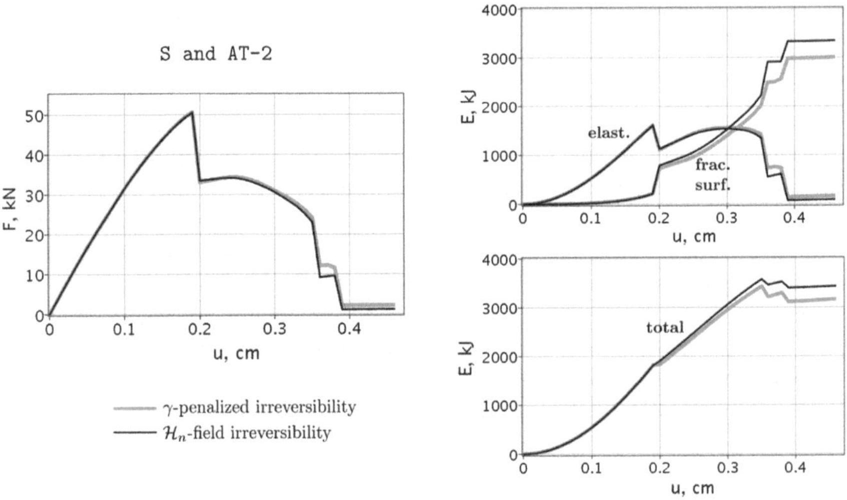


**Fig. 9** Comparison of the load–displacement and energy–displacement curves obtained for the fiber-reinforced matrix traction test using the formulations in (17) and (20) for the S-split and the AT-2 model combination

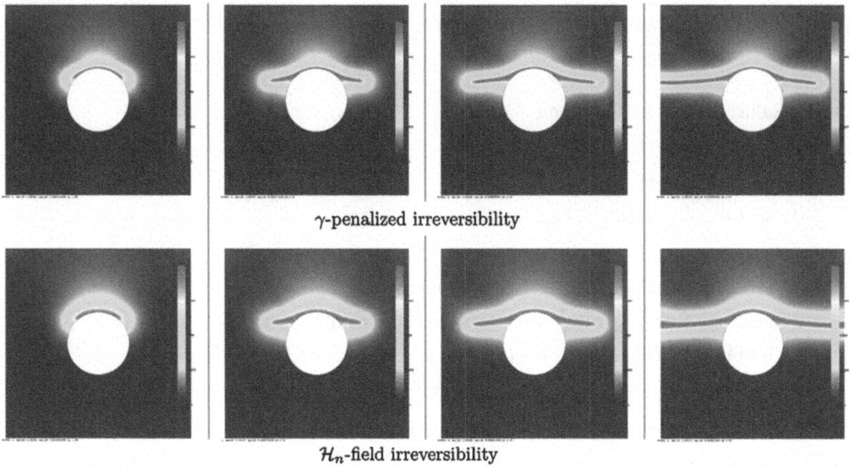


**Fig. 10** Comparison of the crack phase-field profile obtained at the last loading step for the fiber-reinforced matrix traction test using the formulations in (17) and (20) for the S-split and the AT-2 model combination

discussed in the related references. The purpose of the following section is to provide the necessary technical details for both approaches, as well as to illustrate their performance for three benchmark problems.

## *Staggered*

The staggered solution algorithm applied to $\mathcal{E}' = 0$ implies partitioning of this equation into the system $\mathcal{E}'_{\boldsymbol{u}} = 0$ and $\mathcal{E}'_{\alpha} = 0$, as has been already encountered in Section "Treatment of Irreversibility". By alternately fixing $\boldsymbol{u}$ and $\alpha$, the above coupled system is then solved in an iterative manner until convergence is achieved. The algorithm is sketched in Table 2 in a general form, that is, regardless of the system in (14), (17) and (20). Adaptation of step 2 to the corresponding equation in (14), (17) and (20) is straightforward.[3]

Both equations in Table 2 are non-linear: for the first one this is due to the non-linearity of $g(\alpha)\frac{\partial \Psi^+}{\partial \varepsilon}(\varepsilon(\boldsymbol{u})) + \frac{\partial \Psi^-}{\partial \varepsilon}(\varepsilon(\boldsymbol{u})) =: \boldsymbol{\sigma}(\boldsymbol{u}, \alpha)$, whereas for the second one it is due to the Macaulay bracket term $\langle \cdot \rangle_-$. Therefore, the Newton-Raphson procedure is used to iteratively compute $\boldsymbol{u}^k$ and $\alpha^k$ with $\boldsymbol{u}^{k-1}$ and $\alpha^{k-1}$ taken as the corresponding initial guesses, and $\texttt{TOL}_{\text{NR}}$ as the tolerance. Owing to the nested nature of the Newton-Raphson loops, we impose that $\texttt{TOL}_{\text{NR}} < \texttt{TOL}_{\text{Stag}}$.

[3]In particular, in the case of (17), $\mathcal{H}_n$ entering $\widetilde{\mathcal{E}}_\alpha$ must be re-defined: at every $k \geq 1$ we take the cumulative quantity $\mathcal{H}_n := \max_{x \in \Omega}\{\mathcal{H}_{n-1}, \Psi^+(\varepsilon(\boldsymbol{u}^{k-1}))\}$.

**Table 2** Staggered iterative solution process for $\mathcal{E}' = 0$ at loading step $n \geq 1$

Input: loading data $\bar{\boldsymbol{u}}_n$ on $\Gamma_{D,1}$, and
solution $(\boldsymbol{u}_{n-1}, \alpha_{n-1})$ from step $n-1$.

Initialization, $k = 0$:

1. set $(\boldsymbol{u}^0, \alpha^0) := (\boldsymbol{u}_{n-1}, \alpha_{n-1})$.

Staggered iteration $k \geq 1$:

2. given $\boldsymbol{u}^{k-1}$, solve $\mathcal{E}'_\alpha(\boldsymbol{u}^{k-1}, \alpha; \beta) = 0$ for $\alpha$, set $\alpha =: \alpha^k$,
3. given $\alpha^k$, solve $\mathcal{E}'_{\boldsymbol{u}}(\boldsymbol{u}, \alpha^k; v) = 0$ for $\boldsymbol{u}$, set $\boldsymbol{u} =: \boldsymbol{u}^k$,
4. for the obtained pair $(\boldsymbol{u}^k, \alpha^k)$, check
$$\mathrm{Res}^k_{\mathrm{Stag}} := |\mathcal{E}'(\boldsymbol{u}^k, \alpha^k; \boldsymbol{v}, \beta)| \leq \mathtt{TOL}_{\mathrm{Stag}},$$
5. if fulfilled, set $(\boldsymbol{u}^k, \alpha^k) =: (\boldsymbol{u}_n, \alpha_n)$ and $n+1 \to n$;
else $k+1 \to k$.

Output: solution $(\boldsymbol{u}_n, \alpha_n)$.

The definition of $\mathrm{Res}^k_{\mathrm{Stag}}$ in Table 2 used for stopping the staggered process is not unique. As an example, the quantity $||\alpha^k - \alpha^{k-1}||_\infty$ is considered as the residual e.g., in Amor et al. (2009), Pham et al. (2011), Mesgarnejad et al. (2015), Farrell and Maurini (2017) when solving (9) and in Bourdin et al. (2000, 2008), Bourdin (2007a, b) while solving (14). Another option for $\mathrm{Res}^k_{\mathrm{Stag}}$ is a relative change in the normalized energy $\mathcal{E}$. It has been taken as a residual in Ambati et al. (2015) while solving (17), and as an auxiliary convergence-tracing quantity in Gerasimov and De Lorenzis (2016) while considering (14).

Finally, as already mentioned in the introductory part, the problem size of the system in (9), (14), (20) and (17) after finite element discretization is typically very large, since both the phase-field and the deformation localize in bands of width of order $\ell$. Solving the system in a staggered way in this case is computationally very demanding, see e.g., Mesgarnejad et al. (2015), Ambati et al. (2015), Gerasimov and De Lorenzis (2016), Farrell and Maurini (2017) for detailed studies.

*Remark 4.1* In recent work of Farrell and Maurini (2017), various strategies for accelerating the partitioned solution schemes are discussed. It is shown that a better efficiency of the staggered approach presented above can be gained.

## *Monolithic (Newton-Raphson with Line-Search)*

The lack of convexity of the governing energy functional $\mathcal{E}$ in (5) typically results in iterative convergence issues while solving $\mathcal{E}' = 0$ monolithically, that is, for both

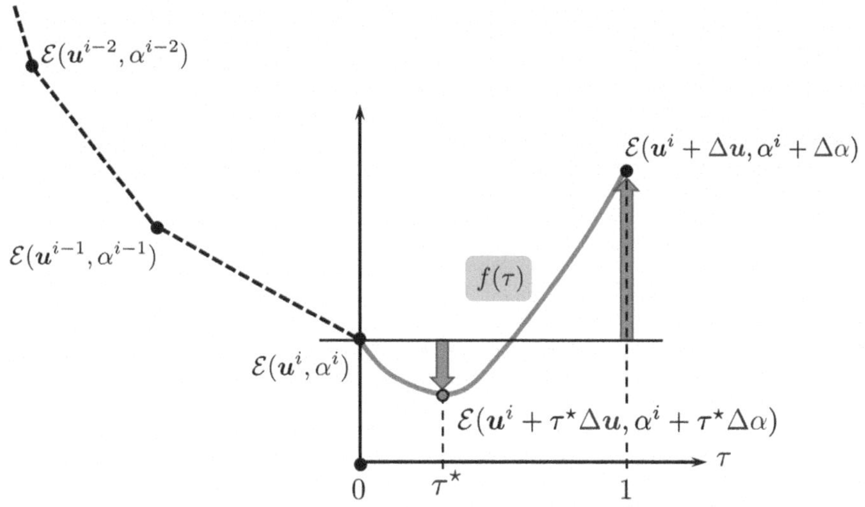


$$\tau^\star = \arg\min \left\{ f(\tau) := \mathcal{E}(\boldsymbol{u}^i + \tau\Delta\boldsymbol{u}, \alpha^i + \tau\Delta\alpha),\ \tau \in [0, 1] \right\}$$

**Fig. 11** The divergence $\mathcal{E}(\boldsymbol{u}^i + \Delta\boldsymbol{u}, \alpha^i + \Delta\alpha) > \mathcal{E}(\boldsymbol{u}^i, \alpha^i)$, with $(\Delta\boldsymbol{u}, \Delta\alpha)$ as the known increment, of the Newton-Raphson solution process is detected; a properly determined line-search parameter $\tau^\star$ provides a converging update Gerasimov and De Lorenzis (2016)

arguments $(\boldsymbol{u}, \alpha)$ using, e.g., the Newton-Raphson procedure. With the known guess $(\boldsymbol{u}^{i-1}, \alpha^{i-1})$ and the computed increment $(\Delta\boldsymbol{u}, \Delta\alpha)$, the divergence of the iterative process is manifested by the increase of the energy for the solution update $(\boldsymbol{u}^{i-1} + \Delta\boldsymbol{u}, \alpha^{i-1} + \Delta\alpha)$. More precisely, it is $\mathcal{E}(\boldsymbol{u}^{i-1} + \Delta\boldsymbol{u}, \alpha^{i-1} + \Delta\alpha) > \mathcal{E}(\boldsymbol{u}^{i-1}, \alpha^{i-1})$. A significant increase of the magnitude of the residual for the updated solution is typically detected as well. Once occurred, the trend will typically continue at the following Newton-Raphson iterations (no self-stabilization is observed), eventually producing no meaningful solution.

In Gerasimov and De Lorenzis (2016), to cope with the aforementioned convergence issues, a specific line-search procedure is devised and validated. The idea is sketched in Fig. 11: once an increase of energy is detected for the direct solution update $(\boldsymbol{u}^i + \Delta\boldsymbol{u}, \alpha^i + \Delta\alpha)$, one rescales the search increment $(\Delta\boldsymbol{u}, \Delta\alpha)$ by a multiplier $\tau \in (0, 1)$ called line-search parameter in order to arrive at a converged energy level. With the optimal value $\tau^\star$ computed, the updated $(\boldsymbol{u}^i + \tau^\star\Delta\boldsymbol{u}, \alpha^i + \tau^\star\Delta\alpha)$ is then taken as the guess for the next Newton-Raphson iteration.

The resulting monolithic procedure for solving $\mathcal{E}' = 0$ based on the combined use of the Newton-Raphson method and the line-search procedure is depicted in Table 3. The items 3(a) and 3(b) in the table constitute the line-search procedure.

*Remark 4.2* Recently, in Kopanicakova and Krause (2019), a monolithic solution procedure based on the trust region method has been proposed. The efficiency and robustness of the approach has been demonstrated. This represents an alternative to

**Table 3** Monolithic iterative solution process for $\mathcal{E}' = 0$ at loading step $n \geq 1$ using Newton-Raphson equipped with line-search

---

Input: loading data $\bar{\boldsymbol{u}}_n$ on $\Gamma_{D,1}$, and
solution $(\boldsymbol{u}_{n-1}, \alpha_{n-1})$ from step $n-1$.

Initialization, $i = 0$:

1. set $(\boldsymbol{u}^0, \alpha^0) := (\boldsymbol{u}_{n-1}, \alpha_{n-1})$.

Newton-Raphson iteration $i \geq 1$:

2. given $(\boldsymbol{u}^{i-1}, \alpha^{i-1})$, solve $\mathcal{E}'(\boldsymbol{u}^{i-1} + \Delta\boldsymbol{u}, \alpha^{i-1} + \Delta\alpha; \boldsymbol{v}, \beta) = 0$ for $(\Delta\boldsymbol{u}, \Delta\alpha)$,
3. **if** for the solution update divergence is detected, i.e.,
$$\mathcal{E}(\boldsymbol{u}^{i-1} + \Delta\boldsymbol{u}, \alpha^{i-1} + \Delta\alpha) > \mathcal{E}(\boldsymbol{u}^{i-1}, \alpha^{i-1}),$$
(a) find the line-search parameter $\tau^\star \in (0, 1)$ such that
$$\mathcal{E}(\boldsymbol{u}^{i-1} + \tau^\star\Delta\boldsymbol{u}, \alpha^{i-1} + \tau^\star\Delta\alpha) < \mathcal{E}(\boldsymbol{u}^{i-1}, \alpha^{i-1}),$$
(b) set $(\boldsymbol{u}^{i-1} + \tau^\star\Delta\boldsymbol{u}, \alpha^{i-1} + \tau^\star\Delta\alpha) =: (\boldsymbol{u}^i, \alpha^i)$,
**else** set $(\boldsymbol{u}^{i-1} + \Delta\boldsymbol{u}, \alpha^{i-1} + \Delta\alpha) =: (\boldsymbol{u}^i, \alpha^i)$;
4. for the obtained pair $(\boldsymbol{u}^i, \alpha^i)$, check
$$\text{Res}^i_{\text{NR}} := |\mathcal{E}'(\boldsymbol{u}^i, \alpha^i; \boldsymbol{v}, \beta)| \leq \texttt{TOL}_{\text{NR}},$$
5. if fulfilled, set $(\boldsymbol{u}^i, \alpha^i) =: (\boldsymbol{u}_n, \alpha_n)$ and $n + 1 \to n$;
else $i + 1 \to i$.

Output: solution $(\boldsymbol{u}_n, \alpha_n)$.

---

the monolithic approach using the Newton-Raphson method with line-search presented above. The comparison of the two approaches is the matter of forthcoming research.

## *Examples*

We compare the performance of the two solution schemes illustrated above by considering three benchmark problems. The first problem is the SEN under shear already depicted in the left plot of Fig. 6 and explored in the context of the different irreversibility approaches. The second and the third one are the plate with hole under tension, see, e.g., Ambati et al. (2015), Gerasimov and De Lorenzis (2016), and the $L$-shaped panel experiment Mesgarnejad et al. (2015), Ambati et al. (2015), Wick (2017a), Gerasimov and De Lorenzis (2016), Gerasimov and De Lorenzis (2019), see Fig. 12. In all cases, the results are obtained for the combination of the `S`-split and the `AT`-2 model. They are taken from the recent publication Gerasimov and De Lorenzis

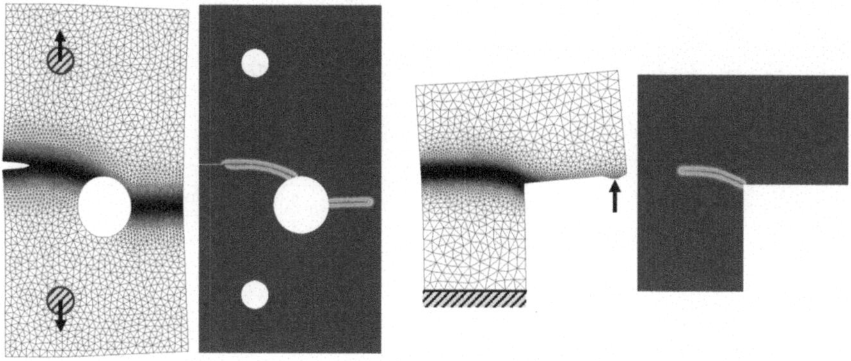

**Fig. 12** Problem setup (sketchy) and the computed failure patterns for the notched plate with hole under tension (left) and the $L$-shaped specimen experiment (right)

(2016). In the plots we designate our monolithic solution scheme using Newton-Raphson with line-search as `MON-ls`. It is assumed that $\texttt{TOL}_{\text{Stag}} = \texttt{TOL}_{\text{NR}} =: \texttt{Tol}$ with $\texttt{TOL}_{\text{Stag}}$ and $\texttt{TOL}_{\text{NR}}$ as in Tables 2 and 3, respectively.

For the three problems, Figs. 13, 14 and 15 report and compare the load-displacement curves computed by the two approaches, as well as the cumulative computational time and the ratio in this time. It can be observed that with the same loading setup and under comparable accuracy requirements expressed through the corresponding tolerance `Tol` the monolithic scheme appears far superior to the staggered one, at least for these benchmark examples. For the detailed study of the energy and the residual convergence at each loading step, we refer the interested reader to Gerasimov and De Lorenzis (2016). There it is shown in particular that even though the monolithic process typically starts with higher residual and energy levels, it converges to the same level of residual and (approximately) the same minimum value of energy faster than the staggered one.

## Conclusions

We have addressed the main challenging computational aspects of phase-field modeling of brittle fracture, focusing on the treatment of the phase-field irreversibility constraint and on the choice of the iterative solution strategy for the non-convex minimization problem stemming from the formulation. For treatment of irreversibility, we have introduced and discussed the available options for enforcing the constraint, in terms of equivalence of the final unconstrained formulation to the original constrained one and of computational efficiency. For solution of the non-convex minimization problem, we have also considered various available options and critically assessed the efficiency and robustness of two of them. Numerical examples on well-known benchmark problems were used for illustrative purposes. At the computational level,

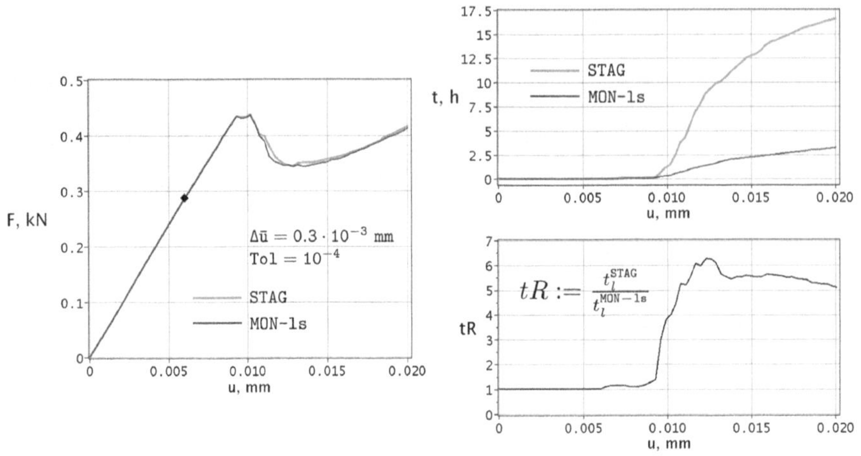


Fig. 13 SEN shear test: comparisons of the load–displacement curves and of the time–displacement diagrams representing the cumulative computational cost for the staggered and monolithic (Newton-Raphson with line-search) approaches

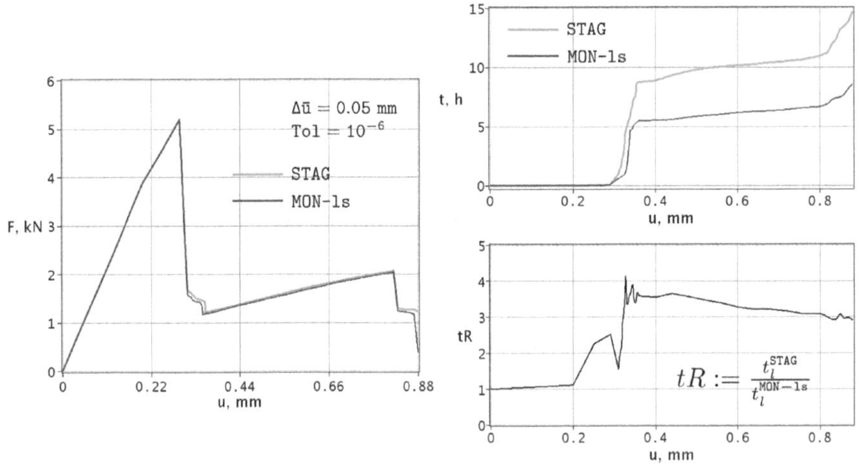


Fig. 14 Notched plate with hole: comparisons of the load–displacement curves and of the time–displacement diagrams representing the cumulative computational cost for the staggered and monolithic (Newton-Raphson with line-search) approaches

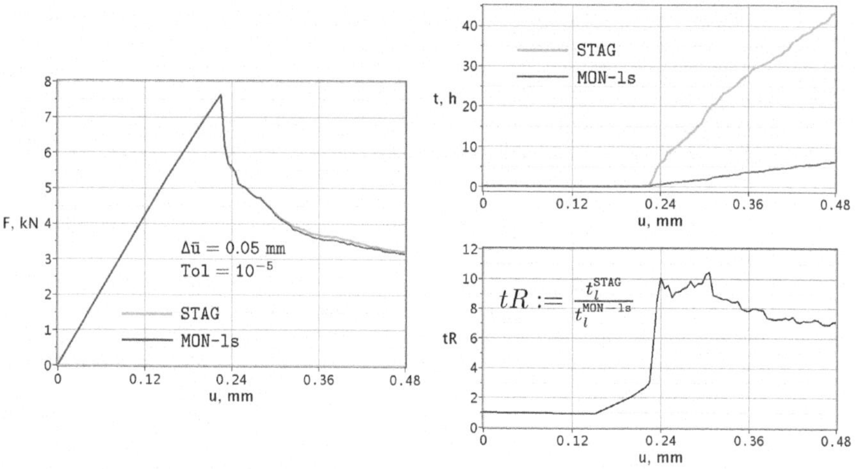


**Fig. 15** $L$-shaped specimen experiment: comparisons of the load–displacement curves and of the time–displacement diagrams representing the cumulative computational cost for the staggered and monolithic (Newton-Raphson with line-search) approaches

finding an optimal compromise between robustness and efficiency seems to be a major and still open issue for phase-field computation of brittle fracture problems. Despite the progress achieved with line-search and trust region methods to improve robustness of the monolithic approach, and with accelerated staggered schemes to improve efficiency of the alternate minimization approach, there seems to be still room for substantial improvement. Another interesting aspect, where theory and numerics closely interact, is the issue of solution non-uniqueness stemming from the non-convexity of the energy functional to be minimized. The occurrence of multiple solutions has been occasionally reported in the literature, e.g., in Bourdin et al. (2000, 2008), Bourdin (2007a), Amor et al. (2009), Burke et al. (2010a), Artina et al. (2014), Artina et al. (2015). However, to the best of our knowledge no attempt has been made so far to investigate and characterize these multiple solutions. Finally, mesh adaptivity has been only briefly mentioned in this chapter but certainly deserves further attention as a key technology for reducing the computational cost.

## References

Alessi, R., Ambati, M., Gerasimov, T., Vidoli, S., & De Lorenzis, L. (2018). Comparison of phase-field models of fracture coupled with plasticity. In *Advances in computational plasticity* (pp. 1–21). https://doi.org/10.1007/978-3-319-60885-3-1.

Alessi, R., & Freddi, F. (2017). Phase-field modelling of failure in hybrid laminates. *Composite Structures*, *181*, 9–25.

Alessi, R., Marigo, J. J., & Vidoli, S. (2015). Gradient damage models coupled with plasticity: Variational formulation and main properties. *Mechanics of Materials*, *80*(Part B), 351–367.

Ambati, M., Gerasimov, T., & De Lorenzis, L. (2015). A review on phase-field models of brittle fracture and a new fast hybrid formulation. *Computational Mechanics*, *55*(2), 383–405.

Ambati, M., Gerasimov, T., & De Lorenzis, L. (2015). Phase-field modeling of ductile fracture. *Computational Mechanics*, *55*(5), 1017–1040.

Ambati, M., & De Lorenzis, L. (2016). Phase-field modeling of brittle and ductile fracture in shells with isogeometric NURBS-based solid-shell elements. *Computer Methods in Applied Mechanics and Engineering*, *312*, 351–373.

Ambrosio, L., & Tortorelli, V. M. (1990). Approximation of functional depending on jumps by elliptic functional via Gamma-convergence. *Communications on Pure and Applied Mathematics*, *43*(8), 999–1036.

Amiria, F., Millán, D., Shen, Y., Rabczuk, T., & Arroyo, M. (2014). Phase-field modeling of fracture in linear thin shells. *Theoretical and Applied Fracture Mechanics*, *69*, 102–109.

Amor, H., Marigo, J.-J., & Maurini, C. (2009). Regularized formulation of the variational brittle fracture with unilateral contact: Numerical experiments. *Journal of the Mechanics and Physics of Solids*, *57*, 1209–1229.

Artina, M., Fornasier, M., Micheletti, S., & Perotto, S. (2015). Anisotropic mesh adaptation for crack detection in brittle materials. *SIAM Journal on Scientific Computing*, *37*(4), 633–659.

Artina, M., Micheletti, S., Perotto, S., & Fornasier, M. (2014). Anisotropic adaptive meshes for brittle fractures: Parameter sensitivity. In A. Abdulle, S. Deparis, D. Kressner, F. Nobile, & M. Picasso (Eds.), *Numerical mathematics and advanced applications* (pp. 293–302). Berlin, Heidelberg: Springer.

Bleyer, J., & Alessi, R. (2018). Phase-field modeling of anisotropic brittle fracture including several damage mechanisms. *Computer Methods in Applied Mechanics and Engineering*, *336*, 213–236.

Borden, M. (2012). Isogeometric Analysis of phase-field models for dynamic brittle and ductile fracture. Ph.D.-thesis.

Borden, M. J., Hughes, T. J. R., Landis, C. M., Anvari, A., & Lee, I. J. (2016). A phase-field formulation for fracture in ductile materials: Finite deformation balance law derivation, plastic degradation, and stress triaxiality effects. *Computer Methods in Applied Mechanics and Engineering*, *312*, 130–166.

Borden, M. J., Hughes, T. J. R., Landis, C. M., & Verhoosel, C. V. (2014). A higher-order phase-field model for brittle fracture: Formulation and analysis within the isogeometric analysis framework. *Computer Methods in Applied Mechanics and Engineering*, *273*, 100–118.

Bourdin, B. (2007). The variational formulation of brittle fracture: Numerical implementation and extensions. In R. de B. A. Combescure & T. Belytschko (Eds.), *IUTAM Symposium on Discretization Methods for Evolving Discontinuities* (pp. 381–393). Berlin: Springer.

Bourdin, B., Chukwodozie, C., & Yoshioka, K. (2012). A variational approach to the numerical simulation of hydraulic fracture. In *Proceedings of the 2012 SPE Annual Technical Conference and Exhibition*. SPE 159154.

Bourdin, B. (2007). Numerical implementation of the variational formulation for quasi-static brittle fracture. *Interfaces and Free Boundaries*, *9*, 411–430.

Bourdin, B., Francfort, G. A., & Marigo, J.-J. (2000). Numerical experiments in revisited brittle fracture. *Journal of the Mechanics and Physics of Solids*, *48*(4), 797–826.

Bourdin, B., Francfort, G. A., & Marigo, J.-J. (2008). The variational approach to fracture. *Journal of Elasticity*, *91*(1–3), 5–148.

Bourdin, B., Marigo, J.-J., Maurini, C., & Sicsic, P. (2014). Morphogenesis and propagation of complex cracks induced by thermal shocks. *Physical Review Letters*, *112*, 014301.

Braides, A. (1998). *Approximation of free-discontinuity problems*, Lecture notes in mathematics. Berlin: Springer.

Burke, S., Ortner, C., & Süli, E. (2010). An adaptive finite element approximation of a generalised Ambrosio–Tortorelli functional, OXMOS preprint No. 29, Mathematical Institute, University of Oxford.

Burke, S., Ortner, C., & Süli, E. (2010). An adaptive finite element approximation of a variational model of brittle fracture. *SIAM Journal on Numerical Analysis*, *48*(3), 980–1012.

Burke, S., Ortner, C., & Süli, E. (2013). An adaptive finite element approximation of a generalized Ambrosio-Tortorelli functional. *Mathematical Models and Methods in Applied Sciences*, *23*(9), 1663–1697.

Cajuhi, T., Sanavia, L., & De Lorenzis, L. (2018). Phase-field modeling of fracture in variably saturated porous media. *Computational Mechanics*, *61*(3), 299–318.

Chambolle, A. (2004). An approximation result for special functions with bounded deformation. *Journal of Mathematics Pures Applications*, *83*, 929–954.

Chambolle, A., Conti, S., & Francfort, G. A. (2018). Approximation of a brittle fracture energy with a constraint of non-interpenetration. *Archive for Rational Mechanics and Analysis*, *228*(3), 867–889.

Del Piero, G., Lancioni, G., & March, R. (2007). A variational model for fracture mechanics: Numerical experiments. *Journal of the Mechanics and Physics of Solids*, *55*, 2513–2537.

Duda, F. P., Ciarbonetti, A., Sanchez, P. J., & Huespe, A. E. (2015). A phase-field/gradient damage model for brittle fracture in elastic-plastic solids. *International Journal of Plasticity*, *65*, 269–296.

Farrell, P. E., & Maurini, C. (2017). Linear and nonlinear solvers for variational phase-field models of brittle fracture. *International Journal for Numerical Methods in Engineering*, *109*, 648–667.

Francfort, G. A., & Marigo, J.-J. (1998). Revisiting brittle fractures as an energy minimization problem. *Journal of the Mechanics and Physics of Solids*, *46*, 1319–1342.

Freddi, F., & Royer-Carfagni, G. (2009). A variational model for cleavage and shear fracture. In *Proceedings of the XIX AIMETA Symposium*, 715–716 (abstract), 1–12 (in extenso on CD-ROM), Ancona, September 11–13, 2009.

Freddi, F., & Royer-Carfagni, G. (2010). Regularized variational theories of fracture: A unified approach. *Journal of the Mechanics and Physics of Solids*, *58*, 1154–1174.

Gerasimov, T., & De Lorenzis, L. (2016). A line search assisted monolithic approach for phase-field computing of brittle fracture. *Computer Methods in Applied Mechanics and Engineering*, *312*, 276–303.

Gerasimov, T., & De Lorenzis, L. (2019). On penalization in variational phase-field models of brittle fracture. *Computer Methods in Applied Mechanics and Engineering*, *354*, 990–1026.

Gerasimov, T., Noii, N., Allix, O., & De Lorenzis, L. (2018). A non-intrusive global/local approach applied to phase-field modeling of brittle fracture. *Advanced Modeling and Simulation in Engineering Sciences*, *5*, 14.

Glowinski, R., Lions, J. L., & Trémolières, R. (1981). *Numerical analysis of variational inequalities*. Amsterdam: North-Holland.

Griffith, A. A. (1921). The phenomena of rupture and flow in solids. *Philosophical Transactions of the Royal Society of London*, *221*, 163–198.

Hecht, F., Leharic, A., & Pironneau, O. (2019). FreeFem++: Language for finite element method and Partial Differential Equations (PDE), Université Pierre et Marie, Laboratoire Jacques-Louis Lions, http://www.freefem.org/ff++/.

Heister, T., Wheeler, M. F., & Wick, T. (2015). A primal-dual active active set method and predictor-corrector mesh adaptivity for computing fracture propagation using a phase-field approach. *Computer Methods in Applied Mechanics and Engineering*, *290*, 466–495.

Hintermüller, M., Ito, K., & Kunisch, K. (2002). The primal-dual active set strategy as a semismooth Newton method. *SIAM Journal on Optimization*, *13*, 865–888.

Kiendl, J., Ambati, M., De Lorenzis, L., Gomez, H., & Reali, A. (2016). Phase-field description of brittle fracture in plates and shells. *Computer Methods in Applied Mechanics and Engineering*, *312*, 374–394.

Kinderlehrer, D., & Stampacchia, G. (1980). *An introduction to variational inequalities and their applications*. Cambridge: Academic Press.

Klinsmann, M., Rosato, D., Kamlah, M., & McMeeking, R. M. (2015). An assessment of the phase field formulation for crack growth. *Computer Methods in Applied Mechanics and Engineering*, *294*, 313–330.

Kopanicakova, A., & Krause, R. (2019). Recursive multilevel trust region method with application to fully monolithic phase-field models of brittle fracture. arXiv:1903.00379v1.

Kuhn, C., & Müller, R. (2010). A continuum phase field model for fracture. *Engineering Fracture Mechanics*, *77*, 3625–3634.

Kuhn, C., Schlüter, A., & Müller, R. (2015). On degradation functions in phase field fracture models. *Computational Materials Science*, *108*, 374–384.

Lancioni, G., & Royer-Carfagni, G. (2009). The variational approach to fracture mechanics: A practical application to the French Panthéon in Paris. *Journal of Elasticity*, *95*, 1–30.

León Baldelli, A. A., Babadjian, J.-F., Bourdin, B., Henao, D., & Maurini, C. (2014). A variational model for fracture and debonding of thin films under in-plane loadings. *Journal of the Mechanics and Physics of Solids*, *70*, 320–348.

Li, T. (2016). Gradient-damage modeling of dynamic brittle fracture: Variational principles and numerical simulations. Ph.D.-thesis.

Li, B., & Maurini, C. (2019). Crack kinking in a variational phase-field model of brittle fracture with strongly anisotropic surface energy, *submitted to Journal of the Mechanics and Physics of Solids*.

Li, B., Peco, C., Millán, D., Arias, I., & Arroyo, M. (2015). Phase-field modeling and simulation of fracture in brittle materials with strongly anisotropic surface energy. *International Journal for Numerical Methods in Engineering*, *102*, 711–727.

Marigo, J.-J., Maurini, C., & Pham, K. (2016). An overview of the modelling of fracture by gradient damage models. *Meccanica*, *51*, 3107–3128.

Mesgarnejad, A., Bourdin, B., & Khonsari, M. M. (2013). A variational approach to the fracture of brittle thin films subject to out-of-plane loading. *Journal of the Mechanics and Physics of Solids*, *61*(11), 2360–2379.

Mesgarnejad, A., Bourdin, B., & Khonsari, M. M. (2015). Validation simulations for the variational approach to fracture. *Computer Methods in Applied Mechanics and Engineering*, *290*, 420–437.

Miehe, C., Aldakheel, F., & Raina, A. (2016). Phase field modeling of ductile fracture at finite strains: A variational gradient-extended plasticity-damage theory. *International Journal of Plasticity*, *84*, 1–32.

Miehe, C., Hofacker, M., Schänzel, L., & Aldakheel, F. (2015). Phase field modeling of fracture in multi-physics problems. Part II. Coupled brittle-to-ductile failure criteria and crack propagation in thermo-elastic-plastic solids. *Computer Methods in Applied Mechanics and Engineering*, *294*, 486–522.

Miehe, C., Hofacker, M., & Welschinger, F. (2010). A phase field model for rate-independent crack propagation: Robust algorithmic implementation based on operator splits. *Computer Methods in Applied Mechanics and Engineering*, *199*, 2765–2778.

Miehe, C., & Mauthe, S. (2016). Phase field modeling of fracture in multi-physics problems. Part III. Crack driving forces in hydro-poro-elasticity and hydraulic fracturing of fluid-saturated porous media. *Computer Methods in Applied Mechanics and Engineering*, *304*, 619–655.

Miehe, C., Schänzel, L., & Ulmer, H. (2015). Phase field modeling of fracture in multi-physics problems. Part I. Balance of crack surface and failure criteria for brittle crack propagation in thermo-elastic solids. *Computer Methods in Applied Mechanics and Engineering*, *294*, 449–485.

Miehe, C., Welschinger, F., & Hofacker, M. (2010). Thermodynamically consistent phase-field models of fracture: Variational principles and multi-field FE implementations. *International Journal for Numerical Methods in Engineering*, *83*, 1273–1311.

Mikelić, A., Wheeler, M. F., & Wick, T. (2015). A quasi-static phase-field approach to pressurized fractures. *Nonlinearity*, *28*, 1371–1399.

Mikelić, A., Wheeler, M. F., & Wick, T. (2015). A phase-field method for propagating fluid-filled fractures coupled to a surrounding porous medium. *SIAM Multiscale Modeling and Simulation*, *13*, 367–398.

Mikelić, A., Wheeler, M. F., & Wick, T. (2015). Phase-field modeling of a fluid-driven fracture in a poroelastic medium. *Computational Geosciences*, *19*, 1171–1195.

Nagaraja, S., Elhaddad, M., Ambati, M., Kollmannsberger, S., De Lorenzis, L., & Rank, E. (2018). Phase-field modeling of brittle fracture with multi-level *hp*-FEM and the finite cell method. *Computational Mechanics*, *63*, 1283–1300.

Nguyen, T. T., Réthoré, J., & Baietto, M.-C. (2017). Phase field modelling of anisotropic crack propagation. *European Journal of Mechanics - A/Solids*, *65*, 279–288.

Pham, K., Amor, H., Marigo, J.-J., & Maurini, C. (2011). Gradient damage models and their use to approximate brittle fracture. *International Journal of Damage Mechanics*, *20*(4), 618–652.

Reinoso, J., Paggi, M., & Linder, C. (2017). Phase field modeling of brittle fracture for enhanced assumed strain shells at large deformations: Formulation and finite element implementation. *Computational Mechanics*, *59*(6), 981–1001.

Sargadoa, J. M., Keilegavlen, E., Berre, I., & Nordbotten, J. M. (2018). High-accuracy phase-field models for brittle fracture based on a new family of degradation functions. *Journal of Physics and Chemistry of Solids*, *111*, 458–489.

Sicsic, P., Marigo, J.-J., & Maurini, C. (2013). Initiation of a periodic array of cracks in the thermal shock problem: A gradient damage modeling. *Journal of the Mechanics and Physics of Solids*, *63*, 256–284.

Strobl, M., & Seelig, T. (2016). On constitutive assumptions in phase field approaches to brittle fracture. *Procedia Structural Integrity*, *2*, 3705–371.

Tanné, E., Li, T., Bourdin, B., Marigo, J.-J., & Maurini, C. (2018). Crack nucleation in variational phase-field models of brittle fracture. *Journal of the Mechanics and Physics of Solids*, *110*, 80–99.

Teichtmeister, S., Kienle, D., Aldakheel, F., & Keip, M.-A. (2017). Phase field modeling of fracture in anisotropic brittle solids. *International Journal of Non-Linear Mechanics*, *97*, 1–21.

Vignollet, J., May, S., de Borst, R., & Verhoosel, C. V. (2014). Phase-field models for brittle and cohesive fracture. *Meccanica*, *49*, 2587–2601.

Weinberg, K., & Hesch, C. (2017). A high-order finite deformation phase-field approach to fracture. *Continuum Mechanics and Thermodynamics*, *29*(4), 935–945.

Wheeler, M. F., Wick, T., & Wollner, W. (2014). An augmented-Lagangrian method for the phase-field approach for pressurized fractures. *Computer Methods in Applied Mechanics and Engineering*, *271*, 69–85.

Wick, T. (2016). Goal functional evaluations for phase-field fracture using PU-based DWR mesh adaptivity. *Computational Mechanics*, *57*, 1017–1035.

Wick, T. (2017). An error-oriented Newton/inexact augmented Lagrangian approach for fully monolithic phase-field fracture propagation. *SIAM Journal on Scientific Computing*, *39*(4), 589–6017.

Wick, T. (2017). Modified Newton methods for solving fully monolithic phase-field quasi-static brittle fracture propagation. *Computer Methods in Applied Mechanics and Engineering*, *325*, 577–611.

Wilson, Z. A., & Landis, C. H. (2016). Phase-field modeling of hydraulic fracture. *Journal of the Mechanics and Physics of Solids*, *96*, 264–290.

Wu, J.-Y., Nguyen, V.P., Zhou, H., & Huang, Y. (2019). A variationally consistent phase-field anisotropic damage model for fracture (submitted).

Wu, T., & De Lorenzis, L. (2016). A phase-field approach to fracture coupled with diffusion. *Computer Methods in Applied Mechanics and Engineering*, *312*, 196–223.

Wu, J.-Y., & Nguyen, V. P. (2018). A length scale insensitive phase-field damage model for brittle fracture. *Journal of the Mechanics and Physics of Solids*, *119*, 20–42.

Zhang, X., Sloan, S. W., Vignes, C., & Sheng, D. (2017). A modification of the phase-field model for mixed mode crack propagation in rock-like materials. *Computer Methods in Applied Mechanics and Engineering*, *322*, 123–136.

# Practical Computational Fluid Dynamics with the Finite Volume Method

**Hrvoje Jasak and Tessa Uroić**

## Nomenclature

| | |
|---|---|
| $F$ | Face flux ($m^3/s$) |
| $R$ | Right hand side in a linear equation |
| $S$ | General source term |
| $V_P$ | Cell volume ($m^3$) |
| $\mathbf{u}$ | Velocity vector (m/s) |
| $\mathbf{H}(\mathbf{u})$ | Off-diagonal of matrix $\mathbf{A_u}$ multiplied by velocity, and associated r.h.s. contributions |
| $\mathbf{A}, \mathbf{B}, \mathbf{C}, \mathbf{D}$ | Arbitrary coefficient matrices |
| $\mathbf{A_u}$ | Convection-diffusion coefficient matrix |
| $\mathbf{D_u}$ | Diagonal of matrix $\mathbf{A_u}$ |
| $\mathbf{LU_u}$ | Off-diagonal of matrix $\mathbf{A_u}$ |
| $\mathbf{M}$ | Preconditioning matrix |
| $\mathbf{S}$ | Iteration matrix in fixed point algorithms |
| $\mathbf{b}$ | Arbitrary right hand side vector |
| $\mathbf{k}$ | Non-orthogonal vector component in orthogonal correction |
| $\mathbf{p}$ | Search direction in Conjugate Gradients |
| $\mathbf{r}$ | Residual vector |
| $\mathbf{x}, \mathbf{y}$ | Unknown vectors |
| $\mathbf{x}_P$ | Cell centroid (m) |
| $\mathbf{d}_f$ | Distance between cell centroids (m) |
| $\mathbf{s}_f$ | Surface normal face area vector ($m^2$) |

H. Jasak (✉)
Wikki Ltd., Studio 459 China Works, 100 Black Prince Road, London SE1 7SJ, UK
e-mail: hrvoje.jasak@fsb.hr

H. Jasak · T. Uroić
Faculty of Mechanical Engineering and Naval Architecture,
University of Zagreb Croatia, Ivana Lučića 5,10002 Zagreb, Croatia



L. De Lorenzis and A. Düster (eds.), *Modeling in Engineering Using Innovative Numerical Methods for Solids and Fluids*, CISM International Centre for Mechanical Sciences 599,
https://doi.org/10.1007/978-3-030-37518-8_4

| | |
|---|---|
| $a_N$ | Off-diagonal matrix element |
| $a_P$ | Diagonal matrix element |
| $a_{ii}$ | Diagonal matrix element |
| $a_{ij}$ | Off-diagonal matrix element |
| $f$ | Cell face centre |
| $p$ | Kinematic pressure ($m^2/s^2$) |
| $p^*$ | Intermediate pressure field ($m^2/s^2$) |
| $t$ | Time-step (s) |

## Calligraphy Letters

| | |
|---|---|
| $\mathcal{K}$ | Krylov subspace |

## Greek Letters

| | |
|---|---|
| $\alpha$ | Length of the step in Conjugate Gradients |
| $\alpha_{\mathbf{u}}$ | Underrelaxation factor for momentum equation |
| $\alpha_P$ | Underrelaxation factor for pressure field |
| $\beta$ | Projection operator in Conjugate Gradients |
| $\epsilon$ | Residual tolerance for measuring convergence |
| $\gamma$ | Diffusion coefficient |
| $\lambda$ | Limiter value in orthogonal correction |
| $\Delta$ | Orthogonal vector component in orthogonal correction |
| $\nu$ | Kinematic viscosity ($m^2/s$) |
| $\phi$ | General scalar variable |
| $\rho$ | Density ($kg/m^3$) |

## Superscripts

| | |
|---|---|
| $(k)$ | Current time step or iteration |
| $\top$ | Transpose operation |
| C | Value on the coarse level of multigrid |
| F | Value on the fine level of multigrid |
| n | Value of (new) solution in current iteration |
| o | Value of (old) solution from previous iteration |

## Subscripts

| | |
|---|---|
| $f$ | Value which belongs to a face |
| $N$ | A value which belongs to neighbouring cells |
| $P$ | A value which belongs to the cell center |
| $p, q$ | Matrix dimensions |

## Introduction

The Finite Volume Method (FVM) of discretisation is the dominant solution method for continuum mechanics problems in heat transfer, fluid flow and related phenomena in the early 21st century. It is based directly on underlying conservation principles, provides consistent second order accuracy in space and time which is sufficient for many "fluid flow related" problems, it is versatile in terms of handling complex and coupled problems, it operates naturally on unstructured or mixed element meshes and it is compatible with today's High Performance Computing (HPC) assets, using the domain decomposition approach.

Clearly, the FVM also has significant drawbacks, compared to other discretisation methods. For example, penalty–based Petrov–Galerkin Finite Element Method (FEM) is proven to be optimal for linear elasticity problems, and the Discontinuous Galerkin (DG) method provides a natural and consistent extension to high–order discretisation. However, a combination of its simplicity and ability of practitioners to manipulate the properties of the discretisation matrices make the FVM capable of dealing with industrial and academic needs.

This is best demonstrated by the size of largest conventional FVM simulation. To the author's knowledge, the largest simulation case executed with a general purpose CFD code used 100 billion ($10^{11}$) control volumes and was executed on a parallel computer with 98,304 nodes (Phuc et al. 2016), using the free surface Volume–of–Fluid (VOF) model. The number of degrees of freedom in this simulation is 500 billion (velocity, pressure, turbulence variables), which is several orders of magnitude larger than largest FEM simulations in structural mechanics.

A further benefit of the modern FVM is the ability of the method to obey the basic constraints of the continuum model, described via Partial Differential Equations (PDE). Examples include:

- Strong conservation, where the amount of mass, species concentration or energy in the system is preserved to the level of machine accuracy (as opposed to the level of discretisation error);
- Enforcing of variable boundedness, where the physical bounds on variables such as temperature (greater than 0 K), turbulence kinetic energy (always positive) or concentration (bounded between 0 and 1) are enforced not only in the converged solution, but also during the iterative procedure;
- Stabilised boundary conditions, where the properties of differential operators are enforced in the implementation and linearization of boundary conditions; and others.

Perhaps the most important characteristic of the modern FVM is the way discretisation practitioners manipulate the form of discretisation to achieve beneficial properties of the discretisation matrix. Matrix sparsity, diagonal dominance, manipulation of the source terms, symmetric positive definite matrices for elliptic problems, control of the sign and eigen–spectrum of matrix elements, control of matrix band and condition number all feature prominently in the discretisation process. Such matrix manipulation allows the modern FVM to rely almost exclusively on iterative solvers for linear systems of equations. A direct consequence is the ability to handle huge problems and operate on massively parallel computing platforms.

Second–order FVM in its "classical" form also comes with significant deficiencies. Perhaps the most serious is its loss of accuracy and stability on highly non–orthogonal meshes for elliptic problems and difficulties of handling strongly coupled sets of PDEs. While the problems can be considered to be "under control", better discretisation shall certainly emerge in the future.

In this text, we shall present the "classical" Finite Volume Method of second order accuracy in space and time, with support for polyhedral meshes. The derivation originates from the work at Imperial College, starting in late 1960s and reaching its current form in the late 20th century. For practitioners looking to reproduce the examples, or use the formulation as a starting point, this particular form of second order FVM is implemented in OpenFOAM (Weller et al. 1998).

## *Governing Equations*

Most physics–based simulations (as opposed to e.g. simulation of stock prices in financial markets) are based on the principle of conservation. In fact, historical methods of approaching fluid flow and heat transfer are based on the analysis of the system under consideration through a control volume view. Here, the system is isolated from the environment via boundary conditions and integral equations for conservation of mass, momentum and energy allow us to draw insights into its behaviour using a single control volume.

In numerical solution methods based on the FVM we shall extend this principle by applying basic conservation laws on smaller chunks of space, posing conservation equations for each of them and looking at their inter–dependence. The first step is, of course to state the conservation equations for a control volume.

Clearly, the conservation of mass momentum and energy follow a simple law, stated below as the scalar transport equation. Scalar transport equation in the standard form will be our model for discretisation. Conservation laws, governing the continuum mechanics adhere to the standard form.

The scalar transport Eq. (1) stated below results from the Reynolds transport theorem and a gradient–based diffusion transport, with source and sink terms usually associated with aspects of physical modelling.

$$\underbrace{\frac{\partial \phi}{\partial t}}_{\text{temporal derivative}} + \underbrace{\nabla \bullet (\phi \mathbf{u})}_{\text{convection term}} - \underbrace{\nabla \bullet (\gamma \nabla \phi)}_{\text{diffusion term}} = \underbrace{q_v}_{\text{source term}} . \tag{1}$$

Here, $\phi$ is the place–holder for the variable under consideration, $\mathbf{u}$ is the velocity field, $\gamma$ is the diffusivity field and $q_v - q_v(\phi)$ is the volumetric source/sink contribution to the balance of $\phi$.

Physical meaning can be associated with each term of (1). The temporal derivative represents inertia of the system. The convection term represents the convective transport by a prescrived velocity field, which can also be described as a "coordinate transformation": a feature is carried by the given velocity field through space, giving this term a hyperbolic nature. The diffusion term results from the gradient transport hypothesis and is elliptic in nature. Finally, sources and sinks account for local (non–transport) effects: local volumetric production or destruction of $\phi$.

In the framework of many control volumes filling the domain of interest mentioned above, individual control volumes (or cells) shall exchange the information via cell faces in the form of convective or diffusive transport, while locally capturing the inertial, source or sink effects.

The FVM discretisation described in this text shall primarily address the numerical solution for (1). In order to address the issue of inter–equation coupling and handling of non–linearity, a simplest flow model shall be used: incompressible laminar transient Navier–Stokes equations for a Newtonian fluid. Governing equations read as follows:

- Momentum equation:

$$\frac{\partial \mathbf{u}}{\partial t} + \nabla\mathbf{.}(\mathbf{u}\mathbf{u}) - \nabla\mathbf{.}\,(\nu\nabla\mathbf{u}) = -\nabla p, \tag{2}$$

  where $\mathbf{u}$ is the velocity field, $p$ is the pressure field and $\nu$ is the kinematic viscosity;

- and Continuity equation:

$$\nabla\mathbf{.u} = 0. \tag{3}$$

A number of more complex models consist effectively of the Navier–Stokes equations and additional equations capturing various phenomena, such as the free surface flow (position of the interface in the Volume–of–Fluid approach is tracked using the advection equation for phase fraction $\alpha$) or various models for turbulent flow, where turbulence is characterised by transport equations for its characteristic properties, such as eddy viscosity turbulence models, or Reynolds stress transport models. Such models can be simulated in a segregated approach (see Section "Pressure–Velocity Coupling"), with individual equations be discretised along the same guidelines.

## *Numerical Solution of Continuum Problems*

A solution to a set of partial differential equations can be calculated analytically or numerically. Analytical solution is represented as a continuous function of space and time and is usually provided in a general form which is then adapted to a set of boundary and initial conditions. Evaluation of the solution implies the evaluation of this known analytical function at certain space and time coordinates. However, the analytical solution is unavailable for most problems of engineering interest.

While the power of analytical solution is impressive, its practical use may not be straightforward. Consider a problem of calculating form drag on a vehicle, assuming that the analytical solution is available. The drag force $F_p$ is the integral of the pressure along the vehicle's body:

$$F_p = \oint_S \mathbf{s} p. \tag{4}$$

Assuming that a closed solution is available, $F_p$ would be calculated by analytical integration over a given analytical surface. However, there are few cars whose body shape is known as a closed analytical function: this is not how they are designed. A practical solution to the problem would involve splitting up the surface into a large number of surface elements, assuming the pressure distribution across each element, and evaluating the integral as a sum of forces of individual elements. To achieve this, the value of the analytical function is extracted at a number of positions in space: this is a striaghtforward task. As the number of evaluation elements on the surface increases, this numerically evaluated force would approach its analytical evaluation.

Extending the analogy, one can claim that for engineering use, it is sufficient to know the field value of a continuous field $p$ at a large number of points, instead of possessing the (ultimately more powerful) closed form of the analytical solution.

Such a collection of evaluation points, on a surface or in space, where the value of the variable is known, comprise the *computational mesh*. Position of evaluation points, also called mesh resolution, depends on the shape of the computational domain and the complexity of the solution function.

Looking at Fig. 1, one can postulate a number of rules about a good numerical representation of the function. For example, looking at a sinusoidal curve, one can state that having less than 4 evaluation points points per period of the sinusoidal function may give a misleading picture of the solution. Having a small number of points or assuming a constant value of the solution variable across the computational cell may lead to an erroneous idea of the true shape of the function. However, as the number of sampling points increases to the level where the solution looks "smooth"

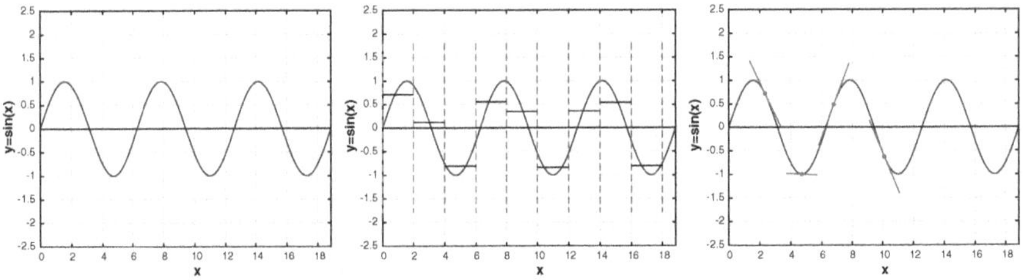


**Fig. 1** Analytical or numerical: target function (left); constant–per–interval sampling; evaluation of derivatives from a numerical approximation

on the mesh, the power of discrete representation of a continuous function becomes unquestionable.

Taking the numerical solution further, one can easily evaluate not only the value of an (analytically unknown) function from a set of evaluation points, but also its derivatives, using intuitive right–minus–left–over–distance rules.

## *Closure*

In what follows, we shall start with a set of governing partial differential equations in analytical form, whose solution is sought over a domain of arbitrary shape. We shall define the domain of interest as a computational mesh in space and time interval of interest as a set of discrete time–steps. Having obtained such numerical description of a computational domain, we shall proceed to assemble a solution procedure for sets of partial differential equations stated above. The solution itself shall be in numerical form, consisting of values of field $\phi$ at a set of predefined points in space–time.

Along the way, we will address the methods of mesh generation, postulate the shape of continuous function $\phi$ between the computational points, perform spatial and temporal integration of various differential operators in (1) and explore ways of obtaining the field values in computational points.

We shall also consider solution methods for large sets of linear algebraic equations that occur from discretisation, consider the effect of boundary conditions and issues of nonlinearity and inter–equation coupling. This is the journey of Finite Volume discretisation.

## Mesh Generation and Mesh Handling

Solution of partial differential equations in continuum mechanics is sought over a region of space: a spatial domain; and a time interval. It is therefore necessary to describe the spatial domain for a computer simulation in an appropriate manner. In 1–D or simple 2–D problems, this may be trivial; however 3–D space and complex curved boundaries bring significant challenges.

Apart from capturing the solution domain, two further issues need to be addressed. First, outer boundary of the spatial domain carries boundary conditions, needed to close the problem. Problem definition and consequently case setup are simplified when the outside boundary is split into patches. For example, in the flow simulation around a vehicle in a wind tunnel it is practical to split the outer boundary into: wind tunnel inlet and outlet; bottom wall (which may be moving), far–field walls and the car body itself. In such a setup it is easier to define boundary conditions for each of the solution variables on a per–patch basis.

Second, for a numerical solution of a PDE using the FVM, it is necessary to define points in the spatial domain where the solution is sought. This is not a trivial task: it defines the sampling density of the continuous function that will be used to capture the solution. It is known in advance that the primary region of interest is the vehicle body, rather than the far–field of the tunnel; hence computational points should be clustered close to it. Functional shape of the solution function will be more complex close to the body (measured in terms of solution gradients): features such as boundary layers, recirculation zones, stagnation points and vortices induced by the body are of primary interest. This judgement comes from the practitioner's knowledge of the problem and expected characteristics of the solution and are captured in the creation of the computational mesh.

The primary role of a computational mesh is to describe the domain shape (and regions of interest) to the computer. When building the mesh, a number of questions should be considered:

- How to capture the geometrical shape of the boundary for the computer?
- How to generate a surface mesh for the boundary?
- How to fill the volume with computational points? and
- How to cluster the points in regions of interest?

The challenges will be addressed in the ramainder of the text in this section.

## *Guidance on Mesh Generation*

A computational mesh is a description of the spatial domain in the simulation: it defines the external shape of the domain as well the regions of highest interest, where mesh resolution is increased. Generation of a high–quality computational mesh remains the current bottle–neck in CFD simulations. Fully automatic mesh generators are improving and are routinely used. At the same time, requirements on rapid and high–quality meshing and massively increased mesh size are becoming a problem.

Mesh size varies considerably with the choice of physical model, purpose of simulation, required accuracy, available computing power and user experience. As a guidance to a novice, some of the routinely used mesh sizes at the time of writing are:

- 100–50,000 cells: Small mesh for model experimentation and quick games with fast turn–around;
- 10,000–1 million cells: 2–D geometry. However, low–*Re* turbulent simulations may require more, due to near–wall mesh resolution requirements;
- 50,000 to several million cells: 3–D geometry;
- 100,000 to 10–50 million cells: Complex geometry, 3D, industrial size meshes. Mesh size varies considerably depending on geometry and physics, steady/transient flow etc.;
- 1–20 million cells: Large Eddy Simulation (LES) 3–D, transient. LES requires very long transient runs and averaging (20–50k time steps), which keeps the mesh resolution down;

- 20–200 million cells: Full car aerodynamics, Formula 1 for routine use;
- 100 billion cells: Largest engineering CFD simulation with a general–purpose CFD code (Phuc et al. 2016). The simulation was executed on a supercomputer using 100,000 cores.

On very large meshes, problems with the current generation of CFD software becomes a limiting factor: only a few massively parallel mesh generation algorithms are available, data file read/write, post–processing of results, as well as high hardware and software prices.

Typical geometrical data formats which are used for mesh generation are:

- 2–D boundary shape: e.g. airfoils. Usually a detailed map of $x - y$ locations on the surface. Sometimes defined as spline curve data;
- Stereo Lithographic Surface (STL): a surface is represented by a set of triangular facets. Resolution can be automatically adjusted to capture the surface curvature or control points. Creation of STL usually available from CAD packages;
- Native CAD description: Initial Graphics Exchange Specification (IGES), solid model etc. In most cases, the surface is represented by Non–Uniform Rational B–Splines (NURBs) or approximated by quadric surfaces. Typically, both are too expensive for the manipulations required in mesh generation and either avoided or simplified.

Even though CAD files are the most accurate representation of the geometry, CAD description is very rarely built specifically for CFD—in most cases, CAD surfaces are assembled from various sources, with varying quality and imperfect matching. Surface clean–up is usually done by hand, time–consuming and not trivial. Taking these facts into account, the procedure of mesh generation can be divided into a number of steps, as follows.

**Geometry clean–up**. Very rarely is the CAD description built specifically for CFD—in most cases, CAD surfaces are assembled from various sources, with varying quality and imperfect matching. Surface clean–up is time–consuming and not trivial.

**Feature removal**. CAD description or STL surface may contain a level of detail too fine to be captured by the desired mesh size, causing trouble with 3–D mesh generation. Feature removal creates an approximation of the original geometry with the desired level of detail.

**Surface mesh generation**. In cases where the surface description is not discrete, a surface mesh may be created first. An STL surface is already a mesh. However, it may be necessary to additionally split the surface for easier imposition of boundary conditions: inlet, outlet, symmetry plane etc. Surface mesh is usually triangular or quadrilateral and there are potential issues with capturing surface curvature: surface mesh will be considered "sufficiently fine".

**Volume mesh generation**. The role of the volume mesh is to capture the 3–D geometry The cells should not overlap and should completely fill the computational domain. Additionally, some convexness criteria (FVM) or a library of pre–defined cell shapes

(FEM) is included. Computational mesh defines the location and distribution of solution points (vertices, cells, etc.). Thus, filling the domain with the mesh is not sufficient—ideally, some aspects of the solution should be taken into account. Thus, *a*–priori knowledge of the solution is useful in mesh generation in order to locate the regions where high mesh resolution ("fine mesh") is needed to capture critical parts of the solution: shocks, boundary layers and similar. The quality of the mesh critical for a good solution and is not measured only in mesh resolution and it depends on the discretisation method.

## *Mesh Types*

Since mesh generation is the bottle–neck, the solvers are generalised to be extremely flexible regarding meshing, thus simplifying the most difficult part of the simulation process. Certain numerical solution techniques require specific mesh types, e.g. Cartesian meshes are used for high–order finite difference method and the supported mesh structure may severely limit the use of a chosen discretisation method. The following mesh types are distinguished:

- Cartesian mesh
- Structured body–fitted mesh
- Multi–block mesh
- Unstructured shape–consistent meshes
- Tetrahedral and hybrid tet–hex mesh
- Overset and Chimera meshes
- Polyhedral meshes.

A Cartesian mesh is a trivial regular $\Delta x \times \Delta y \times \Delta z$ mesh used for rectangular domains. While this mesh type is interesting for theoretical purposes, it is limited in practical applications.

Body–fitted meshes, Fig. 2, originate from the non–orthogonal curvilinear coordinate system approach where the case–specific coordinate system is created to fit the boundary. The mesh is hexahedral and regularly connected and realistic geometry can be captured but with insufficient control over local mesh resolution.

Multi–block mesh is created as a combination of multiple body–fitted blocks, Fig. 3. All blocks and cells are still hexahedral. It provides more control over mesh grading and local resolution and significantly more complex geometries can be captured. Automatic smoothing and mesh optimisation tools are extensively used while the initial block decomposition is still done by hand which makes mesh generation in 3–D for relatively complex shapes hard and time–consuming since block interfaces need to match.

From the numerical simulation point of view, a major step forward was done with unstructured meshes. Instead of being calculated, the mesh connectivity is stored. Loose definition of connectivity allows more freedom: hexahedral and degenerate hexahedral meshes: prisms, pyramids, wedges etc. allow easier meshing. Geometries

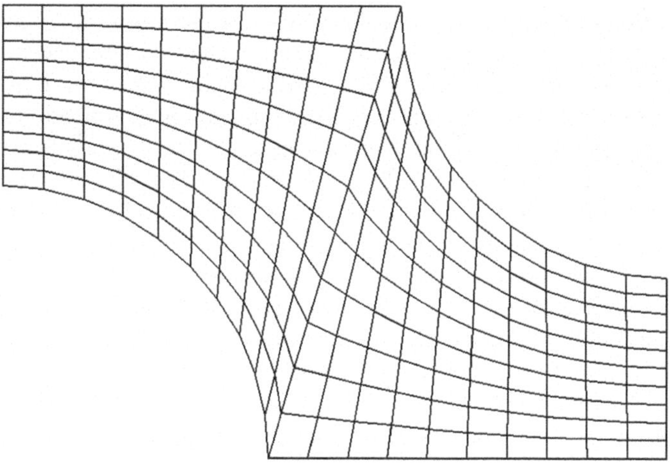

**Fig. 2** An example of a 2–D body fitted mesh

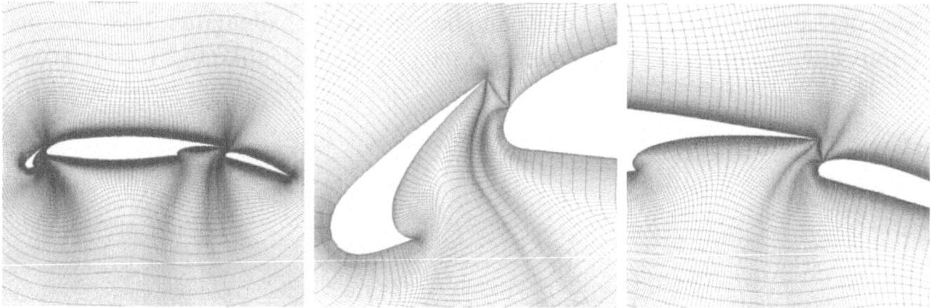

**Fig. 3** Multi–block structured mesh

of industrial interest can now be tackled with a detailed description, which satisfies what the design engineer needs. Still, at this stage, all meshes are hand–built. Manual vertex–by–vertex and cell–by–cell mesh generation for a complex 3–D geometry may take 2–3 months.

From a numerics point of view, tetrahedral meshes are not optimal as they do not provide ability of aligning with solution gradient. However, such meshes can be generated automatically for a spatial domain of complex shape, using mathematical tessalation algorithms. This is a great improvement over manual generation of unstructured or body–structured meshes. If a solver can support tetrahedral meshes, mesh generation time for complex geometry reduces from weeks to hours.

Advantages of tetrahedral meshing are great savings in mesh generation effort, faster turn–around of simulations and geometrical variation and mesh sensitivity studies can now be performed on realistic geometries. However, tetrahedra are particularly poor in boundary layers close to walls, which led to hybrid meshes. A hybrid mesh is built by creating a layer of hexahedral cells next to the wall. The rest of the domain is filled with tetrahedra. A combined tet–hex mesh is a great improvement in quality and flexibility. On the negative side, cell count for a tetrahedral mesh of equivalent resolution is higher than for hexahedra. A part of the price is also paid in lower accuracy of the solver on tetrahedra due to limited neighbourhood connectivity.

Tetrahedral meshes can be automatically generated using the advancing front method or Delaunay triangulation, Fig. 4. Advancing front algorithm starts from the boundary surface triangulation and inserts tetrahedra from the live front using priority lists. In Delaunay triangulation the initial mesh is created by triangulating the boundary. New points are added in a way which improves the quality of the most distorted triangles and creates a convex hull around each point.

As an illustration of relative quality of a hexahedral and tetrahedral mesh, Fig. 5 shows two meshes for the coolant system in a block of an Internal Combusion (IC) engine. Experienced CFD professionals claim that a "pretty mesh" allows for high–quality solutions: visual inspection can therefore give some guidance on solution quality.

In spite of the availability of automatic generation techniques, tetrahedral meshes are often not of sufficient quality for industrial use. On the other hand, automatic hexahedral mesh generation has proven to be extremely challenging. Finite Volume discretisation is not actually dependent on the cell shape. Unlike FEM, there are no

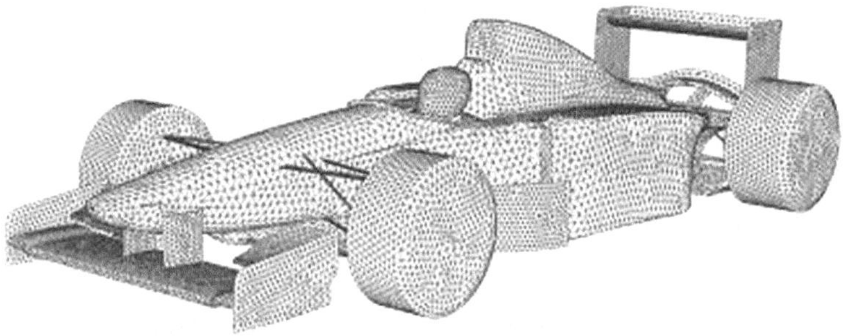

**Fig. 4** A triangular surface mesh of a Formula 1 car in 1996

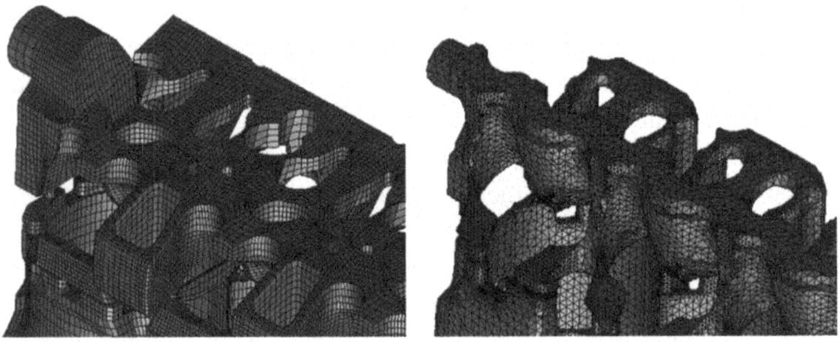

**Fig. 5** Comparison of a hexahedral and tetrahedral mesh for two models of coolant passages in a block of an internal combustion engine

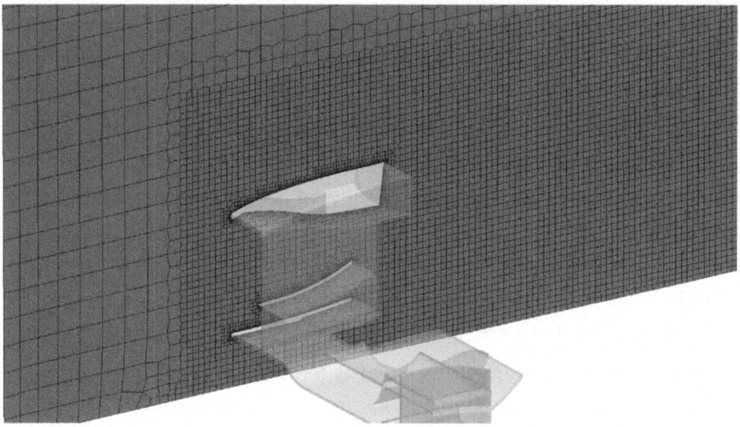

**Fig. 6** A mesh around a Formula 1 front wing produced by cut cell mesh generation technology

pre–defined shape functions and transformation tensors, which brings the possibility of polyhedral mesh support. Polyhedral meshes are considerably better than tetrahedral, can be manipulated to be predominantly hexahedral, orthogonal and regular and can be created automatically. For polyhedra, Finite Volume discretisation algorithm is reformulated (optimised) into loops over cells and faces. While creating a polyhedral mesh, the Delaunay triangulation algorithm introduces points on proximity rules and a dual mesh of convex polyhedra is created and can be extracted. The interaction of the tessalated mesh and the boundary needs to be recovered after polyhedral mesh assembly while local control of mesh size is achieved in the same way as in tetrahedral meshes.

The strategy for mesh generation is straightforward: filling space with non–overlapping cells. Even close to boundaries, it is easy to build high quality layered structure. Problematic parts of mesh generation are related to interaction of advancing generation surfaces or boundary interaction in complex corners of regions where the mesh resolution dos not match the level of detail on the boundary description. Cut cell technology creates a rough background mesh, either uniform hexahedral or capturing major features of the geometry, Fig. 6. The mesh inside of the domain is kept and the one interacting with the boundary surface is adjusted or cut by the surface. In some cases, the background mesh resolution can be automatically adjusted around the surface to match the local resolution requirements Such meshes are good quality and generation is fast. Prismatic boundary layers may also be added and in some cases, background mesh adjustment or concave cell corrections are required.

## *Adaptive Mesh Refinement*

Mesh structure specifies where the computational points are located, while the discretisation practice postulates the shape of the solution between the computational points, which is the main source of discretisation error. A sensible meshing strategy requires high resolution in regions of interest instead of uniformly distributing points in the domain. This implies some knowledge of the solution during mesh generation. The same can be achieved in an iterative way:

1. Create initial mesh and initial solution;
2. Examine the solution from the point of view of accuracy or resolution in "regions of interest";
3. Based on the available solution, adjust mesh resolution in order to improve the solution in the selected parts of the domain;
4. Repeat until sufficient accuracy is achieved or computer resources are exhausted.

Performing mesh improvement by hand is tedious and time–consuming. For an automatic procedure, Fig. 7, two questions need to be answered: "Where to refine the mesh (adjust resolution)" and "How to change the mesh to achieve the required accuracy". Error indicators are usually located in the regions of interest, see Fig. 8. For example: magnitude of the second pressure gradient, Mach number distribution etc. Error estimates, apart from the spatial information (error distribution), provide guidance on the absolute error level. Traditionally, mesh adaptation was a part of the CFD solver instead of the mesh generator. In cases where the refinement algorithm resorts to cell splitting, we may end up with a faceted surface representation instead of a smooth surface, which compromises the results. The remedy lies in geometrical description of the boundary and it needs to be available from the solver instead of trying to recover the data from the original (coarse) mesh. A step further is related to the specification of boundary conditions. In, for example, wind tunnel simulations, the velocity and turbulence at the inlet plane in shown from the measured data and interpolated onto the inlet patch of the mesh. Ideally, the boundary condition should be associated with space or with the boundary description in order to avoid problems with interpolation. This leads to issues of CAD integration, which is beyond our scope.

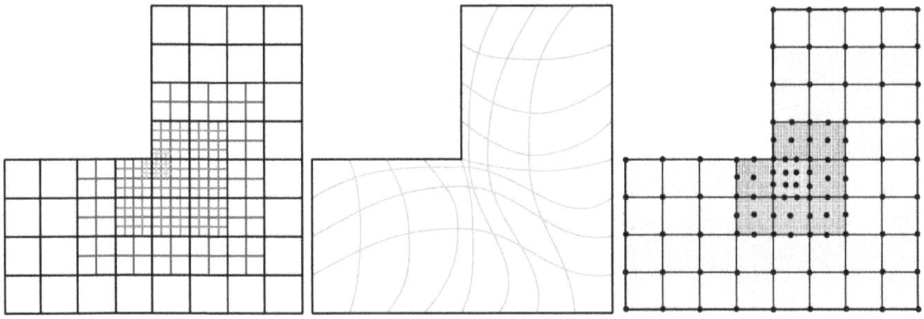

**Fig. 7** Adaptive mesh refinement. *h*–refinement: introducing new computational points in regions of interest. *r*–refinement: re-organise the existing points such that more points fall into the region of interest. *p*–refinement: enriching the space of shape functions in order to capture the solution more closely

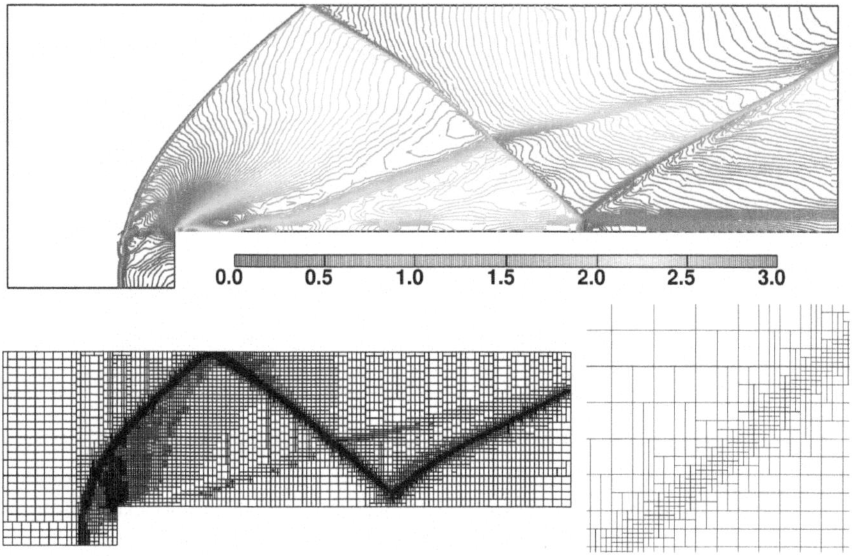


**Fig. 8** Mesh adaptivity on shocked flows

## *FVM Mesh Quality Metrics*

Suitability of a computational mesh for numerical simulation is closely connected with the underlying discretisation method. For example, a cell away from "ideal" isotropic shape is not necessarily bad, if it is aligned with local solution gradients in an appropriate way. Some of the mesh quality measures in FVM are: cell aspect ratio, face non–orthogonality, face skewness, cell distortion from ideal shape, etc.

Cell aspect ratio is defined as ratio of longest to shortest edge length. In many cases, anisotropic cells are desirable to align the cell with the highest solution gradient.

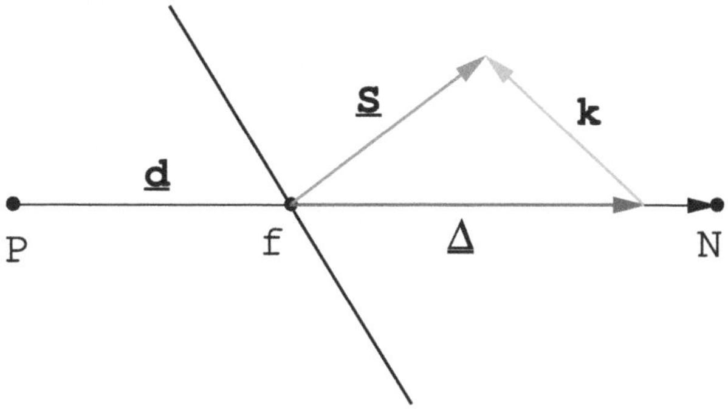


**Fig. 9** Face non–orthogonality

**Fig. 10** Face skewness

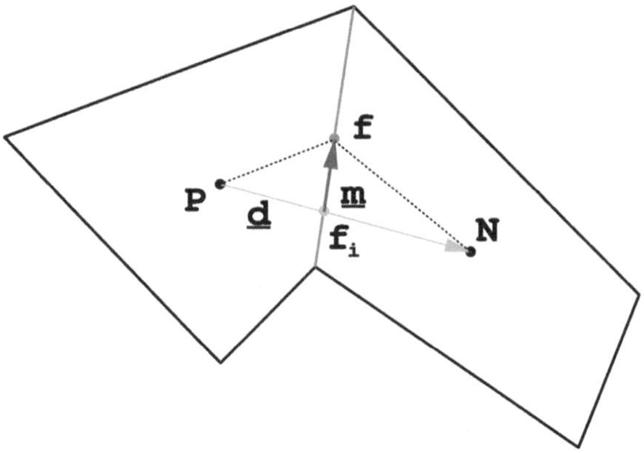


Face non–orthogonality, Fig. 9, is defined as the angle between the face normal and the vector connecting two cell centres which share the face. Non–orthogonality of 70−90° increases the solution cost and reduces accuracy due to diffusion discretisation in FVM. Mesh with non–orthogonality over 90° is not valid.

Face skewness is defined as the distance between face centroid and face integration point and it reduces the order of face integration but without stability implications. It is shown in Fig. 10.

A distorted cell remains weakly convex: all cell–face pyramids need to be of positive volume. Negative cell volumes (inverted cells) are usually fatal. Sensitivity to convexity errors depends on the details of discretisation method.

## *Dynamic Mesh Simulations*

In many cases, the shape of the simulation domain may vary during the simulation time of interest, either in a prescribed manner or as function of the solution. Examples include simulation of flow in a cylinder of an Internal Combustion (IC) engine, where boundary motion is known before the start of a simulation; or cases of fluid–solid interaction, where domain shape is a function of solution variables. In both scenarios, domain shape and computational mesh needs to be adjusted to the desired shape.

Shape change may be achieved in two ways, Jasak (2009):

- Moving deforming mesh, where the point, face and cell connectivity remains fixed, and shape change is achieved via changes in point positions only;
- Topological mesh changes, where both point position and mesh connectivity changes.

In moving deforming meshes, point motion can be described a–priori (if boundary deformation is known ahead of simulation), or by means of automatic mesh motion algorithms. Here, boundary motion is prescribed (e.g. from a fluid–solid interaction coupling) and positions of internal points are calculated by solving a mesh motion equation Jasak and Tuković (2007), Jasak (2009).

Examples of topological include sliding meshes, cell layering, attach/detach boundaries, local remeshing and adaptive mesh refinement. In good implementations, the solution can be transferred between meshes using the space conservation law Ferziger and Perić (1995).

### *Closure*

In this section we have presented the challenges and requirements for building an acceptable computational mesh. It all begins at the geometrical data format which will be used as a basis for mesh generation. Several format types have been discussed as well as the steps necessary to prepare the geometry for generation of the corresponding surface and volume mesh. When dealing with complex geometries, different types of cells are often required to adequatly capture the objects, which can lead to difficulties in terms of discretisation procedure and obtaining a high–quality solution. Thus, we have also presented geometric metrics which are used to assess the quality of the mesh. An overview of refined techniques such as moving meshes and meshes with adaptive local refinement was also given. Once the spatial domain has been discretised, i.e. represented with a computational mesh, discretisation of equations using the Finite Volume Method can be conducted, which is presented in the following section.

## Finite Volume Discretisation

In the previous section methodology for discretisation of the spatial domain was presented. In this section we shall describe the discretisation of equations using the Finite Volume Method. The process shall be explained for a generic transport equation of a scalar property, term by term. Treatment of numerical boundary conditions will be also covered. We shall also comment on the contribution of each discretised term to the linear system.

### *Generic Scalar Transport Equation*

Scalar transport equation in the standard form will be our model for discretisation, since conservation laws, governing the continuum mechanics adhere to the standard form. Standard form is not the only one available: modelled equations may be more complex or some source/sink terms can be recognised as transport. This leads to other forms, but the basics of their discretisation remain the same.

Moving away from physical conservation laws, almost identical equations can be found in other areas, for example financial modelling. The common factor for all equations under consideration is the same set of operators: temporal derivative, gradient, divergence, Laplacian, curl, as well as various source and sink terms.

The standard form of the transport equation reads:

$$\underbrace{\frac{\partial \phi}{\partial t}}_{\text{temporal derivative}} + \underbrace{\nabla\cdot(\phi \mathbf{u})}_{\text{convection term}} - \underbrace{\nabla\cdot(\gamma \nabla \phi)}_{\text{diffusion term}} = \underbrace{q_v}_{\text{source term}} . \tag{5}$$

Temporal derivative represents the inertia of the system. Convection term represents the convective transport by the prescribed velocity field (coordinate transformation). The term is hyperbolic in nature: information comes from the vicinity, defined by the direction of the convection velocity. Diffusion term represents gradient transport. This is an elliptic term: every point in the domain feels the influence of every other point instantaneously. Sources and sinks account for non–transport effects, i.e. local volume production and destruction of $\phi$.

## Numerical Discretisation

Generic transport equation can only rarely be solved analytically and we need to resort to numerical methods. Discretisation is a process of representing the differential equation we wish to solve by a set of algebraic expressions of equivalent properties, typically written in a form of a matrix.

The discretisation will be assembled on a per–operator basis. Space and time will be described via computational mesh for the spatial domain and time–steps which cover the time interval. A generic scalar variable $\phi$ varies in space and time and it will be represented as a discrete field data. Equation operators will be integrated over a cell, and spatial and temporal variation of $\phi$ will be used to interpret the operator in discrete terms.

The solution variable will be stored in cell centroid which is a characteristic of the collocated cell–centred finite volume method. Boundary data, corresponding to prescribed boundary conditions, will be stored on face centres of boundary faces.

For some purposes, e.g. face flux, different type of data is required—in this case it will be a field over all faces in the mesh. Spatial variation can be used for interpolation in general since, e.g. post–processing tools typically use point–based data.

A convex polyhedral cell is shown in Fig. 11. The faces of the cell are convex polygons. Point $P$ is the computational point located at cell centroid $\mathbf{x}_P$. The definition of the centroid reads:

$$\int_{V_P} (\mathbf{x} - \mathbf{x}_P)\, dV = \mathbf{0}. \tag{6}$$

Cell volume is denoted by $V_P$. For the cell, there is one neighbouring cell across each face. Neighbour cell and cell centre will be marked with $N$. The face centre $f$ is defined in the equivalent manner, using the centroid rule:

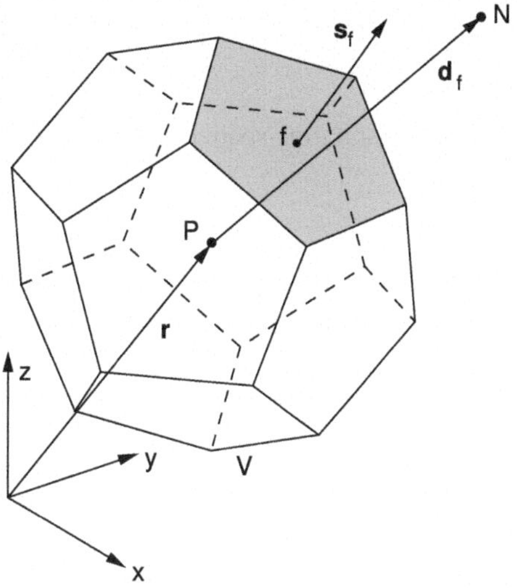


**Fig. 11** Convex polyhedral cell

$$\int_{S_f} (\mathbf{x} - \mathbf{x}_f)\, dS = \mathbf{0}. \tag{7}$$

Delta vector for the face $f$ is defined as the vector between centroids of cells $P$ and $N$, straddling the face:

$$\mathbf{d}_f = \overline{PN}. \tag{8}$$

Face area vector $\mathbf{s}_f$ is a surface normal vector whose magnitude is equal to the area of the face. The face is numerically never flat, so the face centroid and area are calculated from the integral:

$$\mathbf{s}_f = \int_{S_f} \mathbf{n}\, dS, \tag{9}$$

where $\mathbf{n}$ is the face normal vector. The fact that the face centroid does not necessarily lay on the plane of the face is not worrying: we are dealing with surface–integrated quantities. However, we shall require the cell centroid to lay within the cell. In practice, cell volume and face area are calculated by decomposition of a face into triangles and cell into pyramids/tetrahedra.

Faces in a mesh can be internal faces, shared by two cells or boundary faces, adjacent to one cell only and pointing outwards of the computational domain. When operating on a single cell it is assumed that all face area vectors $\mathbf{s}_f$ point outwards of cell $P$.

Discretisation is based on the integral form of the transport equation over each cell (control volume):

$$\int_V \frac{\partial \phi}{\partial t}\, dV + \oint_S \phi\, (\mathbf{n.u})\, dS - \oint_S \gamma\, (\mathbf{n.}\nabla\phi)\, dS = \int_V Q_v\, dV. \tag{10}$$

When postulating spatial variation of $\phi$, we assume second order discretisation in space:

$$\phi(\mathbf{x}) = \phi_P + (\mathbf{x} - \mathbf{x}_P)\mathbf{.}(\nabla\phi)_P. \tag{11}$$

This expression is given for each individual cell. Here, $\phi_P = \phi(\mathbf{x}_P)$.
Linear variation in time is also assumed:

$$\phi(t + \Delta t) = \phi^t + \Delta t \left(\frac{\partial \phi}{\partial t}\right)^t, \tag{12}$$

where $\phi^t = \phi(t)$. In FVM, we have specified the "shape function" without reference to the actual cell shape (tetrahedron, prism, brick, wedge). The variation is always linear. Thus, doing polyhedral Finite Volume should be straightforward.
Volume integrals are evaluated as:

$$\begin{aligned}
\int_V \phi\, dV &= \int_V [\phi_P + (\mathbf{x} - \mathbf{x}_P)\mathbf{.}(\nabla\phi)_P]\, dV \\
&= \phi_P \int_V dV + (\nabla\phi)_P\mathbf{.} \int_V (\mathbf{x} - \mathbf{x}_P) dV \\
&= \phi_P V_P.
\end{aligned} \tag{13}$$

Surface integrals are split into a sum over faces and evaluated as:

$$\begin{aligned}
\oint_S \mathbf{n}\, \phi\, dS &= \sum_f \int_{S_f} \mathbf{n}\phi_f\, dS_f = \sum_f \int_{S_f} \mathbf{n}[\phi_f + (\mathbf{x} - \mathbf{x}_f)\mathbf{.}(\nabla\phi)_f]\, dS_f \\
&= \sum_f \mathbf{s}_f \phi_f.
\end{aligned} \tag{14}$$

Assumption of linear variation of $\phi$ and selection of $P$ and $f$ in the centroid results in second–order discretisation. Gauss' theorem is a tool which will be used for handling the volume integrals of divergence and gradient operators:

$$\int_{V_P} \nabla\mathbf{.a}\, dV = \oint_{\partial V_P} d\mathbf{s.a}. \tag{15}$$

$$\int_{V_P} \nabla\phi\, dV = \oint_{\partial V_P} d\mathbf{s}\, \phi. \tag{16}$$

Note that the face area vector operates from the same side as the gradient operator and it fits with the definition of the gradient for a vector field. In the rest of the analysis, we shall look at the problem face by face. A diagram of a face is given below for 2–D. Working with vectors will ensure no changes are required when switching from 2–D to 3–D.

When assembling the terms from the discretisation method it should be considered whether the solution in a point depends on the values around it. For each computational point, a linear equation will be created by the process of discretisation:

$$a_P x_P + \sum_N a_N x_N = b. \tag{17}$$

Index $P$ denotes an influence of $x$ on itself, while $N$ indicates the influence from neighbouring cells. The influences are stored in matrix elements $a_P$ and $a_N$, while other contributions go into r.h.s. vector $b$.

Thus, the procedure begins by splitting the space into cells and time into time steps. A discrete description of a continuous field variable is assembled and spatial and temporal variation of the solution for second–order discretisation is postulated. Expressions for evaluation of volume and surface integrals are generated and all is used to assemble the discretisation of the differential operators:

1. Rate of change term,
2. Gradient operator,
3. Convection operator,
4. Diffusion operators,
5. Source and sink terms.

## Rate of Change

Time derivative captures the rate–of–change of $\phi$ and only the volume integral must be handled. First, time–step size $\Delta t$ must be defined and then $t_{new} = t_{old} + \Delta t$, while time levels of the variable are defined as $\phi^n$ and $\phi^o$:

$$\phi^o = \phi(t = t_{old}), \tag{18}$$

$$\phi^n = \phi(t = t_{new}). \tag{19}$$

The first and second order approximation of the temporal derivative are:

$$\frac{\partial \phi}{\partial t} = \frac{\phi^n - \phi^o}{\Delta t} \tag{20}$$

$$\frac{\partial \phi}{\partial t} = \frac{\frac{3}{2}\phi^n - 2\phi^o + \frac{1}{2}\phi^{oo}}{\Delta t}. \tag{21}$$

Thus, with the volume integral:

$$\int_V \frac{\partial \phi}{\partial t} dV = \frac{\phi^n - \phi^o}{\Delta t} V_P \tag{22}$$

$$\int_V \frac{\partial \phi}{\partial t} dV = \frac{\frac{3}{2}\phi^n - 2\phi^o + \frac{1}{2}\phi^{oo}}{\Delta t} V_P \tag{23}$$

Since $\frac{\partial \phi}{\partial t}$ in cell $P$ depends on $\phi_P$, the matrix will only have a diagonal contribution and a source

- Diagonal element: $a_P = \frac{V_P}{\Delta t}$,
- Source contribution: $b_P = \frac{V_P \phi^o}{\Delta t}$.

## *Gradient*

A gradient of a given field is evaluated using the Gauss theorem:

$$\int_{V_P} \nabla \phi \, dV = \oint_{\partial V_P} d\mathbf{s} \, \phi, \tag{24}$$

where the surface integral in discretised form is split into a sum of face integrals:

$$\oint_S \mathbf{n} \phi \, dS = \sum_f \mathbf{s}_f \phi_f. \tag{25}$$

It is necessary to evaluate the face value of $\phi$. Consistently with second–order discretisation, linear variation between $P$ and $N$ is assumed:

$$\phi_f = f_x \phi_P + (1 - f_x)\phi_N, \tag{26}$$

where $f_x = \overline{fN}/\overline{PN}$. Gradient evaluation is almost exclusively done explicitly, using the known values of field variables.

It is also possible to use a least–square formulation of the gradient. Consider a cell centre $P$ and a cluster of points $N$ around it and fit a plane:

$$e_N = \phi_N - (\phi_P + \mathbf{d}_N \mathbf{\cdot} (\nabla \phi)_P). \tag{27}$$

Minimising the weighted error $e_P$:

$$e_P^2 = \sum_N (w_N e_N)^2 \text{ where } w_N = \frac{1}{|\mathbf{d}_N|} \tag{28}$$

yields a second–order least–square form of gradient:

$$(\nabla\phi)_P = \sum_N w_N^2\, \mathbf{G}^{-1}.\mathbf{d}_N(\phi_N - \phi_P). \tag{29}$$

**G** is a 3 × 3 symmetric matrix:

$$\mathbf{G} = \sum_N \mathbf{d}_N \mathbf{d}_N. \tag{30}$$

Gradient reconstruction may lead to local over– or under–shoots in the reconstructed field:

$$\min_N(\phi_N) \le \phi_P + \mathbf{d}_N.(\nabla\phi)_P \le \max_N(\phi_N) \tag{31}$$

This is important for bounded variables, especially when gradients are used in further discretisation or coupling terms. The remedy is that a minimum and maximum neighbourhood value is calculated and a gradient limiter is applied to preserve bounds in cell centres, Fig. 12.

## *Convection*

The convection term captures the transport by convective velocity. It is split into a sum of face integrals to transform it from differential to integral form:

$$\int_V \nabla.(\phi\mathbf{u})\, dV = \oint_S \phi(\mathbf{n}.\mathbf{u}) dS. \tag{32}$$

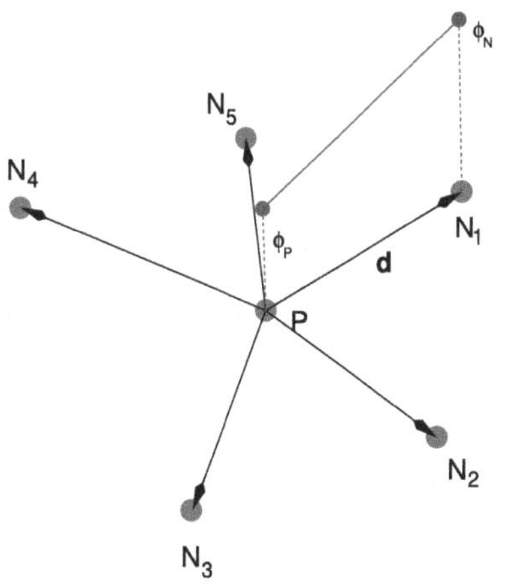


**Fig. 12** Calculation of gradient limiter

**Fig. 13** Face interpolation

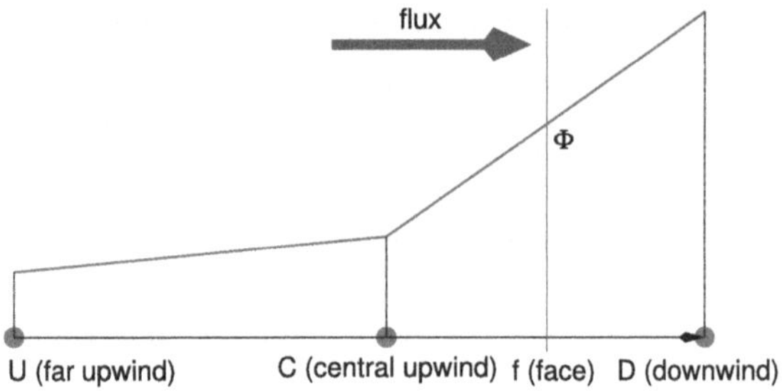


Integration follows the same path as before:

$$\int_V \nabla.(\phi \mathbf{u})\, dV = \oint_S \phi(\mathbf{n}.\mathbf{u}) dS = \sum_f \phi_f (\mathbf{s}_f.\mathbf{u}_f) = \sum_f F\, \phi_f, \tag{33}$$

where $\phi_f$ is the face value of $\phi$ and:

$$F = \mathbf{s}_f.\mathbf{u}_f \tag{34}$$

is the face flux, which is a measure of the flow through the face. In order to close the system, we need a way of evaluating $\phi_f$ from the cell values $\phi_P$ and $\phi_N$ which is done by face interpolation, Fig. 13.

The simplest face interpolation is central differencing:

$$\phi_f = f_x \phi_P + (1 - f_x)\phi_N. \tag{35}$$

It is second–order accurate, but causes oscillations. A more stable scheme is upwind differencing. The transportive property of the term is taken into account, i.e. the information to the face comes from upstream. There are no oscillations, but the solution is smeared, due to the numerical diffusion error:

$$\phi_f = pos(F)\, \phi_P + neg(F)\, \phi_N \;\text{ or }\; f_x = pos(F). \tag{36}$$

$pos(.)$ function takes the value of 1 if the argument is greater or equal to zero and 0 otherwise.

There exists a large number of schemes which try to achieve good accuracy without causing oscillations: e.g. TVD, and NVD families: $\phi_f = f(\phi_P, \phi_N, F, \ldots)$.

For unstructured polyhedral meshes, the concept of a "far upwind node" does not exist. However, the basis of the algorithm is comparison of two cell gradients, which can be achieved using the $grad(\phi)_P$ and $grad(\phi)_N$ around the face, without the need for further addressing Jasak et al. (1999).

In the convection term, $\phi_f$ depends on the values of $\phi$ in two computational points: $P$ and $N$. Note that $F$ can be positive or negative, depending on the flow direction with respect to the face normal. Therefore, the solution in $P$ will depend on the solution in $N$ and vice versa, which means there exists an off–diagonal element $a_N$ in the matrix.

In the case of central differencing on a uniform mesh, a contribution for a face $f$ is:

- Diagonal element: $a_P = f_x\, F$; for all faces $a_P = \sum_N f_x\, F$.
- Off–diagonal element: $a_N = (1 - f_x)F$.
- Source contribution: in our case, nothing. However, some other schemes may have additional (gradient–based) correction terms.

In general, the $P$–to–$N$ element will be different from the $N$–to–$P$ element: the matrix is asymmetric.

In case of upwind differencing, a contribution for face $f$ is:

- Diagonal element: $a_P = \max(F, 0)$.
- Off–diagonal element: $a_N = \min(F, 0)$.

## *Diffusion*

Diffusion term captures the gradient transport. Integration of the term is done using the established procedure:

$$\int_V \nabla\mathbf{.}(\gamma\nabla\phi)\, dV = \oint_S \gamma(\mathbf{n}\mathbf{.}\nabla\phi) dS = \sum_f \int_{S_f} \gamma(\mathbf{n}\mathbf{.}\nabla\phi)\, dS$$
$$= \sum_f \gamma_f\, \mathbf{s}_f\mathbf{.}(\nabla\phi)_f, \tag{37}$$

where $\gamma_f$ is evaluated from cell values using central differencing. Evaluation of the face–normal gradient is done with respect to the alignment of face area normal vector $\mathbf{s}$ and vector connecting the two cell centres $\mathbf{d}_f = \overline{PN}$. If $\mathbf{s}$ and $\mathbf{d}_f = \overline{PN}$ are aligned, difference across the face is used:

$$\mathbf{s}_f\mathbf{.}(\nabla\phi)_f = |\mathbf{s}_f| \frac{\phi_N - \phi_P}{|\mathbf{d}_f|}. \tag{38}$$

This is the component of the gradient in the direction of the $\mathbf{d}_f$ vector. For non–orthogonal meshes, a correction term may be necessary.

For an orthogonal mesh, a contribution for a face $f$ is:

- Diagonal value: $a_P = -\gamma_f \frac{|\mathbf{s}_f|}{|\mathbf{d}_f|}$. For all faces $a_P = -\sum_N \gamma_f \frac{|\mathbf{s}_f|}{|\mathbf{d}_f|}$.
- Off–diagonal value: $a_N = \gamma_f \frac{|\mathbf{s}_f|}{|\mathbf{d}_f|}$.
- Source contribution: for orthogonal meshes, nothing. Non–orthogonal correction will produce a source.

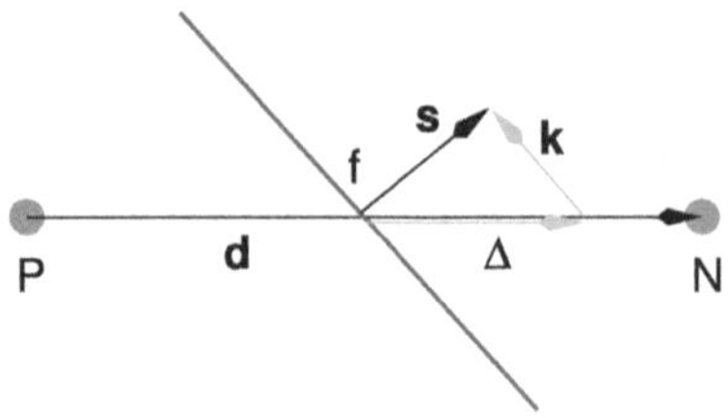


**Fig. 14** Non–orthogonal correction in diffusion term

The $P$–to–$N$ and $N$–to–$P$ elements are identical, which creates a symmetric contribution to the matrix. This is an important characteristic of the diffusion operator.

For non–orthogonal meshes, a correction is added to compensate for the angle between the face area and $\overline{PN}$ vectors.

The decomposition of face gradient into "orthogonal component" $\Delta$ and "non–orthogonal correction" $\mathbf{k}$, depends on mesh quality: mesh non–orthogonality is measured as an angle between $\overline{PN}$ and $\mathbf{s}_f$, Fig. 14. Mathematically, a Laplacian is a perfect operator: smooth, bounded, self–adjoint. Its discretisation yields a symmetric matrix. In contrast, non–orthogonal correction is explicit, unbounded and unsigned. Often on low quality meshes, non–orthogonal correction is limited: the explicit part ($\mathbf{k}$) is clipped to be smaller than its implicit counterpart, based on the current solution:

$$\lambda \frac{|\mathbf{s}_f|}{|\mathbf{d}_f|}(\phi_N - \phi_P) > \mathbf{k}_f \bullet \nabla(\phi)_f, \tag{39}$$

and $\lambda$ is the limiter value.

## *Source and Sink Terms*

Source and sink terms are local in nature:

$$\int_V S \, dV = S V_P. \tag{40}$$

In general, $S$ may be a function of space and time, the solution itself or other variables and can be quite complex. In complex physics cases, the source term can carry the main interaction in the system, e.g. in complex chemistry mechanisms. We shall for the moment consider only a simple case. Typically, linearisation with respect to $\phi$ is performed to promote stability and boundedness:

$$S(\phi) = S_u - S_p \phi$$

where $S_p = \frac{\partial S(\phi)}{\partial \phi}$ and for cases where $S_p > 0$ (sink), treated separately. The contribution to the linear system is limited to diagonal and r.h.s. since source and sink terms do not depend on the neighbourhood:

- Diagonal value created for $S_p < 0$: "boosting diagonal dominance".
- Explicit source contribution: $S_u$.

## *Numerical Boundary Conditions*

Boundary conditions will contribute to the discretisation through the prescribed boundary behaviour. Boundary condition is specified for the whole equation but we shall study them term by term to simplify the problem.

**Dirichlet Condition**

Dirichlet (fixed value) boundary condition specifies the value of the variable on the boundary: $\phi_f = \phi_b$.

In convection term there is a fixed contribution $F\,\phi_b$, i.e. source contribution only. For diffusion term it is necessary to evaluate the near–boundary gradient:

$$\mathbf{n}{\bullet}(\nabla\phi)_b = \frac{\phi_b - \phi_P}{|\mathbf{d}_b|}. \tag{41}$$

This produces a source and a diagonal contribution.

**Neumann Condition**

A generalised form of the Von Neumann (fixed gradient) boundary condition specifies the near–wall gradient $\mathbf{n}{\bullet}(\nabla\phi)_b = g_b$.

In the convection term the boundary value of $\phi$ is evaluated from the internal value and the known gradient:

$$\phi_b = \phi_P + \mathbf{d}_b{\bullet}(\nabla\phi)_b = \phi_P + |\mathbf{d}_b| g_b \tag{42}$$

The evaluated boundary value is used as the face value, which creates a source and a diagonal contribution.

In the diffusion term boundary–normal $g_b$ gradient can be used directly. There is only a source contribution.

Mixed (Robin) boundary condition is a combination of Dirichlet and von Neumann boundary conditions: $\alpha$ times Dirichlet plus $(1 - \alpha)$ times Neumann.

Besides the basic numerical boundary conditions, geometric and coupled conditions are used: symmetry plane condition enforces using the mirror–image of internal solution. Cyclic and periodic boundary conditions couple near–boundary cells to cells on another boundary.

## *Time Advancement*

There are two basic types of time advancement schemes: implicit and explicit schemes. Properties of the algorithm critically depend on this choice, but both are useful under given circumstances. Temporal accuracy depends on the choice of scheme and time step size.

For steady–state simulations, if the equations are linear, the system can be solved with a single linear solver call (provided the discretisation is linear as well). For non–linear equations or special discretisation practices, relaxation methods are used, which exhibit characteristics of time integration, i.e. we are free to redefine the meaning of time.

We have established that an equation can be written in the following form for every control volume:

$$a_P x_P + \sum_N a_N x_N = R, \tag{43}$$

where $N$ denotes the neighbourhood of a computational point $P$. Every time $x_P$ depends on itself, the contribution is added into $a_P$, while when $x_N$ depends on itself, contribution is added into $a_N$. Other contributions go into $R$. For example, in the time derivative, $x$ depends on a value from previous iteration, which goes into $R$. In convection, $x_f$ depends on $x_P$ and $x_N$, where $f$ denotes a face. In diffusion, $\mathbf{s}_f \bullet (\nabla x)_f$ depends on $x_P$ and $x_N$.

There are two solution advancement methods: explicit and implicit.

In the explicit method, $x_P^{\mathrm{n}}$ depends on the neighbour values $x_N^{\mathrm{o}}$ from previous iteration. The expressions are evaluated using the currently available $x$ and the new $x$ is obtained from the time term. Thus, the solution is calculated cell–by–cell, by using the available $x^{\mathrm{o}}$:

$$x_P^{\mathrm{n}} = \frac{R - \sum_N a_N x_N^{\mathrm{o}}}{a_P}. \tag{44}$$

Explicit schemes carry a Courant number ($Co$) limitation, where for stability reasons convective information cannot progress to the mesh more than one cell at a time.

$$Co = \frac{U \, \Delta t}{\Delta x}. \tag{45}$$

Here, $U$ is the face–normal velocity, usually calculated from the face flux:

$$U = \frac{F}{|\mathbf{S}_f|}, \tag{46}$$

$\Delta x$ is the representative cell size, such as distance between two adjacent cell centres and $\Delta t$ is the time–step size. The Courant number poses a particularly inconvenient stability limit: as the computational mesh becomes finer, the maximum stable time–step size also reduces accordingly. Even worse, the global stability limit is related to a maximum local $Co$ number, meaning that locally refined cells, e.g. in boundary layers may cause a massive increase in computational effort.

Courant number limit is the major limitation of explicit methods: information can only propagate at the order of cell size, otherwise the algorithm is unstable. It is quick and efficient, no additional storage is needed but it is very bad for elliptic behaviour.

In the implicit method, $x_P^n$ depends on the neighbour values $x_N^n$ from current iteration. Each term is expressed in matrix form and the resulting linear system is solved. The new solution takes into account the new values in the complete domain which is ideal for elliptic problems. Implicitness removes the Courant number limitation, which allows larger time–steps. However, there is substantial additional storage for matrix elements.

The solution is calculated as:

$$x_P^n = \frac{R - \sum_N a_N x_N^n}{a_P}. \tag{47}$$

Thus, each cell value of $x$ for the "new" level depends on others and all equations need to be solved simultaneously.

First order time integration presented here is often not accurate enough. Runge–Kutta schemes Ferziger and Perić (2002) achieve higher order of accuracy by introducing multiple stages of evaluation within a single time–step. The basic idea of Runge–Kutta methods is to evaluate the variable at several instances in the interval between $t$ and $t + \Delta t$ and then combine them to obtain a high order approximation of the variable.

### *Closure*

In this section discretisation of a generic scalar transport equation using the Finite Volume Method was presented. Discretisation procedure was conducted for each term separately, to examine the properties of the underlying matrix, i.e. contribution of discrete terms to diagonal and off–diagonal elements. The effect of basic numerical boundary conditions on each term was also presented, as well as the effect of sub–optimal computational mesh. The methods presented in this section can be directly applied to other types of transport equations, e.g. Navier–Stokes equations. Algorithms for solving the discrete pressure–velocity system will be presented in the following section.

## Pressure–Velocity Coupling

In the previous section FVM discretisation of a generic scalar transport equation was derived for each transport mechanism. In this section, strategies for resolving the coupling between pressure and velocity fields in a transient, incompressible, single–phase, laminar flow will be presented.

## *Governing Equations*

Equations describing transient, incompressible ($\rho =$ const.), laminar and single–phase flow are the momentum and continuity equation:

$$\underbrace{\frac{\partial \mathbf{u}}{\partial t}}_{\text{rate of change}} + \underbrace{\nabla_{\bullet}(\mathbf{uu})}_{\text{convection}} - \underbrace{\nabla_{\bullet}(\nu \nabla \mathbf{u})}_{\text{diffusion}} = - \underbrace{\nabla p}_{\text{pressure gradient}} \quad . \tag{48}$$

$$\underbrace{\nabla_{\bullet}\mathbf{u}}_{\text{velocity divergence}} = 0, \tag{49}$$

Here, $\mathbf{u}$ is the velocity field, $p$ is kinematic pressure field ($p = P/\rho$) and $\nu$ is kinematic viscosity. Looking at the number of equations and unknowns, the system is well–posed: 1 vector and 1 scalar equation.

There exists linear coupling between Eqs. (48) and (49). $\mathbf{u}$ is a vector variable governed by the (vector) momentum equation. Continuity equation imposes a constraint on velocity, i.e. velocity field should be divergence free. This is an example of a scalar constraint on a vector variable, as $\nabla_{\bullet}\mathbf{u}$ is a scalar.

The convection term in the momentum equation is non–linear. The term will be linearised by reusing the available (previous) values of the velocity field:

$$\nabla_{\bullet}(\mathbf{uu}) \approx \nabla_{\bullet}(\mathbf{u}^{\mathrm{o}}\mathbf{u}^{\mathrm{n}}), \tag{50}$$

where $\mathbf{u}^{\mathrm{o}}$ is the currently available solution or an initial guess, $\mathbf{u}^{\mathrm{n}}$ is the "new" solution. An iterative solution technique will be used to resolve the non–linear coupling and the algorithm cycles until $\mathbf{u}^{\mathrm{o}} = \mathbf{u}^{\mathrm{n}}$.

The system of Eqs. 48 and 49 can be written in matrix form:

$$\begin{bmatrix} \mathbf{A_u} & \nabla(\bullet) \\ \nabla_{\bullet}(\bullet) & 0 \end{bmatrix} \begin{bmatrix} \mathbf{u} \\ p \end{bmatrix} = \begin{bmatrix} 0 \\ 0 \end{bmatrix}, \tag{51}$$

where $\mathbf{A_u}$ is the linear combination of the temporal term, linearised convection and diffusion and ($\bullet$) denotes the position of the unknown variable. For clarity, effects of boundary conditions (although nominally either source or diagonal contribution) are included in the relevant operator matrices.

## *Pressure Equation as a Schur Complement*

The operators in the block system could be considered both as differential and discrete operators. Pressure field does not appear in the continuity equation and there is a zero term on the diagonal of the block matrix, meaning that the system is a saddle point. A

Schur complement (Zhang 2005) will be used to precondition the system and enable the use of standard iterative linear algorithms.

Consider a general block matrix system **M**, consisting of 4 block matrices, **A**, **B**, **C** and **D**, which are respectively $p \times p$, $p \times q$, $q \times p$ and $q \times q$ matrices and **A** is invertible:

$$\begin{bmatrix} \mathbf{A} & \mathbf{B} \\ \mathbf{C} & \mathbf{D} \end{bmatrix}. \tag{52}$$

This structure will arise naturally when trying to solve a block system of equations:

$$\mathbf{Ax} + \mathbf{By} = \mathbf{a}, \tag{53}$$

$$\mathbf{Cx} + \mathbf{Dy} = \mathbf{b}. \tag{54}$$

The Schur complement arises when trying to eliminate $x$ from the system using partial Gaussian elimination by multiplying the first row with $A^{-1}$:

$$\mathbf{A}^{-1}\mathbf{Ax} + \mathbf{A}^{-1}\mathbf{By} = \mathbf{A}^{-1}\mathbf{a} \tag{55}$$

and

$$\mathbf{x} = \mathbf{A}^{-1}\mathbf{a} - \mathbf{A}^{-1}\mathbf{By} \tag{56}$$

Substituting Eq. (56) into Eq. (54) yields:

$$\underbrace{(\mathbf{D} - \mathbf{CA}^{-1}\mathbf{B})}_{\text{Schur complement}}\mathbf{y} = \mathbf{b} - \mathbf{CA}^{-1}\mathbf{a}. \tag{57}$$

The same procedure can be repeated on the block form of the pressure–velocity system, attempting to assemble a pressure equation. Formally, this leads to the following form of the pressure equation:

$$[\nabla\cdot(\bullet)][\mathbf{A_u}^{-1}][\nabla(\bullet)][p] = 0. \tag{58}$$

Here, $\mathbf{A_u}^{-1}$ represent the inverse of the momentum matrix in the discretised form, which acts as diffusivity in the Laplace equation for the pressure. While $\mathbf{A_u}$ is a sparse matrix, its inverse is likely to be dense. Discretised form of the divergence and gradient operator are sparse and well–behaved. However, a triple product with $\mathbf{A_u}^{-1}$ would result in a dense matrix, making it expensive to solve.

This can be remedied be decomposing the momentum matrix before the triple product into the diagonal part off–diagonal matrix:

$$[\mathbf{A_u}] = [\mathbf{D_u}] + [\mathbf{LU_u}], \tag{59}$$

where $\mathbf{D_u}$ only contains diagonal entries. $\mathbf{D_u}$ is easy to invert and will preserve the sparsity pattern in the triple product. Revisiting the momentum equation before the formation of the Schur complement and moving the off–diagonal component of $\mathbf{A_u}$ onto r.h.s. yields:

$$\begin{bmatrix} [\mathbf{D_u}] & [\nabla(\bullet)] \\ [\nabla.(\bullet)] & [0] \end{bmatrix} \begin{bmatrix} \mathbf{u} \\ p \end{bmatrix} = \begin{bmatrix} -[\mathbf{LU_u}][\mathbf{u}] \\ 0 \end{bmatrix}. \tag{60}$$

A revised formulation of the pressure equation via the Schur complement yields:

$$[\nabla.(\bullet)][\mathbf{D_u}^{-1}][\nabla(\bullet)][p] = [\nabla.(\bullet)][\mathbf{D_u}^{-1}][\mathbf{LU_u}][\mathbf{u}]. \tag{61}$$

In both cases, matrix $\mathbf{D_u}^{-1}$ is simple to assemble. It follows that the pressure equation is a Poisson equation with the diagonal part of the discretised momentum acting as diffusivity and the divergence of the velocity on the r.h.s.

## *Derivation of the Pressure Equation*

We shall now rewrite the procedure using discrete form to formally derive an equation for the pressure. This may be done in several ways: formally correct involves a Schur complement (Elman et al. 2005) of the block pressure–velocity system.

We start by discretising the momentum equation using the techniques described for a scalar transport equation. For the purposes of derivation, the pressure gradient term will remain in the differential form. For each control volume, the discretised momentum equation yields:

$$a_P^{\mathbf{u}} \mathbf{u}_P + \sum_N a_N^{\mathbf{u}} \mathbf{u}_N = \mathbf{r} - \nabla p. \tag{62}$$

For simplicity, we shall introduce the $\mathbf{H(u)}$ operator (Jasak 1996), containing the off–diagonal part of the momentum matrix and any associated r.h.s. contributions:

$$\mathbf{H(u)} = \mathbf{r} - \sum_N a_N^{\mathbf{u}} \mathbf{u}_N \tag{63}$$

Using the above, it follows:

$$a_P^{\mathbf{u}} \mathbf{u}_P = \mathbf{H(u)} - \nabla p \tag{64}$$

and

$$\mathbf{u}_P = (a_P^{\mathbf{u}})^{-1}(\mathbf{H(u)} - \nabla p). \tag{65}$$

Substituting the expression for $\mathbf{u}_P$ into the incompressible continuity equation $\nabla.\mathbf{u} = 0$ yields

$$\nabla.\big[(a_P^{\mathbf{u}})^{-1} \nabla p\big] = \nabla.((a_P^{\mathbf{u}})^{-1}\mathbf{H(u)}). \tag{66}$$

This is the form of the pressure equation for incompressible fluid. Note the implied decomposition of the momentum matrix into the diagonal and off–diagonal contribution, where $a_P^{\mathbf{u}}$ is an element in $\mathbf{D_u}$ matrix and $\mathbf{H(u)}$ is the product $\mathbf{LU_u u}$.

The pressure equation is derived from continuity which implies divergence–free velocity. The discretised form of the continuity equation is:

$$\nabla{\bf .u} = \sum_f {\bf s}_f{\bf .u} = \sum_f F, \tag{67}$$

where $F$ is the face flux:

$$F = {\bf s}_f{\bf .u}. \tag{68}$$

The conservative face flux should be created from the solution of the pressure equation. Substituting expression for $\mathbf{u}$ into the flux equation, it follows:

$$F = -(a_P^{\mathbf{u}})^{-1}{\bf s}_f{\bf .}\nabla p + (a_P^{\mathbf{u}})^{-1}{\bf s}_f{\bf .H(u)}. \tag{69}$$

A part of the above, $(a_P^{\mathbf{u}})^{-1}{\bf s}_f{\bf .}\nabla p$ appears during the discretisation of the Laplacian, for each face:

$$(a_P^{\mathbf{u}})^{-1}{\bf s}_f{\bf .}\nabla p = (a_P^{\mathbf{u}})^{-1}\frac{|{\bf s}_f|}{|{\bf d}|}(p_N - p_P) = a_N^p(p_N - p_P). \tag{70}$$

Here, $a_N^p = (a_P^{\mathbf{u}})^{-1}\frac{|{\bf s}_f|}{|{\bf d}|}$ is equal to the off–diagonal matrix element in the pressure Laplacian. Note that in order for the face flux to be conservative, assembly of the flux must be completely consistent with the assembly of the pressure equation, e.g. it must include non–orthogonal correction.

## *Segregated Solution Algorithm: SIMPLE*

The earliest pressure–velocity coupling algorithm is SIMPLE (Semi–Implicit Algorithm for Pressure–Linked Equations), conceived by Patankar and Spalding in 1972 at the Imperial College London (Patankar and Spalding 1972). It comprises the following sequence of operations:

1. Guess the pressure field $p^*$.
2. Solve the momentum equation using the guessed pressure. This step is called momentum predictor:
$$a_P^{\mathbf{u}}\mathbf{u}_P = \mathbf{H(u)} - \nabla p^*.$$
3. Calculate the new pressure based on the velocity field. This is called a pressure correction step:
$$\nabla{\bf .}\left[(a_P^{\mathbf{u}})^{-1}\nabla p\right] = \nabla{\bf .}((a_P^{\mathbf{u}})^{-1}\mathbf{H(u)}).$$

4. Based on the pressure solution, assemble conservative face flux $F$:

$$F = \mathbf{s}_f \mathbf{.H}(\mathbf{u}) - a_N^p (p_N - p_P).$$

5. Repeat to convergence.

The algorithm in its base form produces a series of corrections on $\mathbf{u}$ and $p$. Unfortunately, it will not converge in the presented form. Divergence happens due to the fact that pressure correction contains both the pressure as a physical variable and a component which forces the discrete fluxes to become conservative. In order to achieve convergence, under–relaxation is used:

$$p^{**} = p^* + \alpha_P (p - p^*), \tag{71}$$

and

$$\mathbf{u}^{**} = \mathbf{u}^* + \alpha_{\mathbf{u}} (\mathbf{u} - \mathbf{u}^*), \tag{72}$$

where $p$ and $\mathbf{u}$ are the solution of the pressure and momentum equations and $\mathbf{u}^*$ and $p^*$ represent a series of pressure and velocity approximations. $\alpha_P$ and $\alpha_{\mathbf{u}}$ are the pressure and velocity under–relaxation factors. Note that in practice, momentum under–relaxation is implicit (included in the coefficient matrix) while pressure is under–relaxed explicitly, since the pressure equation is elliptic:

$$\frac{a_P^{\mathbf{u}}}{\alpha_U} \mathbf{u}_P = \mathbf{H}(\mathbf{u}) - \nabla p^* + \frac{1 - \alpha_U}{\alpha_U} a_P^{\mathbf{u}} \mathbf{u}_P^*. \tag{73}$$

Some guidelines for choosing under–relaxation factors are:

$$0 < \alpha_P \leq 1,$$
$$0 < \alpha_U \leq 1,$$
$$\alpha_P + \alpha_U \approx 1,$$

or the standard set (for guidance only):

$$\alpha_P = 0.2,$$
$$\alpha_U = 0.8.$$

Under–relaxation dampens the oscillation in the pressure–velocity coupling and is very efficient in stabilising the algorithm.

To summarise, SIMPLE algorithm prescribes that the momentum predictor will be solved using the available pressure field. The role of pressure in the momentum equation is to ensure that the velocity field is divergence free. After the first momentum solution, the velocity field is not divergence–free since a guessed pressure field was used. Therefore, the pressure field after the first pressure corrector will contain two parts: physical pressure, consistent with the global flow field and a "pressure correction" component, which enforces the continuity and counter–balances the error in the initial pressure guess. Only the first component should be built into the physical pressure field. In SIMPLE, this is handled by severely under–relaxing the pressure.

## *Segregated Solution Algorithm: PISO*

Having 2 under–relaxation factors which balance each other is very inconvenient as the tuning is difficult. The solution was offered by Issa in 1986 in the form of the PISO (Pressure Implicit with Splitting of Operators) algorithm (Issa 1986).

The pressure–velocity system contains two complex coupling terms: non-linear convection term, containing $\mathbf{u} - \mathbf{u}$ coupling and linear pressure–velocity coupling. On low Courant number, i.e. small time–steps, the pressure velocity coupling is much stronger than the non–linear coupling. It is therefore possible to repeat a number of pressure correctors without updating the discretisation of the momentum equation, using the new fluxes. In such a setup, the first pressure corrector will create a conservative velocity field, while the second and following will establish the pressure distribution.

Since multiple pressure correctors are used with a single momentum equation, it is no longer necessary to under–relax the pressure. In steady–state simulations, the system is stabilised by momentum under–relaxation. On the negative side, derivation of PISO is based on the assumption that momentum discretisation may be safely frozen through a series of pressure correctors, which is true only at small time–steps. PISO contains the following sequence of operations:

1. Use the available pressure field $p^*$ from previous corrector or time–step. Conservative fluxes corresponding to $p^*$ are also available.
2. Discretise the momentum equation with the available flux field.
3. Solve the momentum equation using the guessed pressure. This step is called momentum predictor:
$$a_P^{\mathbf{u}} \mathbf{u}_P = \mathbf{H}(\mathbf{u}) - \nabla p^*.$$
4. Calculate the new pressure based on the velocity field. This is called a pressure correction step:
$$\nabla\bullet\left[(a_P^{\mathbf{u}})^{-1} \nabla p\right] = \nabla\bullet((a_P^{\mathbf{u}})^{-1} \mathbf{H}(\mathbf{u})).$$
5. Based on the pressure solution, assemble conservative face flux $F$:
$$F = \mathbf{s}_f \bullet \mathbf{H}(\mathbf{u}) - a_N^p (p_N - p_P).$$
6. Explicitly update cell–centred velocity field with the assembled momentum coefficients:
$$\mathbf{u}_P = (a_P^{\mathbf{u}})^{-1} (\mathbf{H}(\mathbf{u}) - \nabla p).$$

7. Return to pressure correction step if convergence is not reached.
8. Proceed from beginning for a new time–step.

PISO is useful in kinds of simulations where the time–step is controlled by external issues and temporal accuracy is important. In such cases, assumption of slow variation over non-linearity holds and the cost of momentum assembly and solution can be safely avoided, e.g. for Large Eddy simulations. Functional equivalent of the PISO algorithm is also used as a preconditioner in Krylov space saddle–point solvers.

In transient flows there is a natural time–step limit, related to mesh size. In many cases local mesh resolution, e.g. near walls limits the global time–step size to unreasonable level. The limit can be violated locally "without effect on global time accuracy". Some options are:

- Transient SIMPLE–based algorithms where each time-step is considered as a quasi–steady solution, by assuming the time derivative term is a "local inertial source".
- Inertial under–relaxation: in small cells the temporal accuracy will be ruined anyway. It is assumed as it is already the case and "inertial under–relaxation" is added in problematic cells.
- Sub–cycling: in cases where a large time–step or inertial under–relaxation is not an option, e.g. wave propagation, a transport equation can be solved in sub–steps to satisfy the Courant number limit.

However, non–uniform time step size will potentially lead to unboundedness problems. Local time–stepping algorithms can be reformulated to work around this problem.

The weakest point of SIMPLE–type algorithms is equation segregation since a linear problem (pressure and velocity coupling) is addressed by parts. Block–coupled solution of both equations in a single linear system can significantly accelerate convergence as iterations happen only over non–linear terms.

### *Block–Implicit Coupled Solution*

It is possible to formulate a system of equations which will discretise both the pressure and momentum equation in a single matrix and solve them simultaneously. Efficiency of such algorithms is good for steady–state computations but at a cost of considerable increase in storage and choice of linear solver technology. We have previously derived the pressure equation as a Schur complement of the momentum equation. Elman preconditioning of saddle–point systems (Elman et al. 2005) implies replacing the zero term on the diagonal with a Schur complement. Thus, the final form of the block–coupled pressure–velocity system is:

$$\begin{bmatrix} [\mathbf{A_u}] & [\nabla(.)] \\ [\nabla\bullet(.)] & [\nabla\bullet(.)][\mathbf{D_u}^{-1}][\nabla(.)] \end{bmatrix} \begin{bmatrix} \mathbf{u} \\ p \end{bmatrix} = \begin{bmatrix} 0 \\ \nabla\bullet(\mathbf{D_u}^{-1}\nabla(p)) \end{bmatrix}. \tag{74}$$

Here, matrix elements are $4 \times 4$ blocks, consistent with discretisation of operators and boundary conditions of the segregated system. The r.h.s. term in $p$ is the conventional Rhie–Chow correction (Rhie and Chow 1983). The complete $p$–$\mathbf{u}$ matrix is no longer diagonally dominant because of the $[\nabla(\bullet)]$ and $[\nabla.(\bullet)]$ blocks, and these two off–diagonal blocks are a transpose of each other. Flux calculation after block solution is identical to the segregated method. Momentum under–relaxation is needed only for ensuring the convergence of linear solvers, while there is no under–relaxation of pressure equation. The iteration is only performed to handle true non–linearity in convection and turbulence. However, due to matrix properties, stronger solvers and smoothers are needed, e.g. Algebraic Multigrid (AMG).

### *Closure*

In this section algorithms for the solution of pressure–velocity coupling in equations for incompressible, transient, single–phase, laminar flow were presented. The non–linear momentum equation is linearised, which means an iterative procedure must be employed to solve the non–linear coupling. Since there is no equation for pressure, a Schur preconditioning technique is used for derivation as it counteracts the saddle–point problem. Several algorithms based on this preconditioning technique were presented: SIMPLE, PISO and block–coupled implicit technique. The available linear solver technology for the solution of linear coupling will be presented in the following section.

## Linear Equation Solvers

In the previous section, various forms of pressure–velocity coupling algorithms were presented. In this section, numerical solution methods for the resulting linear systems of equations which arise from the FVM discretisation of equations will be outlined. We shall discuss direct solvers and whether they are applicable to CFD problems as well as different types of iterative solvers.

### *Finite Volume Matrix*

Going back to the Finite Volume discretisation, we can observe that an equation can be written in the following form for every computational point (control volume):

$$a_P x_P + \sum_N a_N x_N = R. \tag{75}$$

Together, equations for all cells form a linear system or a matrix:

$$\mathbf{A}\mathbf{x} = \mathbf{b}, \tag{76}$$

where **A** contain matrix elements, **x** is the value of $x_P$ in all cells and **b** is the right-hand-side vector.

Matrix **A** is potentially very big: $n$ cells $\times$ $n$ cells and it is square since the number of equations is equal to the number of unknowns. Also, most elements are equal to zero because the connectivity is always local. Thus, the matrix is sparse which potentially leads to storage savings if a good matrix storage format can be identified. The equations are linearised if they are non–linear by nature, as we typically wish to avoid handling non–linearity at this level due to high cost of non–linear matrix solvers.

## *Matrix Storage Formats*

The matrix can be stored in either dense, compressed row or arrow format. Dense matrix format is used for matrices which are completely or almost full, i.e. do not have many zero elements. All matrix elements are stored, typically in a two–dimensional array: diagonal elements ($a_{ii}$) and off-diagonal elements ($a_{ij}$).

This format is convenient for small matrices and when direct solvers are used. Matrix elements represent a large chunk of memory and efficient operations imply memory management optimisation. Also, it is impossible to say if the matrix is symmetric or not without floating point comparisons.

If the matrix is sparse, as FVM matrices are, only non–zero elements may be stored, which yields considerable savings in memory. In the compressed row format, the elements are typically stored in a single 1-D array (diagonal elements may be separate) and they are ordered row–by–row. Addresing is stored in two arrays: "row start" and "column". The row array records the start and end of each row in the column array. Thus, row $i$ has got elements from `row[i]` to `row[i + 1]`. Size of row arrays is equal to number of rows + 1.

An example of the code for calculating a matrix–vector product using the compressed row format is given:

```
vectorProduct(b, x) //   [b] = [A] [x]
{
    for (int n = 0; n < count; n++)
    {
        for (int ip = row[n]; ip < row[n+1]; ip++)
        {
            b[n] = coeffs[ip]*x[col[ip]];
        }
    }
}
```

Arrow format is an arbitrary sparse format. Elements are stored in 3 arrays: diagonal, upper and lower triangle. While the diagonal addressing is implied, off–diagonal addressing is stored in 2 arrays: "owner" (row index) and "neighbour" (column index) array. The size of addressing arrays is equal to the number of off-diagonal elements. The matrix structure (fill–in) is assumed to be symmetric: presence of $a_{ij}$ implies the presence of $a_{ji}$. Symmetric matrices are easily recognised from this format and if the matrix elements are symmetric, only the upper triangle is stored.

An example of the code for calculating a matrix–vector product using the arrow format is given:

```
vectorProduct(b, x) //    [b] = [A] [x]
{
    for (int n = 0; n < coeffs.size(); n++)
    {
        int c0 = owner(n);
        int c1 = neighbour(n);
        b[c0] = upperCoeffs[n]*x[c1];
        b[c1] = lowerCoeffs[n]*x[c0];
    }
}
```

## *Matrix Properties*

A relationship between the FV mesh and matrix is established through the connectivity of cells and equations. A cell value depends on other cell values only if the two cells share a face. Therefore, a correspondence exists between the off-diagonal matrix elements and the mesh structure. In practice, the matrix is assembled by looping through the mesh. The structure (pattern of elements) of the matrix depends on the numbering of cells.

We have established that FV discretisation produces sparse matrices with only a few non–zero elements per row, depending on the cell type, i.e. the number of faces. A sparse matrix is banded if its non-zero elements are grouped in a stripe around the diagonal. It has a multi–diagonal structure if its non–zero off–diagonal elements form a regular diagonal pattern. A symmetric matrix is equal to its transpose:

$$\mathbf{A} = \mathbf{A}^{\top}. \tag{77}$$

A matrix is positive definite if for every $\mathbf{x} \neq 0$:

$$\mathbf{x}^{\top}\mathbf{A}\mathbf{x} > 0. \tag{78}$$

A matrix is diagonally dominant if in each row the sum of off–diagonal element magnitudes is equal or smaller than the diagonal element:

$$a_{ii} \geq \sum_{j=1}^{N} |a_{ij}| \ ; \ \ j \neq i, \tag{79}$$

and for at least one $i$

$$a_{ii} > \sum_{j=1}^{N} |a_{ij}| \ ; \ \ j \neq i. \tag{80}$$

The exact solution of the linear system $\mathbf{Ax} = \mathbf{b}$ we wish to solve, can be obtained by inverting the matrix $\mathbf{A}$:

$$\mathbf{x} = \mathbf{A}^{-1}\,\mathbf{b}. \tag{81}$$

This is how direct solvers operate: number of operations required for the inversion of $\mathbf{A}$ is fixed and until the inverse is constructed, $\mathbf{x}$ cannot be obtained.

Note that in direct solution algorithm, an initially sparse matrix shall be filled in elements: inverting a sparse matrix $\mathbf{A}$ does not preserve the sparsity pattern in $\mathbf{A}^{-1}$.

## *Definition of Residual*

Iterative solvers start from an approximate solution $\mathbf{x}^{(0)}$ and generate a set of solution estimates $\mathbf{x}^{(k)}$, where $k$ is the iteration counter. Their most important property is that the storage of matrix inverse is not needed—overall storage during the solution process will be only slightly higher (by several vectors of size $n$) than the storage of the sparse matrix.

In iterative solution algorithms, quality of the solution estimate is measured through a residual:

$$\mathbf{r} = \mathbf{b} - \mathbf{Ax}^{(k)}. \tag{82}$$

A residual is a vector showing how far the current estimate $\mathbf{x}^{(k)}$ is from the exact solution $\mathbf{x}$. Note that for correct $\mathbf{x}$, $\mathbf{r}$ will be zero. Since $\mathbf{r}$ defines a value for every equation (row) in $\mathbf{A}$, it is impractical to measure and compare. Thus a residual norm $||\mathbf{r}||$ is used. It can be assembled in many ways, but usually:

$$||\mathbf{r}|| = \sum_{j=1}^{N} |r_j|. \tag{83}$$

In CFD software, the residual norm is normalised further for easier comparison between the equations. Convergence of the iterative solver is usually measured in terms of residual reduction. The linear system of equations is considered to be solved when:

$$\frac{||\mathbf{r}^{(k)}||}{||\mathbf{r}^{(0)}||} < \epsilon, \tag{84}$$

where $\epsilon$ is a user defined tolerance. Good (implicit) numerical simulation software will spend 50–90% of CPU time inverting matrices and performance of linear solvers is absolutely critical for the performance of the entire solution method. Like in the case of mesh generation, the characteristics of a discretisation method and the solution algorithm are coupled with the linear solver. Only a combination of a discretisation method and a linear solver will result in a useful solver. Typically, properties of discretisation will be set up in a way that allows the choice of an efficient solver.

## *Direct Solvers*

Whe looking for a solution of a linear equation set, it is possible to use a direct solver, which performs a given number of operations, after which a solution is obtained.

Direct solvers are expensive in storage and CPU time but can handle any type of matrix and there are no concerns about matrix properties during discretisation. Thus, direct solvers are typically used for cases where it is difficult to control matrix properties through discretisation: boundary element methods, high-order FEM methods, Hermitian elements, Discontinuous Galerkin etc. Iterative solvers start from an initial solution and perform an a number of operations (unknown in advance), which will result in an improved solution.

Iterative solvers may be variants of the direct solution algorithm with special characteristics. For large problems, iterative solvers are the only option, however, they require matrices with "special" properties. Fortunately, FVM matrices are ideally suited (read: carefully constructed) for use with iterative solvers.

Since the number of operations required for the solution by a direct method is known, intermediate solutions are of no interest. When operating on a large sparse matrix like the one from discretisation methods, the direct solver will create entries for elements that were not previously present. As a consequence, formal matrix storage requirement for a direct solver is a full matrix for a complete system, which is extremely large. Also, the advantage of direct solvers is that they can handle any sort of well-posed linear system. In reality this is compromised by the round-off error, and even though it can be partially taken into account through the details of the solution algorithm, for really bad matrices this cannot be helped.

Gaussian elimination is the easiest direct solver, see (Saad 2000). Elimination is performed by combining row elements until a matrix becomes triangular. The elimination step is followed by backward substitution to obtain the solution. In order to control the discretisation error, equations are chosen for elimination based on the diagonal element, which is called pivoting. Linear combination of matrix rows leads to additional fill–in of elements. The number of operations in direct solvers scales with the number of equations cubed, which is computationally very expensive in CFD.

When handling sparse systems, the fill–in is very problematic since it leads to a large increase in storage size and accounts for the bulk of operations. The modern implementation of direct solvers is based on the window approach. Looking at the structure of the sparse system, it can be established that equation for $x_P$ depends only on a small subset of other nodes: in principle, it should be possible to eliminate the equation for $P$ just by looking at a small subset of the complete matrix. If all equations under elimination have overlapping regions of zero off–diagonal elements, there will be no fill–in in the shared regions of zeros. Thus, instead of operating on the complete matrix, an active window for elimination can be created in a multi–frontal solver (Duff and Reid 1983). The window will sweep over the matrix, adding equations one by one and performing elimination immediately. The window matrix will be dense, but much smaller than the complete matrix. Also, the triangular matrix (needed for back–substitution) can be stored in a sparse format.

The window approach may reduce the cost of direct solvers by several orders of magnitude and it is acceptable for medium–sized systems. The number of operations scales roughly with $N \cdot M^2$, where $N$ is the number of equations and $M$ is the maximum size of the solution window. The first step in the implementation is the control of the window size: the window changes its width dynamically and in the worst case may be the size of the complete matrix. Maximum size of the window depends on the matrix connectivity and ordering of equations. Special optimisation software is used to control the window size. Matrix renumbering and ordering heuristics is regularly used, e.g. numbering of a Cartesian mesh to minimise the matrix band. The most expensive operation in the multi–frontal solver is the calculation of the Schur's complement: the difference between the trivial and optimised operation can be a factor of 1000. In practice, Basic Linear Algebra (BLAS) (National Science Foundation 2014) library is used for such problems. It is a special assembly code implementation for matrix manipulation. The code is optimised by hand and sometimes written specially for processor architecture. It is unlikely that a hand–written code for the same operation achieves more than 10 % efficiency of BLAS and a good implementation can be measured in how much the code spends on operations outside of BLAS.

## *Iterative Solvers*

Performance of iterative solvers depends on the matrix characteristics. The solver operates by incrementally improving the solution, which leads to the concept of error propagation: if the error is augmented in the iterative process, the solver diverges. The easiest way of analysing the error is in terms of eigen–spectrum of the matrix. The general idea of iterative solvers is to replace $\mathbf{A}$ with a matrix that is easy to invert and approximates $\mathbf{A}$ and use this to obtain the new solution.

The most basic iterative solvers are Jacobi and Gauss–Seidel iterations (Saad 2000). Propagation of information in these simple iterations is very slow. Jacobi propagates the "data" one equation at a time. For Gauss–Seidel, the information propagation depends on the matrix ordering and sweep direction. In practice, forward and backward sweeps are alternated.

Consider again a linear problem $\mathbf{Ax} = \mathbf{b}$. A fixed point iterative method is obtained by splitting $\mathbf{A} = \mathbf{M} - \mathbf{N}$:

$$\mathbf{x}^{(k+1)} = \mathbf{S}\mathbf{x}^{(k)} + \mathbf{M}^{-1}\mathbf{b}, \tag{85}$$

where $k$ is the iteration counter, and $\mathbf{S}$ is the iteration matrix:

$$\mathbf{S} = \mathbf{M}^{-1}\mathbf{N}. \tag{86}$$

The solution error $\mathbf{e}$ and error propagation equation are:

$$\mathbf{e}^{(k)} = \mathbf{x}^{(k)} - \mathbf{x}, \tag{87}$$

$$\mathbf{e}^{(k+1)} = \mathbf{S}\mathbf{e}^{(k)}, \tag{88}$$

where $\mathbf{x}$ is the correct solution. Iteration matrix possesses the recursive property:

$$\mathbf{e}^{(k+1)} = \mathbf{S}^{k}\mathbf{e}^{(0)}. \tag{89}$$

Thus, the iterative matrix affects the error vector which should reduce in magnitude each time the matrix is applied to it. The magnitude of error reduction is governed by matrix properties: the method will converge fast if the matrix is strictly diagonally dominant and if all off–diagonal elements are approximately the same. FV matrices stemming from problems with complex physics and unstructured non–uniform meshes can hardly satisfy such conditions, which leads to performance deterioration of fixed point algorithms. A step in an alternate direction was made by transforming a linear system into a minimisation problem by using Conjugate Gradient type solvers.

### *Krylov Subspace Solvers*

Looking at the direct solver, we can imagine that it operates in $N$-dimensional space, where $N$ is the number of equations and searches for a point which minimises the residual. In Gaussian elimination, each direction of the $N$–dimensional space is visited once and eliminated from further consideration.

The idea of Krylov space solvers, see (van der Vorst 2003), is that an approximate solution can be found more efficiently if we look for search directions more intelligently. A residual vector $\mathbf{r}$ at each point gives the "direction" in which the error is minimised and we should search in. Additionally, we would like to always search in a direction orthogonal to all previous search directions, meaning the correct solution will be reached after $N$ iterations.

Even though the idea is good, it is still impractical for large systems. The search procedure is an example of an iterative roughener, i.e. the error term does decrease in magnitude in each iteration, but some components of the error may temporarily grow. In terms of performance, the number of operations in Krylov space solvers scales with $N \cdot log(N)$, where $N$ is the number of unknowns.

The basic Krylov subspace solver, which can be applied onto symmetric, positive–definite systems is the Conjugate Gradient (CG) (Shewchuk 1994). CG solver is an orthogonal projection technique onto the Krylov space $\mathcal{K}(\mathbf{A}, \mathbf{r}^{(0)})$, and comprises the following operations:

1. In the beginning assume $\mathbf{x}^{(0)} = 0$, and then $\mathbf{p}^{(0)} = \mathbf{r}^{(0)} = \mathbf{b} - \mathbf{A}\mathbf{x}^{(0)}$.
2. Calculate the length of the step in direction $\mathbf{p}^{(k)}$:

$$\alpha^{(k)} = \frac{(\mathbf{r}^{(k)})^\top \mathbf{r}^{(k)}}{(\mathbf{p}^{(k)})^\top \mathbf{A}\mathbf{p}^{(k)}}.$$

3. Calculate the new solution:

$$\mathbf{x}^{(k+1)} = \mathbf{x}^{(k)} + \alpha^{(k)}\mathbf{p}^{(k)}.$$

4. Calculate the new residual:

$$\mathbf{r}^{(k+1)} = \mathbf{r}^{(k)} + \alpha^{(k)}\mathbf{A}\mathbf{p}^{(k)}.$$

5. Calculate the projection operator of the new residual onto the previous search direction:

$$\beta^{(k)} = \frac{(\mathbf{r}^{(k+1)})^\top \mathbf{r}^{(k+1)}}{(\mathbf{r}^{(k)})^\top \mathbf{r}^{(k)}}\mathbf{p}^{(k)}.$$

6. Calculate the new search direction by making it **A**–orthogonal to the new residual:

$$\mathbf{p}^{(k+1)} = \mathbf{r}^{(k+1)} + \beta^{(k)}\mathbf{p}^{(k)}.$$

7. Return to step 2 if the convergence criterion is not satisfied.

CG solver seeks the solution in the Krylov space by stepping in direction **p** scaled by step length $\alpha$:

$$\mathbf{x}^{(k+1)} = \mathbf{x}^{(0)} + \alpha_k \mathbf{p}^{(k)}. \tag{90}$$

Direction vectors $\mathbf{p}^{(k)}$ are chosen to be conjugate (**A**–orthogonal):

$$\left(\mathbf{A}\mathbf{p}^{(k)}, \mathbf{p}^{(k)}\right) = 0. \tag{91}$$

Here, symbol $(\bullet, \bullet)$ denotes the scalar product of two vectors. Even though it is guaranteed that CG will converge to correct solution in $N$ iterations, it is still not appropriate for large systems such as those arising from the FVM discretisation. To speed up and stabilise convergence, transformations of the coefficient matrix are used, called preconditioning.

## *Preconditioning*

Preconditioning matrix **M** is a matrix which approximates **A** such that equation

$$\mathbf{M}\mathbf{x} = \mathbf{b}$$

may be inexpensive to solve. Then, the following procedure is applied:

$$\mathbf{A}\mathbf{M}^{-1}\mathbf{u} = \mathbf{b} \tag{92}$$

$$\mathbf{x} = \mathbf{M}^{-1}\mathbf{u}. \tag{93}$$

For example, the additional cost of the preconditioned CG method is equivalent to solving a linear system. Thus, **M** should be chosen in such a way that this system can be solved quickly and efficiently. Some possibilities for choosing a preconditioning matrix are using only the diagonal elements of **A** or using an incomplete lower–uper factorisation of **A**, which is known as the ILU preconditioner (Saad 2000).

## *Algebraic Multigrid*

A return to fixed point methods has been made after mathematical analysis of discretisation showed it makes sense to use coarse–mesh solutions to accelerate the solution process on the fine mesh, through initialisation of fine level solution using the coarse correction. In terms of matrices and linear solvers, the same principle should apply since the matrices come from discretisation. Since it would be impractical to build a series of coarse meshes just to solve a system of linear equations, it can be readily recognised that all the information about the coarse mesh (and therefore the coarse matrix) already exists in the fine mesh. Thus, the same can be done with a linear equation solver which is the basis of Algebraic Multigrid (AMG) (Trottenberg et al. 2001).

Operation of AMG relies on the fact that a high–frequency error is easy to eliminate using the fixed point algorithm. High–frequency error corresponds to those components of the error which can be eliminated using the local information, i.e. by communicating only with the coupled equations. Once the high-frequency error is removed by e.g. Gauss–Seidel solver, iterative convergence slows down. At the same time, the error that looks smooth on the current mesh will behave as oscillatory on a coarser mesh. If the mesh is coarser, the error is both eliminated faster and in fewer iterations. Thus, in multigrid the solution is mapped through a series of coarse levels, each of the levels being responsible for eliminating a "band" of the error.

Communication between fine and coarse level is established through restriction and prolongation matrices, **R** and **P** respectively. Restriction matrix is used for transferring the residual from fine to coarse level:

$$\mathbf{R}\mathbf{r}^{\mathrm{F}} = \mathbf{r}^{\mathrm{C}}, \tag{94}$$

where $\mathbf{r}^{\mathrm{F}}$ is the residual calculated from the fine level approximate solution and $\mathbf{r}^{\mathrm{C}}$ is the value of the residual on coarse level, after restriction. A fine residual, containing the smooth error component, is restricted and used as the right-hand-side of the coarse system. Prolongation matrix is used for interpolating the coarse level correction onto fine level:

$$\mathbf{P}\mathbf{e}^{\mathrm{C}} = \mathbf{e}^{\mathrm{F}}, \tag{95}$$

where $\mathbf{e}^{\mathrm{C}}$ is the correction calculated on the coarse level by solving the residual equation $\mathbf{A}^{\mathrm{C}}\,\mathbf{e}^{\mathrm{C}} = \mathbf{r}^{\mathrm{C}}$, and $\mathbf{e}^{\mathrm{F}}$ is the correction after interpolation onto fine level. Once the coarse system is solved, coarse correction is prolongated to the fine level and added to the solution. Interpolation introduces aliasing errors, which can be efficiently removed by smoothing (using a fixed point method) on the fine level.

Coarse matrix $\mathbf{A}^{\mathrm{C}}$ is constructed through projection:

$$\mathbf{A}^{\mathrm{C}} = \mathbf{R}\mathbf{A}^{\mathrm{F}}\mathbf{P}. \tag{96}$$

Creating a coarse level matrix is roughly equivalent to creation of coarse mesh cells. Two main approaches are Aggregative algebraic multigrid (AAMG) and Selective algebraic multigrid (SAMG). In AAMG (Hutchinson and Raithby 1986), equations are grouped into clusters in a manner similar to grouping fine cells to form a coarse cell. The grouping pattern is based on the strength of off–diagonal elements. In SAMG (Ruge and Stüben 1987), the equations are separated into two groups: the coarse and fine equations. Selection rules specifies that no two coarse points should be connected to each other, creating a maximum possible set. Fine equations form a fine–to–coarse interpolation method (restriction matrix), which is used to form the coarse system.

The bulk of multigrid work is performed by transferring the error and correction through the multigrid levels. Fixed point algorithms only act to remove high–frequency error, i.e. they smooth the error. Smoothing can be applied on each level before the restriction of the residual, which is called pre–smoothing. If it is applied after the coarse correction has been added, it is called post–smoothing. Algorithmically, post-smoothing is more efficient.

Based on the above, AMG can be considered a two–level solver. In practice, the "coarse level" solution is also assembled using multigrid, leading to multi–level systems, which are governed by a cycle. The most important multigrid cycle types are:

- V–cycle (Fig. 15) in which a hierarchy of coarse levels is created, each level (except the coarsest) is visited twice. Residual reduction is performed all the way to the coarsest level and then prolongation and post–smoothing are done on each level in the opposite direction. Mathematically, it is possible to show that the V–cycle is optimal and leads to the solution algorithm where the number of operations scales linearly with the number of unknowns. Other cycles, e.g. W–cycle (Fig. 16) or F–cycle are a variation on the V-cycle, where coarse levels are visited multiple times.

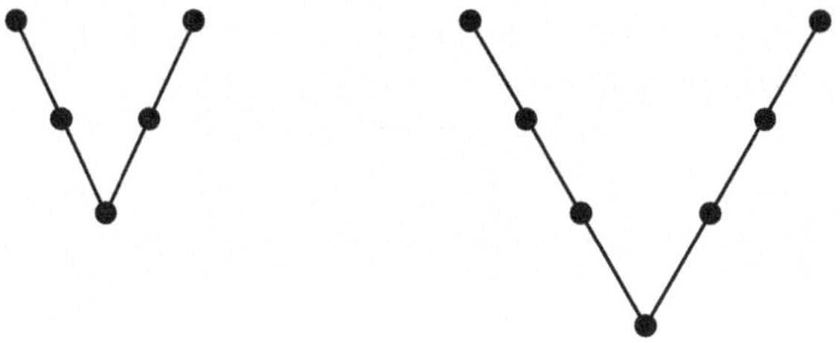

**Fig. 15** Multi–level multigrid V–cycle

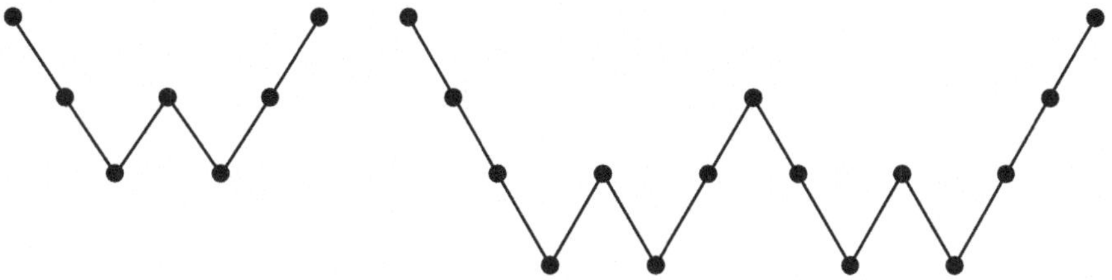

**Fig. 16** Multi–level multigrid W–cycle

- Flex cycle: the creation of coarse levels is done on demand, when the smoother stops converging efficiently.

## *Block–Coupled Solution Algorithms*

For cases of strong coupling between the components of a vector, the components can be solved as a block variable: $(u_x, u_y, u_z)$ will appear as variables in the same linear system. In spite of the fact that the system is much larger, the coupling pattern still exists: components of $\mathbf{u}$ in cell $P$ may be coupled to other components in the same point or to vector components in the neighbouring cell. With this in mind, we can still keep the sparse addressing defined by the mesh: if a variable is a vector, a tensorial diagonal elements couples the vector components in the same cell. A tensorial off–diagonal element couples the components of $\mathbf{u}_P$ to all components of $\mathbf{u}_N$, which covers all possibilities. Important disadvantages of a block coupled system are that the linear system is large as several variables are handled together. Also, different kinds of physics can be present, e.g. the transport–dominated momentum equation and elliptic pressure equation. At matrix level, it is impossible to separate them, which makes the system more difficult to solve.

Irrespective of the level of coupling, the FVM dictates that a cell value will depend only on values in surrounding cells. We still have freedom to organise the matrix by ordering entries for various components of the solution variable $\mathbf{x}$, i.e. global sparsity pattern is still related to mesh connectivity.

An example of a block–coupled system is shown for the pressure–velocity system, see Uroić et al. (2017) and Uroic and Jasak (2018):

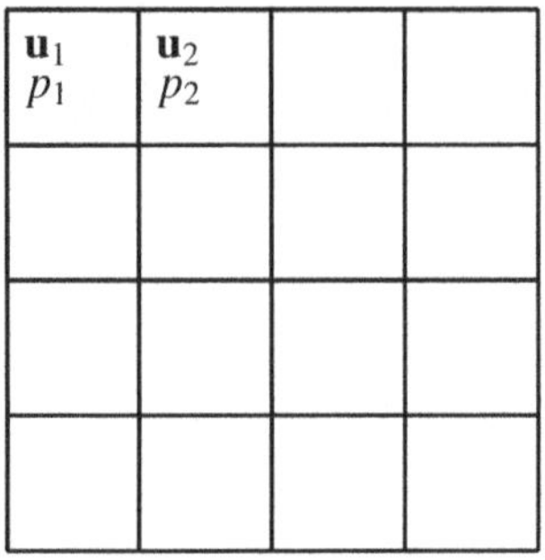


$$\begin{bmatrix} \begin{pmatrix} a_{P\,(\mathbf{u}\,\mathbf{u})} & a_{P\,(\mathbf{u}\,p)} \\ a_{P\,(p\,\mathbf{u})} & a_{P\,(p\,p)} \end{pmatrix} & \begin{pmatrix} a_{N\,(\mathbf{u}\,\mathbf{u})} & a_{N\,(\mathbf{u}\,p)} \\ a_{N\,(p\,\mathbf{u})} & a_{N\,(p\,p)} \end{pmatrix} & \dots \\ \cdot & \begin{pmatrix} a_{P\,(\mathbf{u}\,\mathbf{u})} & a_{P\,(\mathbf{u}\,p)} \\ a_{P\,(p\,\mathbf{u})} & a_{P\,(p\,p)} \end{pmatrix} & \dots \\ \vdots & \vdots & \ddots \end{bmatrix} \begin{bmatrix} \mathbf{u}_1 \\ p_1 \\ \mathbf{u}_2 \\ p_2 \\ \vdots \end{bmatrix} = \begin{bmatrix} b_{\mathbf{u}\,1} \\ b_{p\,1} \\ b_{\mathbf{u}\,2} \\ b_{p\,2} \\ \vdots \end{bmatrix}$$

### *Closure*

In this section algorithms for the solution of linear systems of equations were presented. Historically, fixed point methods were the first iterative algorithms for large, sparse matrices. However, due to their deficiencies which became obvious on non–uniform meshes, other types of solvers were developed. The first class are Krylov subspace solvers which rely in transforming the solution of linear system into a minimisation problem in which the correct solution vector can be found in $N$ iterations, where $N$ is the dimension of the problem. Krylov subspace methods are efficient only when used with matrix preconditioning techniques. Fixed methods were reawakened with the application of algebraic multigrid methods, which efficiently eliminate the components of the error unaffected by fixed point methods.

## Examples

In this section we shall present some applications of CFD using only the basic incompressible, single–phase and turbulent flow equations, which were discussed in the scope of this chapter. Examples from automotive industry will be given as well as applications for turbomachinery and biological flows.

## *CFD for External Aerodynamics*

Numerous automotive components involve fluid flow and require optimisation. This opens a wide area of potential of CFD use in automotive industry. CFD approaches the problem of fluid flow from fundamental equations: there is no problem–specific or industry–specific simplification. A critical step involves complex geometry handling as it is essential to capture real geometrical features of the engineering component under consideration. Traditional applications involve incompressible turbulent flow of Newtonian fluids.

While automotive CFD is mostly perceived in terms of external aerodynamics simulations, reality of industrial CFD use is significantly different. In numbers of users in automotive companies, CFD today is second only to CAD packages and in some areas, CFD replaces experiments. In comparison with CFD, experimental studies are expensive, carry limited information and it is difficult to achieve sufficient turn–over. The biggest obstacle is validation, i.e. confidence in CFD results. CFD is used across the automotive industry, at various levels of sophistication. The impact of simulations and reliance on numerical methods is greatest in areas that were not studied in detail beforehand.

There is considerable use in cases where it is difficult to quantify the results in simple terms like the lift and drag coefficient, such as flow organisation, stability and optimisation or a detailed look at the flow field, especially in complex geometry. Thus, CFD can contribute through parametric study (trends), by reducing experimental work etc. Numerical modelling is particularly useful in understanding the flow or looking for qualitative improvements: e.g. optimisation of vehicle soiling pattern on windows.

Aerodynamic design of a car is important as it plays a significant role in the behaviour of a vehicle. Investigating flow features is useful for optimisation of the aerodynamics which results in reduction of fuel consumption, more comfort (less noise, better ventilation) and improved driving characteristics (stability, handling). In racing car industry, it is important to achieve a large negative lift (downforce) while minimising the drag. Drag usually comes from frontal pressure when a vehicle pushes the air out of the way, and from the vacuum created at the rear when the air molecules are not able to fill the hole left by the vehicle body and it acts in the direction opposite to the velocity vector. The third component is the boundary layer effect, i.e. the friction between the car and the air. Downforce (negative lift) pushes the car into the road, which increases traction. Good traction is extremely important for behaviour of the car in the corners. A Formula 1 car without its front and rear wings would fly off the ground as it reaches the maximum speed, because the higher the velocity the air molecules are travelling, the lower the pressure. Thus, the car would act like an airfoil in a freestream and the force would lift it off the road. The wings, which are essentially inverted airfoils, make sure that the car stays on the ground and even enable it to drive turned upside–down. The best road cars today have the drag coefficient of 0.3, while Formula 1 cars with the wings and open wheels have a minimum of about 0.7. This makes a Formula 1 car slightly more efficient than

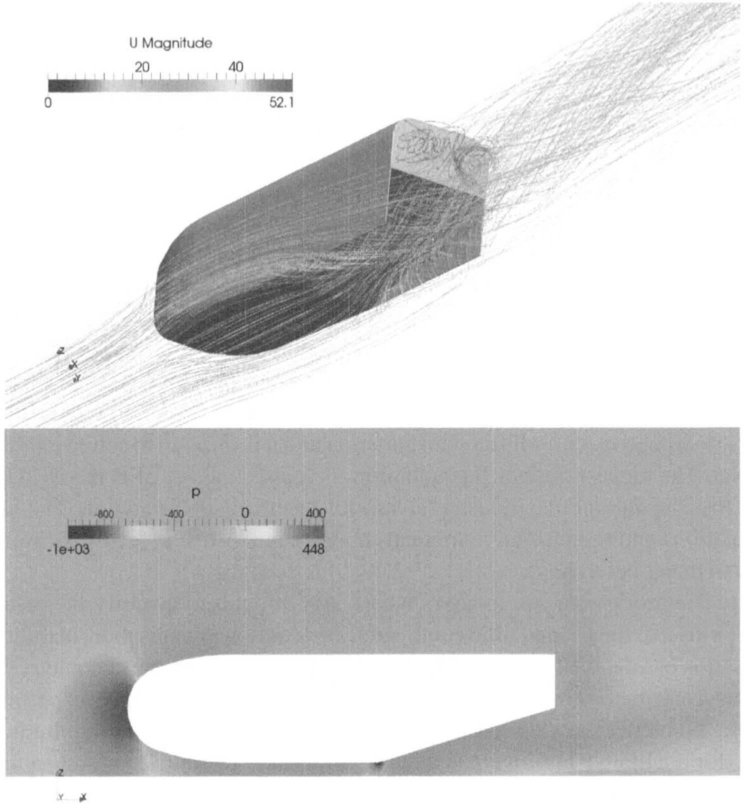


**Fig. 17** Streamlines around the bluff body with a diffuser, coloured by the values of velocity magnitude (top), pressure field around the bluff body (bottom)

a flat plate (drag coefficient = 1), but have in mind the downforce and horsepower. There are two options for estimating the drag and lift acting on a vehicle: wind tunnel measurements and CFD simulation. The financial and time aspect make CFD a better solution. Here, we give an example of isolated aerodynamics components: the diffuser and a rotating wheel. The diffuser is situated at the rear of a racing car undertray and is of great importance for downforce generation. Exposed wheels increase the drag force and also greatly influence the lift characteristics. In general, boundary layer separation is a good indicator of high drag.

A bluff body equipped with an upswept back section that operates in close proximity to the ground is shown in Fig. 17. "A bluff body can be defined as a body of any shape, which experiences complete boundary layer separation before the trailing edge, due to large adverse pressure gradient set up over that part of the body behind the position of maximum thickness. This pressure gradient decelerates the slow moving fluid within the boundary layer near the surface and eventually causes a reversed flow and hence separation." (Fackrell 1974).

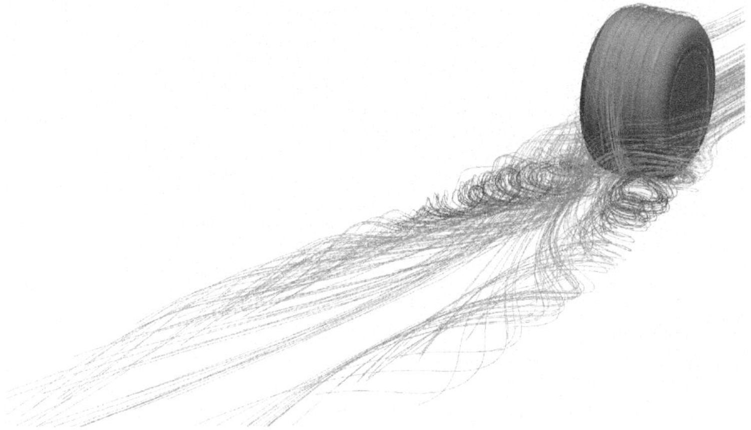

**Fig. 18** Vortices behind a rotating wheel

When placed in ground effect, an expansion/diffuser effect is formed between the upswept surface and the ground. This is used to increase downforce. The experimentally measured lift coefficient is equal to −1.90, (Senior 2002), while the simulated was −1.88 (+1.1%). The measured lift coefficient is 0.49, and the simulated 0.47 (−4.1%). "The flow remains essentially symmetric. The flow velocities accelerating underneath the side-plate have increased due to the reduction in gap through which it flows. The vortex behind is strong and concentrated and the flow separating from the side-plate winds into the vortex" (Senior 2002), visible in Fig. 17.

The second case is a rotating wheel in contact with the ground, with a camber angle equal to 2.4°. The experimentally measured drag coefficient is equal to 0.598. Drag coefficient obtained from the simulation was equal to 0.60 (+0.3%). The flow characteristics can be seen in Fig. 18, and were described in literature (Fackrell 1974):

> The near wake of the tire ... is dominated by two large counter-rotating vortices. Looking from the back of the wind tunnel, the left vortex is larger and more persistent than the right vortex and this is due to the combined effect of the wheel camber angle and strut.

"There is a region of strong downward velocity between the vortex cores in the centerplane of the tire...", evident in Fig. 18.

The third case is a generic car, a crossbreed between BMW 3 and Audi A4 Limousine, developed in TU Munich. Here, the simulated drag coefficient is equal to 0.36, which is high in comparison to experimental data (0.29). This indicates that a mesh refinement study should be conducted. The effects mentioned can be observed for this geometry: high pressure acting on the front of the car and recirculation at the back of the car, Fig. 19.

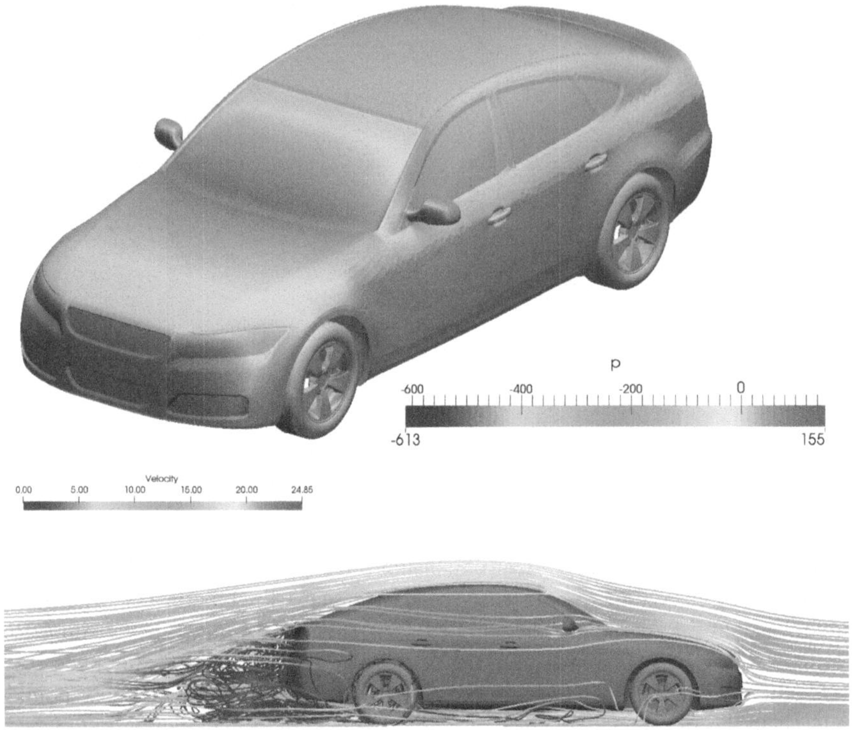


**Fig. 19** Pressure field acting on the surface of a generic car (top). Streamlines coloured by the values of velocity magnitude around the car (bottom)

## *Turbomachinery CFD*

Another industrial example of solving basic pressure–velocity coupling problem are turbomachinery applications. In turbomachinery, CFD can give a good prediction of flow features and integral quantities, such as power, head and efficiency. The first bottle–neck of turbomachinery simulations is the creation of computational mesh. To capture the smallest features of the blades, such as trailing edges and the boundary layer around them, refinement needs to be conducted, i.e. the meshes become very large. Also, rotation of turbomachinery components causes phenomena in the flow field which repeat with the rotational frequency. Thus, a transient problem with a moving mesh has to be solved to capture the correct behaviour of the flow. It was mentioned in Section "Finite Volume Discretisation" that explicit time stepping is limited by the Courant number, which typically makes turbomachinery simulations very slow as there are extremely small cells in the computational mesh. It is possible to simplify the solution procedure by solving the modified steady–state set of equations, accounting for centrifugal and Coriolis forces rather than rotating the rotor mesh, which is known as the Multiple Reference Frame approach (MRF) (Jasak and

Beaudoin 2011). A stationary mesh is used and the results are accurate only for the simulated position of the rotor (frozen rotor approach). Two steady–state examples with incompressible flow are shown: a centrifugal pump and a model of the Francis turbine.

Hydropower is an important contributor to the overall power generation. Water turbines turn the kinetic energy of water into mechanical energy (rotation) which is used to power the generator which generates electricity. Based on the head under which they operate, water turbines can be divided into high, medium and low head. The choice of each type depends on the working conditions, i.e. the available head and flow rate. Francis turbine operates at a medium head and is the most frequent among other notable types (Pelton, Kaplan). The water flows into the turbine through a spiral casing and enters the guide vanes which direct the flow towards the rotor blades. After exiting the rotor, the water enters a draft tube and is ejected into the river through a diffuser. Thus, the inlet into the rotor is radial, while the outlet is axial, which is one of the characteristics of Francis turbines. As established, the choice of the turbine type depends on the available water quantities. However, optimising the chosen type to suit the specific operating conditions is not straightforward. CFD simulations of flow through the turbine can help in exploring local phenomena or even identifying operating conditions which would be unsafe and damaging to the mechanical components, i.e. cavitation inside the rotor. Here, results of the simulation of the best efficiency operating point is shown, Fig. 20 (Norwegian Hydropower Centre 2014). Best efficiency is achieved for minimal flow losses (no recirculation areas or boundary layer separation). Thus, analysis of the flow pattern can reveal potential drawbacks of the design.

A machine complimentary to a water turbine is the centrifugal pump. A pump has a rotating impeller powered by an electric motor which accelerates water. The kinetic energy of water is then transformed into high pressure via the pump casing. Since there exists an adverse pressure gradient (the water flows from an area of low to an area with high pressure), it is more challenging to simulate due to thicker boundary layers and generally more unstable flow conditions. However, the equations and methodology remain the same. An example of a centrifugal pump simulation is shown in Fig. 21.

In addition to investigation of enclosed turbomachines, i.e. turbomachines which have rotational and stationary components, CFD is used to analyse complex flow phenomena which appear in marine propulsion. For example, with an increasing complexity of propulsion systems and a continued need to design faster, more reliable and quieter means of propulsion with greater manoeuvrability, the focus shifted towards Contra-Rotating Propellers (CRP). The main reason for employing such a design is the idea that positioning a secondary propeller behind the main propeller and having it rotate in the opposite direction positively affects the performance of the propulsion system and removes the bulk of the torque transferred from the propeller to the vessel. This is due to the fact that the secondary propeller harvests the additional energy otherwise lost in the rotating flow. An important benefit of CRP sets is better uniformity of flow in downstream wake of the propeller set, resulting in lower noise signatures. Furthermore, if sets with equal and even number of blades are used,

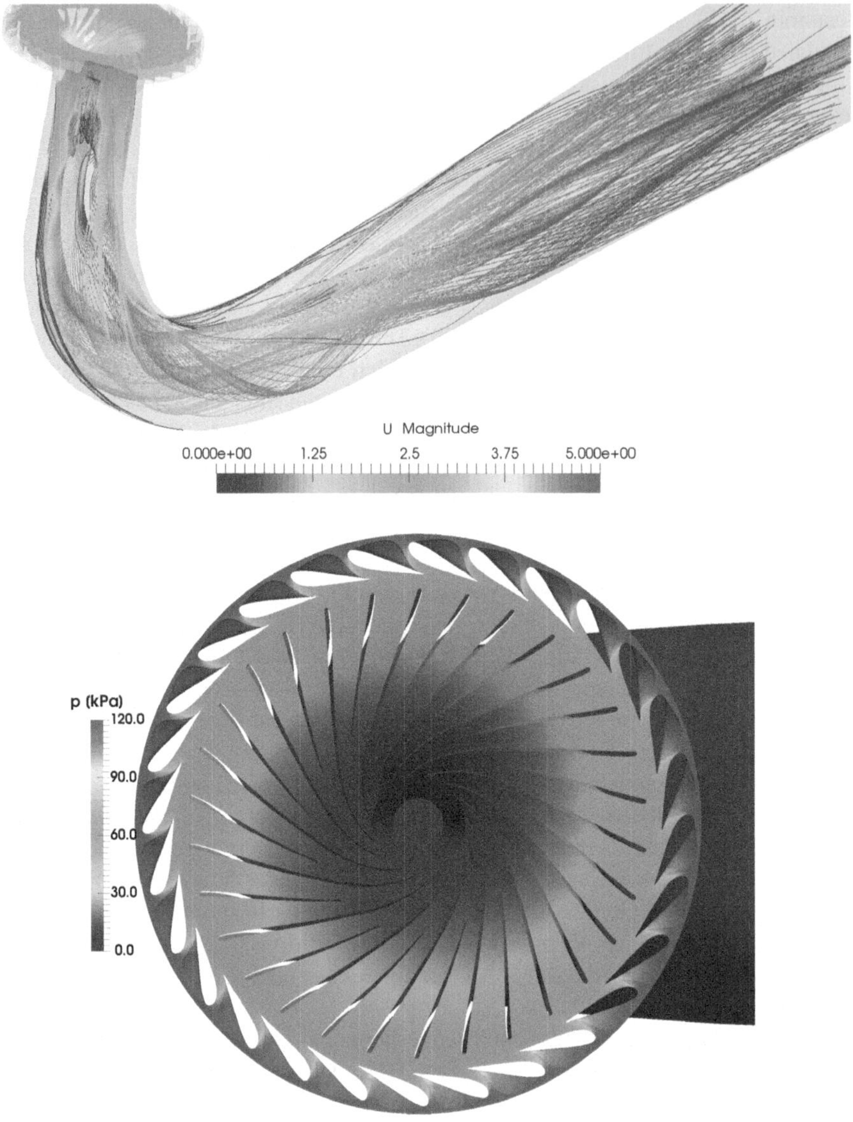


**Fig. 20** Streamlines inside a Francis turbine coloured by velocity magnitude (top). Pressure field acting on the guide vanes and rotor blades (bottom)

fluctuating thrust will be present. If there is an odd number of blades, the fluctuation is smaller, but a sideways force will be present. Classic CRP design recognises two approaches to CRP installation with regard to shaft design: coaxially mounted and single shaft CRP. Coaxially mounted CRP set consists of two propellers fitted on two separate shafts with coaxial axes of rotation. The main (fore) propeller is fitted

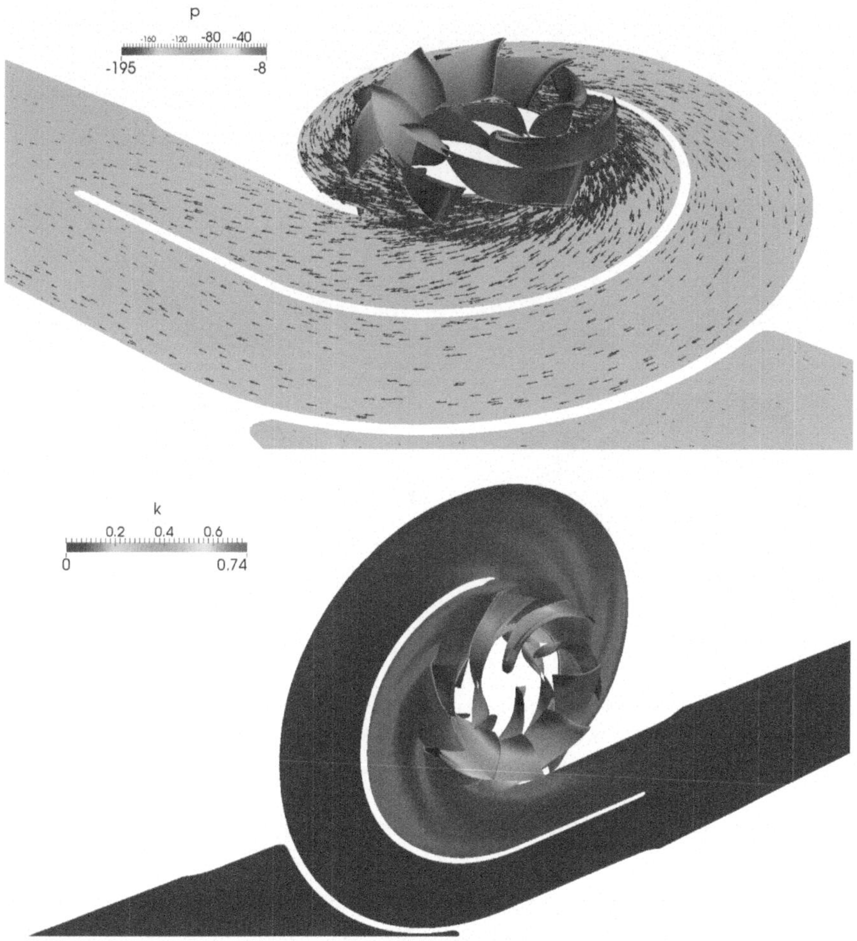


**Fig. 21** Pressure field acting on the pump impeller (top). Turbulent kinetic energy values in the flow inside a centrifugal pump (bottom)

on the main shaft, with the secondary (aft) propeller mounted on a shaft positioned behind the main propeller. Single shaft CRP design features both the fore and the aft propeller on a single shaft. Such design is historically older than the dual–shaft design and was abandoned because of problems with inner shaft lubrication, which lead to the stagnation of further CRP research in general. With recent developments in electromagnetic motors and electric propulsion systems, a renewed interest in CRP design and development has emerged. CFD serves as an irreplacable tool for comparing the performance of different types of propellers, and as a preparation of experimental setup and measurement. An example of CFD analysis of a single shaft CRP is shown in Fig. 22, taken from (Balatinec and Jasak 2019).

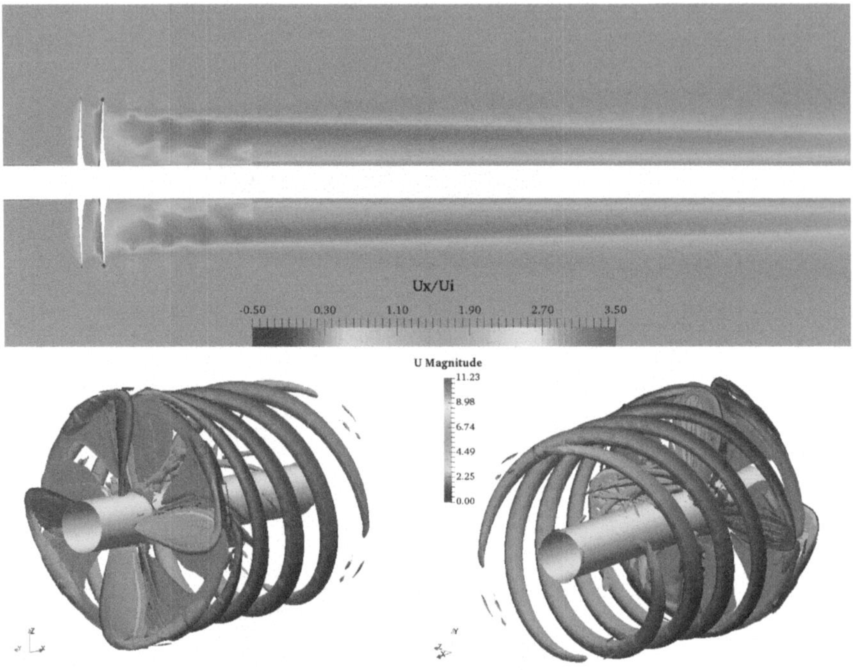


**Fig. 22** Velocity field around contra–rotating propellers normalised against the inlet value of velocity (top). Vortices generated by the rotation of propellers (bottom)

## *Biomedical CFD Simulations*

The most recent application of CFD appeared among medical practitioners, where patient–specific simulations can reveal important informations for an effective surgery. An example of simulating flow in the nasal cavity is presented next. The complex anatomical structure of the nose and the internal soft tissue which gives the form to the nasal cavity, make it extremely difficult to generalise and understand the exact mechanism of nasal breathing. A description of the nasal airflow and the consequential phenomena would contribute to the comprehension and identification of a variety of nasal conditions and selection of an appropriate medical treatment. Here a project (Balatinec 2018) which investigates the idea of creating an atlas of healthy nasal airflows using CFD is presented. The pattern of nasal airflow can be determined by examining a sufficient number of anatomies, in order to capture the significant characteristics of the flow. The challenge of the simulation process is the description of the geometry and the definition of boundary conditions. The computational mesh is created based on computational tomography (CT) scans, Fig. 23. The boundary conditions are defined from in vivo measurements. The airflow is incompressible and turbulent, thus a set of transient incompressible pressure–velocity equations is solved.

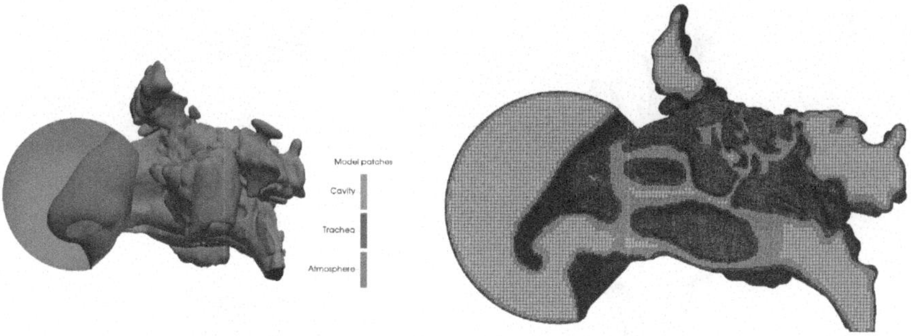


**Fig. 23** CAD model of a nose cavity obtained from CT scans (left). Computational mesh of the nose cavity (Right)

A case of normal inhalation is presented in Fig. 24 where distinct flow features can be observed. Looking at the cross-section of the nasal cavity at several locations and time-steps, it can be seen that the largest velocity magnitude develops along the nasal septum which separates the nasal canals (7.3 m/s at the nasal valve region). In the peripheral part of the nasal cavity (meatus region), the velocity is significantly smaller (3.6 m/s). Weak airflow can be noticed in the sinuses (0.04 m/s) during the second half of the inhalation when the velocity starts to decrease in the first section of the nasal cavity. For the particular patient, there is more obstruction in the right side of the cavity, therefore the velocities are higher in the left side.

Similar conclusions can be observed for the simulation of normal exhalation and quick inhalation, with some simulation-specific phenomena further examined in the project documentation (Balatinec 2018).

## *Closure*

In this section we presented some applications of basic flow equations. In automotive industry, CFD is used mainly for optimisation of vehicle performance as well as improving passenger comfort. There were three examples of aerodynamics simulations in this section: flow around a bluff body equipped with a diffuser, rotating wheel and a generic car. It was shown that, in addition to calculating integral values of interest (forces, power, etc.) CFD provides an insight into flow phenomena which appear at specific working conditions, demonstrated for turbomachinery cases (Francis turbine, centrifugal pump, ship propellers). CFD has even penetrated into medicine as patient–specific simulations provide pieces of information crucial for successful procedures. An example of airflow through a nasal cavity was given.

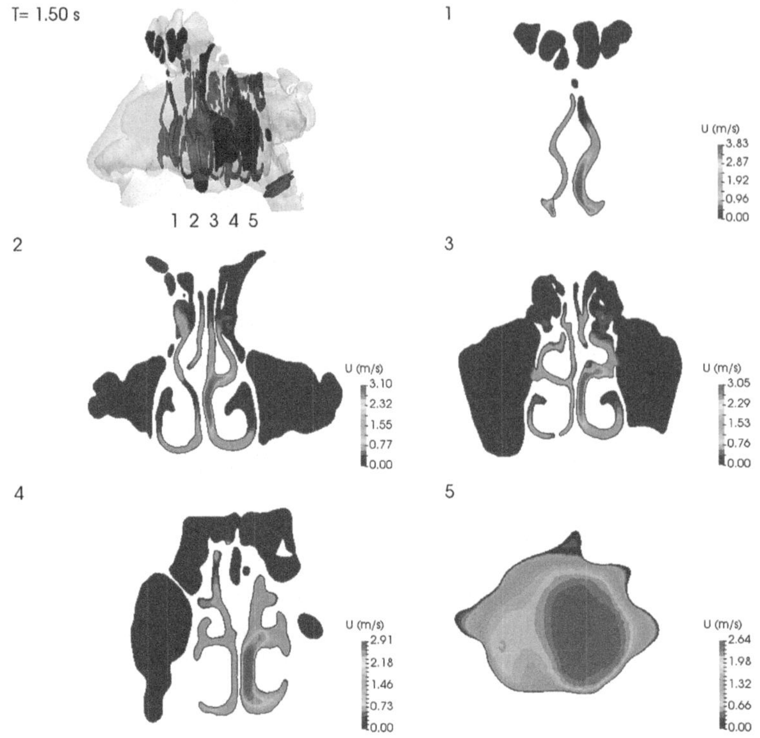


**Fig. 24** Velocity magnitude during the inhalation process on different planes throughout the nose cavity

# References


Balatinec, L. (2018). *Rector's award: Nasal flow simulations*. Faculty of Mechanical Engineering and Naval Architecture: University of Zagreb.

Balatinec, L., & Jasak, H. (2019). CFD evaluation of hydrodynamic performance of contra-rotating propellers (CRP) using OpenFOAM. In *Proceedings of propellers & impellers: Research, design, construction and application 27–28 March 2019, London, UK*. The Royal Institution of Naval Architects, 2019. https://www.rina.org.uk.

Duff, I. S., & Reid, J. K. (1983). The multifrontal solution of indefinite sparse symmetric linear. *ACM Transactions on Mathematical Software (TOMS)*, *9*, 302–325.

Elman, H. C., Silvester, D. J., & Wathen, A. J. (2005). *Finite elements and fast iterative solvers: With applications in incompressible fluid dynamics*. Numerical Mathematics and Scientific Computation: Oxford University Press.

Fackrell, J. E. (1974). *The aerodynamics of an isolated wheel rotating in contact with the ground*. Ph.D. thesis, Faculty of Engineering, University of London.

Ferziger, J. H., & Perić, M. (1995). *Computational methods for fluid dynamics*. Berlin: Springer.

Ferziger, J. H., & Perić, M. (Eds.). (2002). *Computational methods for fluid dynamics*. Berlin: Springer.

Hutchinson, B. R., & Raithby, G. D. (1986). A multigrid method based on the additive correction strategy. *Numerical Heat Transfer*, *9*, 511–537.

Issa, R. (1986). Solution of the implicitly discretized fluid flow equations by operator-splitting. *Journal of Computational Physics*, *62*, 40–65.

Jasak, H. (1996). *Error analysis and estimation for the finite volume method with applications to fluid flows*. Ph.D. thesis, Imperial College of Science, Technology & Medicine, London.

Jasak, H. (2009). *Dynamic mesh handling in OpenFOAM*. Orlando, FL: 48th AIAA Aerospace Sciences Meeting, AIAA Paper, January 2009

Jasak, H., & Tuković, Ž. (2007). Automatic mesh motion for the unstructured finite volume method. *Transactions of FAMENA*, *30*(2), 1–18.

Jasak, H., Weller, H. G., & Gosman, A. D. (1999). High resolution NVD differencing scheme for arbitrarily unstructured meshes. *International Journal for Numerical Methods in Fluids*, *31*, 431–449.

Jasak, H., & Beaudoin, M. (2011). OpenFOAM turbo tools: From general purpose CFD to turbomachinery simulations. In *ASME-JSME-KSME (2011) joint fluids engineering conference: Volume 1, Symposia—Parts A, B, C, and D*. ASME. https://doi.org/10.1115/ajk2011-05015.

National Science Foundation. Basic linear algebra subprograms. (2014). http://www.netlib.org/blas/.

Norwegian Hydropower Centre. (2014). *Experimental study of Francis 99 turbine*. https://www.ntnu.edu/nvks/f99-test-case1.

Patankar, S. V., & Spalding, D. B. (1972). A calculation procedure for heat, mass and momentum transfer in three-dimensional parabolic flows. *International Journal of Heat and Mass Transfer*, *15*, 1787–1806.

Van Phuc, P., Chiba, S., Minami, K. (2016). Large scale transient cfd simulations for buildings using openfoam on a world-s top-class supercomputer. In *2016 North American OpenFOAM user conference*.

Rhie, C. M., & Chow, W. L. (1983). A numerical study of the turbulent flow past an isolated airfoil with trailing edge separation. *AIAA Journal*, *21*, 1525–1532.

Ruge, J. W., & Stüben, K. (1987). Algebraic multigrid. In *SIAM—frontiers in applied mathematics: Multigrid methods*.

Saad, Y. (2000). *Iterative methods for sparse linear systems*. SIAM.

Senior, A. E. (2002). *The aerodynamics of a diffuser equipped bluff body in ground effect*. Ph.D. thesis, School of Engineering Sciences, University of Southampton, Southampton.

Shewchuk, J. R. (1994). *An introduction to the conjugate gradient method without the agonizing pain*. Pittsburgh: School of Computer Science.

Trottenberg, U., Oosterlee, C., & Schüller, A. (2001). *Multigrid*. Academic Press: Elsevier.

Uroic, T., & Jasak, H. (2018). Block-selective algebraic multigrid for implicitly coupled pressure-velocity system. *Computers and Fluids*, *167*, 100–110.

Uroić, T., Jasak, H., Rusche, H. (2017). Implicitly coupled pressure–velocity solver. In H. Jasak, & J. M. Nobrega (Eds.) *OpenFOAM: Selected papers of the 11th Workshop*. Berlin: Springer.

van der Vorst, H. A. (2003). *Iterative Krylov methods for large linear systems*. Cambridge University Press.

Weller, H. G., Tabor, G., Jasak, H., & Fureby, C. (1998). A tensorial approach to computational continuum mechanics using object orientated techniques. *Computers in Physics*, *12*(6), 620–631.

Zhang, F. (Ed.). (2005). *The Schur complement and its applications*. Numerical Methods and Algorithms: Springer.

# Tutorial on Hybridizable Discontinuous Galerkin (HDG) Formulation for Incompressible Flow Problems

**Matteo Giacomini, Ruben Sevilla and Antonio Huerta**

## Introduction

Computational engineering has always been concerned with the solution of equations of mathematical physics. Development of robust, accurate and efficient techniques to approximate solutions of these problems is still an important area of research. In recent years, hybrid discretization methods have gained popularity, in particular, for flow problems. This chapter presents the extension to incompressible flows (Stokes and Navier-Stokes) of the primer presented by Sevilla and Huerta (2016), which was restricted to the Poisson (thermal) problem.

Hybrid methods have been proposed for some time. In fact, already Ciarlet (2002) describes a hybrid method as “any finite element method based on a formulation where one unknown is a function, or some of its derivatives, on the set $\Omega$, and the other unknown is the trace of some of the derivatives of the same function, or the trace of the function itself, along the boundaries of the set”. In fact, Raviart and Thomas (1977) propose a discontinuous Galerkin (DG) technique such that the continuity constraint is eliminated from the finite element space and imposed by means of Lagrange multipliers on the inter-element boundaries. Stemming from this work, several hybrid methods have been developed using both primal (see Egger and Waluga 2012; Oikawa 2015; Di Pietro and Ern 2015) and mixed formulations (see Cockburn and Gopalakrishnan 2004, 2005a, b; Cockburn et al. 2009b). The latter family of techniques is nowadays known as *hybridizable discontinuous Galerkin* (HDG) method. For a literature review on hybrid discretization methods and their recent developments, the interested reader is referred to Cockburn (2017) or Giacomini and Sevilla (2019).

M. Giacomini · A. Huerta (✉)
Laboratori de Calcul Numeric (LaCaN). ETS de Ingenieros de Caminos, Canales y Puertos, Universitat Politecnica de Catalunya—BarcelonaTech, Barcelona, Spain
e-mail: antonio.huerta@upc.es

R. Sevilla
Zienkiewicz Centre for Computational Engineering, College of Engineering, Swansea University, Bay Campus, Swansea SA1 8EN, Wales, UK



L. De Lorenzis and A. Düster (eds.), *Modeling in Engineering Using Innovative Numerical Methods for Solids and Fluids*, CISM International Centre for Mechanical Sciences 599,
https://doi.org/10.1007/978-3-030-37518-8_5

Since its introduction, HDG has been objective of intensive research and has been applied to a large number of problems in different areas, including fluid mechanics (see Cockburn et al. 2009b, 2010; Peraire et al. 2010; Nguyen et al. 2010a, b, 2011b; Giacomini et al. 2018), wave propagation (see Nguyen et al. 2011a, c; Giorgiani et al. 2013a) and solid mechanics (see Soon et al. 2009; Kabaria et al. 2015; Sevilla et al. 2018; Sevilla 2019), to name but a few.

This chapter starts from the HDG method originally proposed by Cockburn et al. (2009b) and, following Sevilla and Huerta (2016), provides a tutorial for the implementation of an HDG formulation for incompressible flow problems. Note that HDG features a mixed formulation and a hybrid variable, which is the trace of the primal one. The *hybridization* (Fraeijs de Veubeke 1965) (also known as *static condensation* for primal formulations, Guyan 1965) allows to reduce the number of the globally-coupled degrees of freedom of the problem (see Cockburn et al. 2009a; Kirby et al. 2011; Giorgiani et al. 2013b; Huerta et al. 2013). Moreover, the superconvergent properties of HDG in elliptic problems allow to define an efficient and inexpensive error indicator to drive degree adaptivity procedures, not feasible in a standard CG approach, see for instance Giorgiani et al. (2013a, 2014) or Sevilla and Huerta (2018).

Moreover, the HDG formulation presented in this chapter approximates the Cauchy stress formulation of incompressible flow equations. In particular, Voigt notation allows to easily enforce the symmetry of second-order tensors pointwise. Optimal convergence of order $k+1$ is obtained for velocity, pressure and strain-rate tensor, even for low-order polynomial approximations. Moreover, the local postprocessing strategy is adapted to construct an approximation of the velocity superconverging with order $k+2$, even for low-order polynomial approximations.

The remainder of this chapter is organized as follows. Section "Incompressible Flows: Problem Statement" introduces the problem statement for incompressible flows. The implementation of HDG for the linearized incompressible Navier-Stokes, known as Oseen equations, is presented in Section "HDG Method for Oseen Flows". Section "Numerical Examples" presents some numerical results to validate the optimal convergence properties of HDG for Stokes, Oseen and incompressible Navier-Stokes equations. Eventually, two Appendices "Appendix: Saddle-Point Structure of the Global Problem" and "Appendix: Implementation Details", present the proof of the symmetry of the matrix of the HDG global problem for Stokes equations and the technical details required for implementation, respectively.

## Incompressible Flows: Problem Statement

In compact form, the steady-state incompressible Navier-Stokes problem in the open bounded computational domain $\Omega \in \mathbb{R}^{\mathtt{n_{sd}}}$ with boundary $\partial\Omega$ and $\mathtt{n_{sd}}$ the number of spatial dimensions, reads

$$\begin{cases} -\boldsymbol{\nabla} \cdot (2\nu\boldsymbol{\nabla}^{\mathrm{s}}\boldsymbol{u} - p\mathbf{I}_{\mathrm{n_{sd}}}) + \boldsymbol{\nabla} \cdot (\boldsymbol{u} \otimes \boldsymbol{u}) = \boldsymbol{s} & \text{in } \Omega, \\ \boldsymbol{\nabla} \cdot \boldsymbol{u} = 0 & \text{in } \Omega, \\ \boldsymbol{u} = \boldsymbol{u}_D & \text{on } \Gamma_D, \\ \big((2\nu\boldsymbol{\nabla}^{\mathrm{s}}\boldsymbol{u} - p\mathbf{I}_{\mathrm{n_{sd}}}) - (\boldsymbol{u} \otimes \boldsymbol{u})\big)\boldsymbol{n} = \boldsymbol{t} & \text{on } \Gamma_N, \end{cases} \tag{1}$$

where the pair $(\boldsymbol{u}, p)$ represents the unknown velocity and pressure fields, $\nu > 0$ is the kinematic viscosity of the fluid, $\boldsymbol{n}$ is the outward unit normal vector to the corresponding boundary, in this case $\Gamma_N$, $\boldsymbol{s}$ is the volumetric source term, and $\boldsymbol{\nabla}^{\mathrm{s}}\boldsymbol{u}$ is the strain-rate tensor, that is, the symmetric part of the gradient of velocity with $\boldsymbol{\nabla}^{\mathrm{s}} := \big(\boldsymbol{\nabla} + \boldsymbol{\nabla}^T\big)/2$. Consequently, $\boldsymbol{\sigma} := 2\nu\boldsymbol{\nabla}^{\mathrm{s}}\boldsymbol{u} - p\mathbf{I}_{\mathrm{n_{sd}}}$ is the Cauchy stress tensor. Recall that $[\boldsymbol{\nabla}\boldsymbol{u}]_{ij} = \partial u_i/\partial x_j$.

The boundary $\partial\Omega$ is composed of two disjoint parts, the Dirichlet portion $\Gamma_D$, where the value $\boldsymbol{u}_D$ of the velocity is imposed, and the Neumann one $\Gamma_N$. Formally, $\partial\Omega = \Gamma_D \cup \Gamma_N$, $\Gamma_D \cap \Gamma_N = \emptyset$. On Neumann boundaries, two typical options are found. On the one hand, material surfaces (i.e. $\boldsymbol{u} \cdot \boldsymbol{n} = 0$ and, thus, $(\boldsymbol{u} \otimes \boldsymbol{u})\boldsymbol{n} = \boldsymbol{0}$) where a traction $\boldsymbol{t}$ is applied. For this reason, many fluid references define Neumann boundary conditions as $(2\nu\boldsymbol{\nabla}^{\mathrm{s}}\boldsymbol{u} - p\mathbf{I}_{\mathrm{n_{sd}}})\,\boldsymbol{n} = \boldsymbol{t}$. On the other hand, artificial boundaries where the convection term cannot be neglected. This is typical when implementing Neumann boundaries in synthetic problems.

*Remark 2.1* (*Cauchy stress vs. velocity-pressure formulation*) It is standard for incompressible flow problems to use the pointwise solenoidal property of the velocity to modify the momentum equations. Then, the Navier-Stokes problem is rewritten as

$$\begin{cases} -\boldsymbol{\nabla} \cdot (\nu\boldsymbol{\nabla}\boldsymbol{u} - p\mathbf{I}_{\mathrm{n_{sd}}}) + \boldsymbol{\nabla} \cdot (\boldsymbol{u} \otimes \boldsymbol{u}) = \boldsymbol{s} & \text{in } \Omega, \\ \boldsymbol{\nabla} \cdot \boldsymbol{u} = 0 & \text{in } \Omega, \\ \boldsymbol{u} = \boldsymbol{u}_D & \text{on } \Gamma_D, \\ \big((\nu\boldsymbol{\nabla}\boldsymbol{u} - p\mathbf{I}_{\mathrm{n_{sd}}}) - (\boldsymbol{u} \otimes \boldsymbol{u})\big)\boldsymbol{n} = \boldsymbol{t} & \text{on } \Gamma_N. \end{cases}$$

However, as noted by Donea and Huerta (2003, Sect. 6.5), from a mechanical viewpoint the two formulations are *not identical* because, in general,

$$\boldsymbol{\sigma}\,\boldsymbol{n} = (2\nu\boldsymbol{\nabla}^{\mathrm{s}}\boldsymbol{u} - p\mathbf{I}_{\mathrm{n_{sd}}})\,\boldsymbol{n} \neq (\nu\boldsymbol{\nabla}\boldsymbol{u} - p\mathbf{I}_{\mathrm{n_{sd}}})\,\boldsymbol{n}.$$

In fact, in a velocity-pressure formulation $\boldsymbol{t}$ is a *pseudo-traction* imposed on the Neumann boundary $\Gamma_N$, unless a Robin-type boundary condition is imposed, namely

$$\big((\nu\boldsymbol{\nabla}\boldsymbol{u} - p\mathbf{I}_{\mathrm{n_{sd}}}) - (\boldsymbol{u} \otimes \boldsymbol{u})\big)\,\boldsymbol{n} = \boldsymbol{t} - (\nu\boldsymbol{\nabla}\boldsymbol{u}^T)\boldsymbol{n},$$

which, obviously, does not correspond to the natural boundary condition associated with the operator in the partial differential equation.

As usual in computational mechanics, when modeling requires it, along the same portion of a boundary it is also typical to impose Dirichlet and Neumann boundary conditions at once but along orthogonal directions. For instance, this is the case of a perfect-slip boundary, which is the limiting case of those discussed in Remark 2.2, and of a fully-developed outflow boundary in Remark 2.3.

*Remark 2.2* (*Slip/friction boundary condition*) Another family of physical boundary conditions that can be considered for the Navier-Stokes equations includes slip/friction boundaries, see Volker (2002), which, in fact, correspond to Robin-type boundary conditions. Slip boundaries are denoted by $\Gamma_S$ and need to verify: $\partial\Omega = \Gamma_D \cup \Gamma_N \cup \Gamma_S$, being $\Gamma_D$, $\Gamma_N$ and $\Gamma_S$ disjoint by pairs. Slip boundary conditions are:

$$\begin{cases} \boldsymbol{u} \cdot \boldsymbol{n} + \alpha \boldsymbol{n} \cdot \boldsymbol{\sigma n} = 0 & \text{on } \Gamma_S, \\ \beta \boldsymbol{u} \cdot \boldsymbol{t}_k + \boldsymbol{t}_k \cdot \boldsymbol{\sigma n} = 0 & \text{for } k = 1, \dots, \mathtt{n_{sd}} - 1, \text{ on } \Gamma_S, \end{cases}$$

where $\alpha$ and $\beta$ are the penetration and friction coefficients respectively, whereas $\boldsymbol{n}$ is the outward unit normal to $\Gamma_S$ and the tangential vectors $\boldsymbol{t}_k$, for $k = 1, \dots, \mathtt{n_{sd}} - 1$, are such that $\{\boldsymbol{n}, \boldsymbol{t}_1, \dots, \boldsymbol{t}_{\mathtt{n_{sd}}-1}\}$ form an orthonormal system of vectors.

Note that symmetry-type boundary conditions can be seen as a particular case of slip boundary conditions. That is, along the plane of symmetry (in 3D and axis of symmetry in 2D) a non-penetration ($\alpha = 0$) and perfect-slip ($\beta = 0$) boundary condition needs to be imposed. Obviously, this plane of symmetry can be oriented arbitrarily in the domain $\Omega$.

*Remark 2.3* (*Outflow boundary condition*) Outflow surfaces are of great interest in the simulation of engineering problems, especially for internal flows. Common choices in the literature are represented by traction-free conditions

$$(2\nu \boldsymbol{\nabla}^{\mathtt{s}} \boldsymbol{u} - p \mathbf{I}_{\mathtt{n_{sd}}}) \boldsymbol{n} = \boldsymbol{0} \tag{2a}$$

or homogeneous Neumann conditions,

$$\big((2\nu \boldsymbol{\nabla}^{\mathtt{s}} \boldsymbol{u} - p \mathbf{I}_{\mathtt{n_{sd}}}) - (\boldsymbol{u} \otimes \boldsymbol{u})\big) \boldsymbol{n} = \boldsymbol{0}. \tag{2b}$$

Nonetheless, as will be shown in Section "Navier-Stokes Flow", these conditions introduce perturbations in the flow near the outlet and do not meet the goal of outflow boundaries, that is, to model a flow exiting the domain undisturbed. Outflow conditions for fully-developed flows can be devised as Robin-type boundary conditions (see van de Vosse et al. 2003) similarly to the perfect-slip boundary conditions described in Remark 2.2. Given an outflow boundary $\Gamma_O$ such that $\partial\Omega = \Gamma_D \cup \Gamma_N \cup \Gamma_S \cup \Gamma_O$, being $\Gamma_D$, $\Gamma_N$, $\Gamma_S$ and $\Gamma_O$ disjoint by pairs, the corresponding conditions are

$$\begin{cases} \boldsymbol{u} \cdot \boldsymbol{t}_k = 0 & \text{for } k = 1, \dots, \mathtt{n_{sd}} - 1, \text{ on } \Gamma_O, \\ \boldsymbol{n} \cdot \boldsymbol{\sigma n} = 0 & \text{on } \Gamma_O, \end{cases} \tag{2c}$$

where $\boldsymbol{n}$ is the outward unit normal to $\Gamma_O$ and the tangential vectors $\boldsymbol{t}_k$, for $k = 1, \ldots, \mathtt{n_{sd}} - 1$, are such that $\{\boldsymbol{n}, \boldsymbol{t}_1, \ldots, \boldsymbol{t}_{\mathtt{n_{sd}}-1}\}$ form an orthonormal system of vectors.

To further simplify the presentation, the linearized incompressible Navier-Stokes equations are considered. The so-called Oseen equations, with linear convection driven by the solenoidal field $\boldsymbol{a}$, are

$$\begin{cases} -\boldsymbol{\nabla} \cdot (2\nu\boldsymbol{\nabla}^{\mathtt{s}}\boldsymbol{u} - p\mathbf{I}_{\mathtt{n_{sd}}}) + \boldsymbol{\nabla}\cdot(\boldsymbol{u} \otimes \boldsymbol{a}) = \boldsymbol{s} & \text{in } \Omega, \\ \boldsymbol{\nabla} \cdot \boldsymbol{u} = 0 & \text{in } \Omega, \\ \boldsymbol{u} = \boldsymbol{u}_D & \text{on } \Gamma_D, \\ \big((2\nu\boldsymbol{\nabla}^{\mathtt{s}}\boldsymbol{u} - p\mathbf{I}_{\mathtt{n_{sd}}}) - (\boldsymbol{u} \otimes \boldsymbol{a})\big)\boldsymbol{n} = \boldsymbol{t} & \text{on } \Gamma_N, \end{cases} \tag{3}$$

where $\boldsymbol{a}$ coincides with $\boldsymbol{u}$ when confronted with Navier-Stokes. In Einstein notation and for $i = 1, \ldots, \mathtt{n_{sd}}$, the strong form of the Oseen equations reads as

$$\begin{cases} -\dfrac{\partial}{\partial x_j}\left(\nu\left(\dfrac{\partial u_i}{\partial x_j} + \dfrac{\partial u_j}{\partial x_i}\right) - p\,\delta_{ij}\right) + \dfrac{\partial}{\partial x_j}(u_i\,a_j) = s_i & \text{in } \Omega, \\ \dfrac{\partial u_j}{\partial x_j} = 0 & \text{in } \Omega, \\ u_i = u_{D,i} & \text{on } \Gamma_D, \\ \nu\left(\dfrac{\partial u_i}{\partial x_j} + \dfrac{\partial u_j}{\partial x_i}\right)n_j - pn_i - u_i a_j n_j = t_i & \text{on } \Gamma_N. \end{cases}$$

Finally, it is important to recall that, when inertial effects are negligible (viz. in micro-fluidics), incompressible flows are modeled by means of the Stokes problem. That is,

$$\begin{cases} -\boldsymbol{\nabla} \cdot (2\nu\boldsymbol{\nabla}^{\mathtt{s}}\boldsymbol{u} - p\mathbf{I}_{\mathtt{n_{sd}}}) = \boldsymbol{s} & \text{in } \Omega, \\ \boldsymbol{\nabla} \cdot \boldsymbol{u} = 0 & \text{in } \Omega, \\ \boldsymbol{u} = \boldsymbol{u}_D & \text{on } \Gamma_D, \\ (2\nu\boldsymbol{\nabla}^{\mathtt{s}}\boldsymbol{u} - p\mathbf{I}_{\mathtt{n_{sd}}})\boldsymbol{n} = \boldsymbol{t} & \text{on } \Gamma_N, \end{cases} \tag{4}$$

which is a second-order elliptic equation. Stokes problem can be viewed as the particular case of the Oseen equations (3), for $\boldsymbol{a} = \boldsymbol{0}$. Thus, in the remainder of this Chapter, the Oseen equations will be detailed. Navier-Stokes and Stokes problems will be recovered after replacing $\boldsymbol{a}$ by $\boldsymbol{u}$ or $\boldsymbol{0}$, respectively.

It is worth noticing that the divergence-free equation in Navier-Stokes, Oseen and Stokes problems induces a compatibility condition on the velocity field, namely

$$\int_{\Gamma_D} \boldsymbol{u}_D \cdot \boldsymbol{n}\, d\Gamma + \int_{\Gamma_N} \boldsymbol{u} \cdot \boldsymbol{n}\, d\Gamma = 0. \tag{5}$$

In particular, if only Dirichlet boundary conditions are considered (i.e. $\Gamma_D = \partial\Omega$), an additional constraint on the pressure needs to be imposed to avoid its indeterminacy. It is common for hybrid formulations (see for instance Cockburn et al. 2009b, 2010; Cockburn and Shi 2014) to impose zero mean pressure on the boundary, namely

$$\int_{\partial\Omega} p \, d\Gamma = 0. \tag{6}$$

## HDG Method for Oseen Flows

### *Functional and Discrete Approximation Setting*

The HDG method relies on a mixed hybrid formulation of the problem under analysis. Thus, assume that $\Omega$ is partitioned in $\mathtt{n_{el}}$ disjoint subdomains $\Omega_e$,

$$\Omega = \bigcup_{e=1}^{\mathtt{n_{el}}} \Omega_e, \quad \text{with } \Omega_i \cap \Omega_j = \emptyset \text{ for } i \neq j,$$

with boundaries $\partial\Omega_e$, which define an internal interface, also known as *internal skeleton*, $\Gamma$

$$\Gamma := \Big[ \bigcup_{e=1}^{\mathtt{n_{el}}} \partial\Omega_e \Big] \setminus \partial\Omega.$$

Moreover, in what follows, the classical $\mathcal{L}_2$ inner products for vector-valued functions are considered for a generic domain $D \subseteq \Omega \subset \mathbb{R}^{\mathtt{n_{sd}}}$ and a generic line/surface $S \subset \Gamma \cup \partial\Omega$, that is,

$$(\boldsymbol{u}, \boldsymbol{w})_D := \int_D \boldsymbol{u} \cdot \boldsymbol{w} \, d\Omega, \qquad \langle \boldsymbol{u}, \boldsymbol{w} \rangle_S := \int_S \boldsymbol{u} \cdot \boldsymbol{w} \, d\Gamma.$$

The $\mathcal{L}_2$ inner product for tensor-valued functions will also be used on $D$, namely

$$(\boldsymbol{L}, \boldsymbol{G})_D := \int_D \boldsymbol{L} : \boldsymbol{G} \, d\Omega, \text{ i.e. in Einstein notation: } \int_D [\boldsymbol{L}]_{ij} [\boldsymbol{G}]_{ij} \, d\Omega.$$

In the following subsections the Sobolev space $\mathcal{H}^1(D)$, $D \subseteq \Omega$, of $\mathcal{L}_2(D)$ functions whose gradient also is an $\mathcal{L}_2(D)$ function, will be used systematically. The space of $\mathcal{L}_2(D)$ symmetric tensors $\mathbb{S}$ of order $\mathtt{n_{sd}}$ with $\mathcal{L}_2(D)$ row-wise divergence is denoted by $[\mathcal{H}(\text{div}; D); \mathbb{S}]$. Moreover, the trace space $\mathcal{H}^{\frac{1}{2}}(\partial D)$, being the space of the restriction of $\mathcal{H}^1(D)$ functions to the boundary $\partial D$, is introduced. Note also that for functions on $S \subset \Gamma \cup \partial\Omega$, the typical $\mathcal{L}_2(S)$ space is employed.

Moreover, the following discrete functional spaces are considered according to the notation introduced in Sevilla and Huerta (2016)

$$\mathcal{V}^h(\Omega) := \left\{ v \in \mathcal{L}_2(\Omega) : v|_{\Omega_e} \in \mathcal{P}^k(\Omega_e)\ \forall \Omega_e\,,\ e = 1, \dots, \mathtt{n_{el}} \right\}, \tag{7a}$$

$$\hat{\mathcal{V}}^h(S) := \left\{ \hat{v} \in \mathcal{L}_2(S) : \hat{v}|_{\Gamma_i} \in \mathcal{P}^k(\Gamma_i)\ \forall \Gamma_i \subset S \subseteq \Gamma \cup \partial\Omega \right\}, \tag{7b}$$

where $\mathcal{P}^k(\Omega_e)$ and $\mathcal{P}^k(\Gamma_i)$ are the spaces of polynomial functions of complete degree at most $k$ in $\Omega_e$ and on $\Gamma_i$, respectively.

## *Strong Forms of the Local and Global Problems*

As noted earlier, the HDG method relies on a mixed hybrid formulation. Following the rationale described by Sevilla and Huerta (2016), the Oseen problem reported in (3) is rewritten as a system of first-order equations as

$$\begin{cases} \boldsymbol{L} + \sqrt{2\nu}\boldsymbol{\nabla}^{\mathtt{s}}\boldsymbol{u} = \boldsymbol{0} & \text{in } \Omega_e, \text{ and for } e = 1, \dots, \mathtt{n_{el}}, \\ \boldsymbol{\nabla}\cdot(\sqrt{2\nu}\boldsymbol{L} + p\mathbf{I}_{\mathtt{n_{sd}}}) + \boldsymbol{\nabla}\cdot(\boldsymbol{u}\otimes\boldsymbol{a}) = \boldsymbol{s} & \text{in } \Omega_e, \text{ and for } e = 1, \dots, \mathtt{n_{el}}, \\ \boldsymbol{\nabla}\cdot\boldsymbol{u} = 0 & \text{in } \Omega_e, \text{ and for } e = 1, \dots, \mathtt{n_{el}}, \\ \boldsymbol{u} = \boldsymbol{u}_D & \text{on } \Gamma_D, \\ \big((\sqrt{2\nu}\boldsymbol{L} + p\mathbf{I}_{\mathtt{n_{sd}}}) + (\boldsymbol{u}\otimes\boldsymbol{a})\big)\boldsymbol{n} = -\boldsymbol{t} & \text{on } \Gamma_N, \\ [\![\boldsymbol{u}\otimes\boldsymbol{n}]\!] = \boldsymbol{0} & \text{on } \Gamma, \\ [\![\big((\sqrt{2\nu}\boldsymbol{L} + p\mathbf{I}_{\mathtt{n_{sd}}}) + (\boldsymbol{u}\otimes\boldsymbol{a})\big)\boldsymbol{n}]\!] = \boldsymbol{0} & \text{on } \Gamma, \end{cases} \tag{8}$$

where $\boldsymbol{L} = -\sqrt{2\nu}\boldsymbol{\nabla}^{\mathtt{s}}\boldsymbol{u}$ is the mixed variable representing the *scaled* strain-rate or second-order velocity deformation tensor. The last two equations are the so-called *transmission conditions*. They impose, respectively, continuity of velocity and normal flux (normal component of the stress—i.e. diffusive flux—and of the convective flux), across the interior faces. The *jump* $[\![\cdot]\!]$ operator has been introduced following the definition by Montlaur et al. (2008), such that, along each portion of the interface $\Gamma$ it sums the values from the element on the left and on the right, say $\Omega_l$ and $\Omega_r$, namely

$$[\![\odot]\!] = \odot_l + \odot_r.$$

Note that the above definition of the jump operator always involves the outward unit normal to a surface, say $[\![\odot\boldsymbol{n}]\!]$. Thus, at the interface between elements $\Omega_l$ and $\Omega_r$, this definition implies $[\![\odot\boldsymbol{n}]\!] = \odot_l\boldsymbol{n}_l + \odot_r\boldsymbol{n}_r$ where $\boldsymbol{n}_l$ and $\boldsymbol{n}_r$ are the outward unit normals to $\partial\Omega_l$ and $\partial\Omega_r$, respectively. Moreover, recall that $\boldsymbol{n}_l = -\boldsymbol{n}_r$ along their interface.

*Remark 3.1* Recall that for incompressible flow problems with purely Dirichlet boundary conditions (i.e. $\Gamma_D = \partial\Omega$) an additional constraint is required to avoid indeterminacy of pressure. A common choice is to impose an arbitrary mean value of the pressure on the boundary (usually zero), that is $\langle p, 1\rangle_{\partial\Omega} = \texttt{Cst}$.

Starting from the mixed formulation on the broken computational domain, see (8), HDG features two stages. First, a set of $\texttt{n}_{\texttt{el}}$ local problems is introduced to define element-by-element $(\boldsymbol{L}_e, \boldsymbol{u}_e, p_e) = (\boldsymbol{L}, \boldsymbol{u}, p)$ for all $\boldsymbol{x} \in \Omega_e \subset \Omega$ in terms of a novel independent variable $\hat{\boldsymbol{u}}$, namely

$$\left\{\begin{aligned} \boldsymbol{L}_e + \sqrt{2\nu}\boldsymbol{\nabla}^{\texttt{s}}\boldsymbol{u}_e &= \boldsymbol{0} && \text{in } \Omega_e,\\ \boldsymbol{\nabla}\cdot(\sqrt{2\nu}\boldsymbol{L}_e + p_e\mathbf{I}_{\texttt{n}_{\texttt{sd}}}) + \boldsymbol{\nabla}\cdot(\boldsymbol{u}_e\otimes\boldsymbol{a}) &= \boldsymbol{s} && \text{in } \Omega_e,\\ \boldsymbol{\nabla}\cdot\boldsymbol{u}_e &= 0 && \text{in } \Omega_e,\\ \boldsymbol{u}_e &= \boldsymbol{u}_D && \text{on } \partial\Omega_e\cap\Gamma_D,\\ \boldsymbol{u}_e &= \hat{\boldsymbol{u}} && \text{on } \partial\Omega_e\setminus\Gamma_D, \end{aligned}\right. \tag{9a}$$

where $\hat{\boldsymbol{u}}$ represents the trace of the velocity on the mesh skeleton $\Gamma \cup \Gamma_N$. Note that (9a) constitutes a purely Dirichlet boundary value problem. As previously observed, an additional constraint needs to be added to remove the indeterminacy of the pressure, namely

$$\langle p_e, 1\rangle_{\partial\Omega_e} = \rho_e, \tag{9b}$$

where $\rho_e$ denotes the scaled mean pressure on the boundary of the element $\Omega_e$. Hence, the local problem defined by (9) provides $(\boldsymbol{L}_e, \boldsymbol{u}_e, p_e)$, for $e = 1, \dots, \texttt{n}_{\texttt{el}}$, in terms of the global unknowns $\hat{\boldsymbol{u}}$ and $\boldsymbol{\rho} = (\rho_1 \dots, \rho_{\texttt{n}_{\texttt{el}}})^T$.

The second stage computes the trace of the velocity $\hat{\boldsymbol{u}}$ and the scaled mean pressure $\boldsymbol{\rho}$ on the element boundaries by solving the global problem accounting for the following transmission conditions and the Neumann boundary conditions

$$\left\{\begin{aligned} \big[\!\big[\big((\sqrt{2\nu}\boldsymbol{L} + p\mathbf{I}_{\texttt{n}_{\texttt{sd}}}) + (\boldsymbol{u}\otimes\boldsymbol{a})\big)\boldsymbol{n}\big]\!\big] &= \boldsymbol{0} && \text{on } \Gamma,\\ \big((\sqrt{2\nu}\boldsymbol{L} + p\mathbf{I}_{\texttt{n}_{\texttt{sd}}}) + (\boldsymbol{u}\otimes\boldsymbol{a})\big)\boldsymbol{n} &= -\boldsymbol{t} && \text{on } \Gamma_N. \end{aligned}\right. \tag{10a}$$

Note that the continuity of velocity on $\Gamma$, $[\![\boldsymbol{u}\otimes\boldsymbol{n}]\!] = \boldsymbol{0}$, is not explicitly written because it is automatically satisfied. This is due to the Dirichlet boundary condition $\boldsymbol{u}_e = \hat{\boldsymbol{u}}$ imposed in every local problem and to the unique definition of the hybrid variable $\hat{\boldsymbol{u}}$ on each edge/face of the mesh skeleton. Moreover, the divergence-free condition in the local problem induces the following compatibility condition for each element $\Omega_e$, $e = 1, \dots, \texttt{n}_{\texttt{el}}$

$$\langle \hat{\boldsymbol{u}}\cdot\boldsymbol{n}_e, 1\rangle_{\partial\Omega_e\setminus\Gamma_D} + \langle \boldsymbol{u}_D\cdot\boldsymbol{n}_e, 1\rangle_{\partial\Omega_e\cap\Gamma_D} = 0, \tag{10b}$$

which is utilized to close the global problem.

*Remark 3.2* (*Neumann local problems*) As detailed by Sevilla and Huerta (2016), an alternative HDG formulation is obtained if the Neumann boundary condition is imposed in the local problem. In this case, the trace of the velocity, $\hat{\boldsymbol{u}}$, is only defined on $\Gamma$ rather than on $\Gamma \cup \Gamma_N$. This leads to a marginally smaller discrete global problem. For the sake of clarity, the standard HDG formulation with Neumann boundary conditions imposed in the global problem is only considered here.

For a complete introduction to HDG, the interested reader is referred to the seminal contribution by Cockburn and Gopalakrishnan (2009b) and to the subsequent series of papers by Cockburn and coworkers (Cockburn et al. 2009b, 2010, 2011; Nguyen et al. 2009a, b, 2010a, b, 2011b; Peraire et al. 2010) where the HDG formulation for flow problems has been theoretically and numerically analyzed.

## Weak Forms of the Local and Global Problems

For each element $\Omega_e$, $e = 1, \ldots, \mathtt{n_{el}}$, the weak formulation of (9) is as follows: given $\boldsymbol{u}_D$ on $\Gamma_D$ and $\hat{\boldsymbol{u}}$ on $\Gamma \cup \Gamma_N$, find $(\boldsymbol{L}_e, \boldsymbol{u}_e, p_e) \in [\mathcal{H}(\text{div}; \Omega_e); \mathbb{S}] \times \left[\mathcal{H}^1(\Omega_e)\right]^{\mathtt{n_{sd}}} \times \mathcal{H}^1(\Omega_e)$ that satisfies

$$
\begin{cases}
-\left(\boldsymbol{G}, \boldsymbol{L}_e\right)_{\Omega_e} + \left(\boldsymbol{\nabla}\cdot(\sqrt{2\nu}\boldsymbol{G}), \boldsymbol{u}_e\right)_{\Omega_e} \\
\qquad\qquad = \langle \boldsymbol{G}\,\boldsymbol{n}_e, \sqrt{2\nu}\,\boldsymbol{u}_D\rangle_{\partial\Omega_e\cap\Gamma_D} + \langle \boldsymbol{G}\,\boldsymbol{n}_e, \sqrt{2\nu}\,\hat{\boldsymbol{u}}\rangle_{\partial\Omega_e\setminus\Gamma_D}, \\
\left(\boldsymbol{w}, \boldsymbol{\nabla}\cdot(\sqrt{2\nu}\boldsymbol{L}_e)\right)_{\Omega_e} + \left(\boldsymbol{w}, \boldsymbol{\nabla} p_e\right)_{\Omega_e} \\
\qquad\qquad + \left\langle \boldsymbol{w}, \widehat{(\sqrt{2\nu}\boldsymbol{L}_e + p_e\mathbf{I}_{\mathtt{n_{sd}}})\,\boldsymbol{n}_e} - (\sqrt{2\nu}\boldsymbol{L}_e + p_e\mathbf{I}_{\mathtt{n_{sd}}})\,\boldsymbol{n}_e \right\rangle_{\partial\Omega_e} \\
\qquad\qquad\qquad - \left(\boldsymbol{\nabla}\boldsymbol{w}, \boldsymbol{u}_e \otimes \boldsymbol{a}\right)_{\Omega_e} + \langle \boldsymbol{w}, (\widehat{\boldsymbol{u}_e \otimes \boldsymbol{a}})\boldsymbol{n}_e\rangle_{\partial\Omega_e} = \left(\boldsymbol{w}, \boldsymbol{s}\right)_{\Omega_e}, \\
\left(\boldsymbol{\nabla} q, \boldsymbol{u}_e\right)_{\Omega_e} = \langle q, \boldsymbol{u}_D \cdot \boldsymbol{n}_e\rangle_{\partial\Omega_e\cap\Gamma_D} + \langle q, \hat{\boldsymbol{u}} \cdot \boldsymbol{n}_e\rangle_{\partial\Omega_e\setminus\Gamma_D}, \\
\langle p_e, 1\rangle_{\partial\Omega_e} = \rho_e,
\end{cases}
$$

for all $(\boldsymbol{G}, \boldsymbol{w}, q) \in [\mathcal{H}(\text{div}; \Omega_e); \mathbb{S}] \times \left[\mathcal{H}^1(\Omega_e)\right]^{\mathtt{n_{sd}}} \times \mathcal{H}^1(\Omega_e)$, where, as defined in Section "Functional and Discrete Approximation Setting", $[\mathcal{H}(\text{div}; \Omega_e); \mathbb{S}]$ is the space of square-integrable symmetric tensors $\mathbb{S}$ of order $\mathtt{n_{sd}}$ on $\Omega_e$ with square-integrable row-wise divergence. Note that the variational form of the momentum equation above is obtained after integrating by parts the diffusive part of the flux twice, whereas the convective term is integrated by parts only once. This, as noted in Sevilla and Huerta (2016), preserves the symmetry of the local problem for the diffusive operator and, consequently, for the Stokes problem ($\boldsymbol{a} = \boldsymbol{0}$).

The numerical trace of the diffusive flux is defined in Cockburn et al. (2009b) and Sevilla and Huerta (2016),

$$\widehat{(\sqrt{2\nu}\boldsymbol{L}_e+p_e\mathbf{I}_{\mathtt{n_{sd}}})\,\boldsymbol{n}_e} := \begin{cases} (\sqrt{2\nu}\boldsymbol{L}_e+p_e\mathbf{I}_{\mathtt{n_{sd}}})\,\boldsymbol{n}_e+\tau^d(\boldsymbol{u}_e-\boldsymbol{u}_D) & \text{on } \partial\Omega_e\cap\Gamma_D, \\ (\sqrt{2\nu}\boldsymbol{L}_e+p_e\mathbf{I}_{\mathtt{n_{sd}}})\,\boldsymbol{n}_e+\tau^d(\boldsymbol{u}_e-\hat{\boldsymbol{u}}) & \text{elsewhere.} \end{cases} \tag{11a}$$

For the convection flux, several alternatives are possible, and they can be written in general form as

$$\widehat{(\boldsymbol{u}_e\otimes\boldsymbol{a})\boldsymbol{n}_e} := \begin{cases} (\tilde{\boldsymbol{u}}\otimes\hat{\boldsymbol{a}})\boldsymbol{n}_e+\tau^a(\boldsymbol{u}_e-\boldsymbol{u}_D) & \text{on } \partial\Omega_e\cap\Gamma_D, \\ (\tilde{\boldsymbol{u}}\otimes\hat{\boldsymbol{a}})\boldsymbol{n}_e+\tau^a(\boldsymbol{u}_e-\hat{\boldsymbol{u}}) & \text{elsewhere,} \end{cases} \tag{11b}$$

where $\hat{\boldsymbol{a}}$ is the trace of the convective field evaluated on the mesh edges/faces (and it is unique, as $\hat{\boldsymbol{u}}$, on the interior faces) whereas $\tilde{\boldsymbol{u}}$ needs to be appropriately defined. The option $\tilde{\boldsymbol{u}} = \boldsymbol{u}_e$ is rarely used because, in general, $\tilde{\boldsymbol{u}}$ can be seen as an intermediate state typically used in exact Riemann solvers (see Hesthaven 2019, Sect. 6.1). Here, as usually done in HDG (see Nguyen et al. 2009a, b, 2011b) $\tilde{\boldsymbol{u}}$ is chosen such that $\tilde{\boldsymbol{u}} = \boldsymbol{u}_D$ on $\Gamma_D$ and $\tilde{\boldsymbol{u}} = \hat{\boldsymbol{u}}$, elsewhere. The coefficients $\tau^d$ and $\tau^a$ are stabilization parameters that play a crucial role on the stability, accuracy and convergence properties of the resulting HDG method, see Remark 3.3.

*Remark 3.3* (*Stabilization paramaters*) The influence of the stabilization parameters on the well-posedness of the HDG method has been studied extensively (see for instance Cockburn and Gopalakrishnan 2009; Cockburn et al. 2008; Nguyen et al. 2010b). As noted in those references, for second-order elliptic problems, $\tau^d$ is of order one *for dimensionless problem*; thus, here $\tau^d$ is proportional to the viscosity, namely

$$\tau^d = \kappa\nu/\ell, \tag{12}$$

where $\ell$ is the characteristic size of the problem under analysis and $\kappa > 0$ is a scaling factor. It is worth noticing that the purely diffusive (i.e. Stokes) problem is not very sensitive to the stabilization parameter $\tau^d$, as will be shown later in the numerical experiments.

Stabilization of the convective term emanates from the extensive literature on DG methods for hyperbolic problems. Obviously, in presence of both diffusion and convection phenomena, an appropriate choice of the stabilization parameters $\tau^d$ and $\tau^a$ is critical to guarantee stability and optimal convergence of the HDG method, as extensively analyzed by Cockburn et al. (2009a) and Cesmelioglu et al. (2013; 2017). Typically for the convection term, a characteristic velocity field of the fluid is considered to determine $\tau^a$, namely

$$\tau^a = \beta\|\boldsymbol{a}\|_2 \quad \text{or} \quad \tau^a = \beta\|\boldsymbol{a}\|_\infty, \tag{13a}$$

where the above norms may be defined either locally on a single element $\Omega_e$ or globally on the domain $\Omega$ and $\beta > 0$ is a positive constant independent on the Reynolds number. Note that alternative choices for the stabilization parameters $\tau^d$ and $\tau^a$ are possible, e.g. depending on the spatial coordinate $\boldsymbol{x}$. More precisely, global definitions of the parameter $\tau^a$ may be obtained by considering the maximum of the expressions reported in (13a), over all the nodes in the computational mesh, that is

$$\tau^a = \beta \max_{\boldsymbol{x} \in \mathcal{N}_\Omega} \|\boldsymbol{a}(\boldsymbol{x})\|_2 \quad \text{or} \quad \tau^a = \beta \max_{\boldsymbol{x} \in \mathcal{N}_\Omega} \|\boldsymbol{a}(\boldsymbol{x})\|_\infty, \tag{13b}$$

where $\mathcal{N}_\Omega$ is the set of nodes of the computational mesh associated with the domain $\Omega$. The possibility of defining different values of the stabilization on each face of $\partial\Omega_e$ is also of great interest, e.g. the expression proposed by Cesmelioglu et al. (2017) for each face $\Gamma_e$, namely

$$\tau^a|_{\Gamma_e} = \beta \max\{\hat{\boldsymbol{a}} \cdot \boldsymbol{n}_e, 0\}. \tag{13c}$$

Nonetheless, $\tau^d + \tau^a - \hat{\boldsymbol{a}} \cdot \boldsymbol{n}_e$ needs to be nonnegative on all the faces of the element and positive at least on one, see Cockburn et al. (2009a). Thus, for any pair $(\tau^d, \tau^a)$, the following admissibility condition is introduced

$$\min_{\boldsymbol{x} \in \partial\Omega_e} \{\tau^d + \tau^a - \hat{\boldsymbol{a}} \cdot \boldsymbol{n}_e\} \geq \gamma > 0.$$

Introducing the definition of the numerical traces in (11) into the momentum equation leads to the weak form of the local problem: for $e = 1, \dots, \mathtt{n_{el}}$, find $(\boldsymbol{L}_e, \boldsymbol{u}_e, p_e) \in [\mathcal{H}(\text{div}; \Omega_e); \mathbb{S}] \times \left[\mathcal{H}^1(\Omega_e)\right]^{\mathtt{n_{sd}}} \times \mathcal{H}^1(\Omega_e)$ that satisfies

$$\begin{cases} -\big(\boldsymbol{G}, \boldsymbol{L}_e\big)_{\Omega_e} + \big(\boldsymbol{\nabla} \cdot (\sqrt{2\nu}\boldsymbol{G}), \boldsymbol{u}_e\big)_{\Omega_e} \\ \qquad = \langle \boldsymbol{G}\,\boldsymbol{n}_e, \sqrt{2\nu}\,\boldsymbol{u}_D \rangle_{\partial\Omega_e \cap \Gamma_D} + \langle \boldsymbol{G}\,\boldsymbol{n}_e, \sqrt{2\nu}\,\hat{\boldsymbol{u}} \rangle_{\partial\Omega_e \setminus \Gamma_D}, \\ \big(\boldsymbol{w}, \boldsymbol{\nabla} \cdot (\sqrt{2\nu}\boldsymbol{L}_e)\big)_{\Omega_e} + \big(\boldsymbol{w}, \boldsymbol{\nabla} p_e\big)_{\Omega_e} - \big(\boldsymbol{\nabla} \boldsymbol{w}, \boldsymbol{u}_e \otimes \boldsymbol{a}\big)_{\Omega_e} + \langle \boldsymbol{w}, \tau \boldsymbol{u}_e \rangle_{\partial\Omega_e} \\ \qquad = \big(\boldsymbol{w}, \boldsymbol{s}\big)_{\Omega_e} \\ \qquad\quad + \langle \boldsymbol{w}, (\tau - \hat{\boldsymbol{a}} \cdot \boldsymbol{n}_e)\boldsymbol{u}_D \rangle_{\partial\Omega_e \cap \Gamma_D} + \langle \boldsymbol{w}, (\tau - \hat{\boldsymbol{a}} \cdot \boldsymbol{n}_e)\hat{\boldsymbol{u}} \rangle_{\partial\Omega_e \setminus \Gamma_D}, \\ \big(\boldsymbol{\nabla} q, \boldsymbol{u}_e\big)_{\Omega_e} = \langle q, \boldsymbol{u}_D \cdot \boldsymbol{n}_e \rangle_{\partial\Omega_e \cap \Gamma_D} + \langle q, \hat{\boldsymbol{u}} \cdot \boldsymbol{n}_e \rangle_{\partial\Omega_e \setminus \Gamma_D}, \\ \langle p_e, 1 \rangle_{\partial\Omega_e} = \rho_e, \end{cases} \tag{14}$$

for all $(\boldsymbol{G}, \boldsymbol{w}, q) \in [\mathcal{H}(\text{div}; \Omega_e); \mathbb{S}] \times \left[\mathcal{H}^1(\Omega_e)\right]^{\mathtt{n_{sd}}} \times \mathcal{H}^1(\Omega_e)$ and where the stabilization parameter is defined as $\tau = \tau^d + \tau^a$. Note that, as expected, the previous problem provides $(\boldsymbol{L}_e, \boldsymbol{u}_e, p_e)$, for $e = 1, \dots, \mathtt{n_{el}}$, in terms of the global unknowns $\hat{\boldsymbol{u}}$ and $\boldsymbol{\rho} = (\rho_1 \dots, \rho_{\mathtt{n_{el}}})^T$.

For the global problem, the weak formulation equivalent to (10) is: find $\hat{\boldsymbol{u}} \in \left[\mathcal{H}^{\frac{1}{2}}(\Gamma \cup \Gamma_N)\right]^{\mathtt{n_{sd}}}$ and $\boldsymbol{\rho} \in \mathbb{R}^{\mathtt{n_{el}}}$ that satisfies

$$\begin{cases} \displaystyle\sum_{e=1}^{\mathtt{n_{el}}} \Big\{ \langle \widehat{\boldsymbol{w}}, \widehat{(\sqrt{2\nu}\boldsymbol{L}_e + p_e \mathbf{I}_{\mathtt{n_{sd}}})\,\boldsymbol{n}_e} + \widehat{(\boldsymbol{u}_e \otimes \boldsymbol{a})\,\boldsymbol{n}_e} \rangle_{\partial\Omega_e \setminus \partial\Omega} \\ \qquad\qquad + \langle \widehat{\boldsymbol{w}}, \widehat{(\sqrt{2\nu}\boldsymbol{L}_e + p_e \mathbf{I}_{\mathtt{n_{sd}}})\,\boldsymbol{n}_e} + \widehat{(\boldsymbol{u}_e \otimes \boldsymbol{a})\,\boldsymbol{n}_e} + \boldsymbol{t} \rangle_{\partial\Omega_e \cap \Gamma_N} \Big\} = 0, \\ \langle \hat{\boldsymbol{u}} \cdot \boldsymbol{n}_e, 1 \rangle_{\partial\Omega_e \setminus \Gamma_D} = -\langle \boldsymbol{u}_D \cdot \boldsymbol{n}_e, 1 \rangle_{\partial\Omega_e \cap \Gamma_D} \qquad \text{for } e = 1, \dots, \mathtt{n_{el}}, \end{cases}$$

for all $\widehat{\boldsymbol{w}} \in \left[\mathcal{L}_2(\Gamma \cup \Gamma_N)\right]^{\mathtt{n_{sd}}}$.

Replacing in the transmission equations the numerical traces defined in (11), the variational form of the global problem is obtained. Namely, find $\hat{\boldsymbol{u}} \in \left[\mathcal{H}^{\frac{1}{2}}(\Gamma \cup \Gamma_N)\right]^{\mathtt{n_{sd}}}$ and $\boldsymbol{\rho} \in \mathbb{R}^{\mathtt{n_{el}}}$ such that, for all $\widehat{\boldsymbol{w}} \in \left[\mathcal{L}_2(\Gamma \cup \Gamma_N)\right]^{\mathtt{n_{sd}}}$, it holds

$$\begin{cases} \displaystyle\sum_{e=1}^{\mathtt{n_{el}}} \Big\{ \langle \widehat{\boldsymbol{w}}, (\sqrt{2\nu}\boldsymbol{L}_e + p_e \mathbf{I}_{\mathtt{n_{sd}}})\,\boldsymbol{n}_e \rangle_{\partial\Omega_e \setminus \Gamma_D} + \langle \widehat{\boldsymbol{w}}, \tau \boldsymbol{u}_e \rangle_{\partial\Omega_e \setminus \Gamma_D} \\ \qquad\qquad - \langle \widehat{\boldsymbol{w}}, \tau \hat{\boldsymbol{u}} \rangle_{\partial\Omega_e \cap \Gamma} - \langle \widehat{\boldsymbol{w}}, (\tau - \hat{\boldsymbol{a}} \cdot \boldsymbol{n}_e) \hat{\boldsymbol{u}} \rangle_{\partial\Omega_e \cap \Gamma_N} \Big\} \\ \qquad\qquad\qquad = -\displaystyle\sum_{e=1}^{\mathtt{n_{el}}} \langle \widehat{\boldsymbol{w}}, \boldsymbol{t} \rangle_{\partial\Omega_e \cap \Gamma_N}, \\ \langle \hat{\boldsymbol{u}} \cdot \boldsymbol{n}_e, 1 \rangle_{\partial\Omega_e \setminus \Gamma_D} = -\langle \boldsymbol{u}_D \cdot \boldsymbol{n}_e, 1 \rangle_{\partial\Omega_e \cap \Gamma_D} \qquad \text{for } e = 1, \dots, \mathtt{n_{el}}. \end{cases} \tag{15}$$

Remark 3.4 shows how the first equation in (15) is obtained exploiting the uniqueness of $\hat{\boldsymbol{u}}$ and $\hat{\boldsymbol{a}}$ on the internal skeleton. Recall that for Stokes it is trivial ($\hat{\boldsymbol{a}} = \mathbf{0}$) and for Navier-Stokes it follows from $\hat{\boldsymbol{a}} = \hat{\boldsymbol{u}}$ on $\Gamma \cup \Gamma_N$.

*Remark 3.4* In order to obtain the first equation in (15) the usual choice $\tilde{\boldsymbol{u}} = \hat{\boldsymbol{u}}$ is employed in the definition of the convection numerical flux, see (11b). Thus,

$$\begin{aligned} &\sum_{e=1}^{\mathtt{n_{el}}} \langle \widehat{\boldsymbol{w}}, \widehat{(\boldsymbol{u}_e \otimes \boldsymbol{a})\boldsymbol{n}_e} \rangle_{\partial\Omega_e \setminus \Gamma_D} \\ &= \sum_{e=1}^{\mathtt{n_{el}}} \langle \widehat{\boldsymbol{w}}, (\hat{\boldsymbol{u}} \otimes \hat{\boldsymbol{a}})\boldsymbol{n}_e + \tau^a (\boldsymbol{u}_e - \hat{\boldsymbol{u}}) \rangle_{\partial\Omega_e \setminus \Gamma_D} \\ &= \sum_{e=1}^{\mathtt{n_{el}}} \langle \widehat{\boldsymbol{w}}, \tau^a \boldsymbol{u}_e - (\tau^a - \hat{\boldsymbol{a}} \cdot \boldsymbol{n}_e) \hat{\boldsymbol{u}} \rangle_{\partial\Omega_e \setminus \Gamma_D} \\ &= \sum_{e=1}^{\mathtt{n_{el}}} \Big\{ \langle \widehat{\boldsymbol{w}}, \tau^a \boldsymbol{u}_e \rangle_{\partial\Omega_e \setminus \Gamma_D} - \langle \widehat{\boldsymbol{w}}, (\tau^a - \hat{\boldsymbol{a}} \cdot \boldsymbol{n}_e) \hat{\boldsymbol{u}} \rangle_{\partial\Omega_e \cap \Gamma} \\ &\quad - \langle \widehat{\boldsymbol{w}}, (\tau^a - \hat{\boldsymbol{a}} \cdot \boldsymbol{n}_e) \hat{\boldsymbol{u}} \rangle_{\partial\Omega_e \cap \Gamma_N} \Big\} \\ &= \sum_{e=1}^{\mathtt{n_{el}}} \Big\{ \langle \widehat{\boldsymbol{w}}, \tau^a \boldsymbol{u}_e \rangle_{\partial\Omega_e \setminus \Gamma_D} - \langle \widehat{\boldsymbol{w}}, \tau^a \hat{\boldsymbol{u}} \rangle_{\partial\Omega_e \cap \Gamma} - \langle \widehat{\boldsymbol{w}}, (\tau^a - \hat{\boldsymbol{a}} \cdot \boldsymbol{n}_e) \hat{\boldsymbol{u}} \rangle_{\partial\Omega_e \cap \Gamma_N} \Big\}, \end{aligned}$$

where the last line follows from the uniqueness of $\widehat{\boldsymbol{w}}$, $\hat{\boldsymbol{u}}$ and $\hat{\boldsymbol{a}}$ on the internal skeleton $\Gamma$ and from the relationship $\boldsymbol{n}_l = -\boldsymbol{n}_r$ between the outward unit normal vectors on an internal edge/face shared by two neighboring elements $\Omega_l$ and $\Omega_r$.

## Discrete Forms and the Resulting Linear System

The discrete functional spaces defined in (7) are used in this section to construct the block matrices involved in the discretization of the HDG local and global problems. Moreover, following the rationale proposed in Giacomini et al. (2018) for the Stokes equations, the HDG-Voigt formulation is introduced to discretize the Cauchy stress form of the incompressible flow equations under analysis. The advantage of this formulation compared with classical HDG approaches lies in the easy implementation of the space $\mathbb{S}$ of symmetric tensors of order $\mathtt{n_{sd}}$, which thus allows to retrieve optimal convergence and superconvergence properties, even for low-order approximations.

### Voigt Notation for Symmetric Second-Order Tensors

A symmetric second-order tensor is written in Voigt notation by storing its diagonal and off-diagonal components in vector form, after an appropriate rearrangement. Exploiting the symmetry, only $\mathtt{m_{sd}} = \mathtt{n_{sd}}(\mathtt{n_{sd}} + 1)/2$ components (i.e. three in 2D and six in 3D) are stored. For the strain-rate tensor $\boldsymbol{\nabla}^{\mathtt{S}}\boldsymbol{u}$, the following column vector $\boldsymbol{e}_{\mathtt{V}} \in \mathbb{R}^{\mathtt{m_{sd}}}$ is obtained

$$\boldsymbol{e}_{\mathtt{V}} := \begin{cases} \begin{bmatrix} e_{11}, & e_{22}, & e_{12} \end{bmatrix}^T & \text{in 2D,} \\ \begin{bmatrix} e_{11}, & e_{22}, & e_{33}, & e_{12}, & e_{13}, & e_{23} \end{bmatrix}^T & \text{in 3D,} \end{cases} \tag{16}$$

where the arrangement proposed by Fish and Belytschko (2007) has been utilized and the components are defined as

$$e_{ij} := \frac{\partial u_i}{\partial x_j} + (1 - \delta_{ij})\frac{\partial u_j}{\partial x_i}, \quad \text{for } i, j = 1, \dots, \mathtt{n_{sd}} \text{ and } i \leq j, \tag{17}$$

being $\delta_{ij}$ the classical Kronecker delta. Moreover, the strain-rate tensor can be written as $\boldsymbol{e}_{\mathtt{V}} = \boldsymbol{\nabla}_{\mathtt{S}}\boldsymbol{u}$ by introducing the $\mathtt{m_{sd}} \times \mathtt{n_{sd}}$ matrix

$$\boldsymbol{\nabla}_{\mathtt{S}} := \begin{cases} \begin{bmatrix} \partial/\partial x_1 & 0 & \partial/\partial x_2 \\ 0 & \partial/\partial x_2 & \partial/\partial x_1 \end{bmatrix}^T & \text{in 2D,} \\ \begin{bmatrix} \partial/\partial x_1 & 0 & 0 & \partial/\partial x_2 & \partial/\partial x_3 & 0 \\ 0 & \partial/\partial x_2 & 0 & \partial/\partial x_1 & 0 & \partial/\partial x_3 \\ 0 & 0 & \partial/\partial x_3 & 0 & \partial/\partial x_1 & \partial/\partial x_2 \end{bmatrix}^T & \text{in 3D.} \end{cases} \tag{18}$$

*Remark 3.5* The components of the strain-rate tensor $\boldsymbol{\nabla}^{\mathtt{s}}\boldsymbol{u}$ can be retrieved from its Voigt counterpart by multiplying the off-diagonal terms $e_{ij},\ i \neq j$ by a factor $1/2$, namely

$$\boldsymbol{\nabla}^{\mathtt{s}}\boldsymbol{u} := \begin{cases} \begin{bmatrix} e_{11} & e_{12}/2 \\ e_{12}/2 & e_{22} \end{bmatrix} & \text{in 2D,} \\ \begin{bmatrix} e_{11} & e_{12}/2 & e_{13}/2 \\ e_{12}/2 & e_{22} & e_{23}/2 \\ e_{13}/2 & e_{23}/2 & e_{33} \end{bmatrix} & \text{in 3D.} \end{cases} \tag{19}$$

Moreover, recall the following definitions introduced in Giacomini et al. (2018) for Voigt notation. First, the vorticity $\boldsymbol{\omega} := \boldsymbol{\nabla} \times \boldsymbol{u}$ is expressed using Voigt notation as $\boldsymbol{\omega} = \boldsymbol{\nabla}_{\mathtt{W}}\boldsymbol{u}$, where the $\mathtt{n_{rr}} \times \mathtt{n_{sd}}$ matrix $\boldsymbol{\nabla}_{\mathtt{W}}$, with $\mathtt{n_{rr}} = \mathtt{n_{sd}}(\mathtt{n_{sd}} - 1)/2$ number of rigid body rotations (i.e. one in 2D and three in 3D), has the form

$$\boldsymbol{\nabla}_{\mathtt{W}} := \begin{cases} \begin{bmatrix} -\partial/\partial x_2, & \partial/\partial x_1 \end{bmatrix} & \text{in 2D,} \\ \begin{bmatrix} 0 & -\partial/\partial x_3 & \partial/\partial x_2 \\ \partial/\partial x_3 & 0 & -\partial/\partial x_1 \\ -\partial/\partial x_2 & \partial/\partial x_1 & 0 \end{bmatrix} & \text{in 3D.} \end{cases} \tag{20}$$

*Remark 3.6* (*Vorticity in 2D*) Recall that in 2D the curl of a vector $\boldsymbol{v} = [v_1, v_2]^T$ is a scalar quantity. By setting $v_3 = 0$, the resulting vector $\boldsymbol{v} = [v_1, v_2, 0]^T$ is embedded in the three dimensional space $\mathbb{R}^3$. Thus, $\boldsymbol{\nabla} \times \boldsymbol{v}$ may be interpreted as a vector pointing along $x_3$ and with magnitude equal to $-(\partial v_1/\partial x_2) + (\partial v_2/\partial x_1)$.

Stokes' law is expressed in Voigt notation as $\boldsymbol{\sigma}_{\mathtt{V}} = \mathbf{D}\boldsymbol{\nabla}_{\mathtt{S}}\boldsymbol{u} - \mathbf{E}p$, where the vector $\mathbf{E} \in \mathbb{R}^{\mathtt{m_{sd}}}$ and the matrix $\mathbf{D} \in \mathbb{R}^{\mathtt{m_{sd}} \times \mathtt{m_{sd}}}$ are defined as

$$\mathbf{E} := \begin{cases} \begin{bmatrix} 1, & 1, & 0 \end{bmatrix}^T & \text{in 2D,} \\ \begin{bmatrix} 1, & 1, & 1, & 0, & 0, & 0 \end{bmatrix}^T & \text{in 3D.} \end{cases} \quad \mathbf{D} := \begin{cases} \begin{bmatrix} 2\nu\mathbf{I}_{\mathtt{n_{sd}}} & \mathbf{0}_{\mathtt{n_{sd}}\times 1} \\ \mathbf{0}^T_{\mathtt{n_{sd}}\times 1} & \nu \end{bmatrix} & \text{in 2D,} \\ \begin{bmatrix} 2\nu\mathbf{I}_{\mathtt{n_{sd}}} & \mathbf{0}_{\mathtt{n_{sd}}} \\ \mathbf{0}_{\mathtt{n_{sd}}} & \nu\mathbf{I}_{\mathtt{n_{sd}}} \end{bmatrix} & \text{in 3D.} \end{cases} \tag{21}$$

Similarly, traction boundary conditions on $\Gamma_N$ are rewritten as $\mathbf{N}^T\boldsymbol{\sigma}_{\mathtt{V}} = \boldsymbol{t}$, where $\mathbf{N}$ is the $\mathtt{m_{sd}} \times \mathtt{n_{sd}}$ matrix

$$\mathbf{N} := \begin{cases} \begin{bmatrix} n_1 & 0 & n_2 \\ 0 & n_2 & n_1 \end{bmatrix}^T & \text{in 2D,} \\ \begin{bmatrix} n_1 & 0 & 0 & n_2 & n_3 & 0 \\ 0 & n_2 & 0 & n_1 & 0 & n_3 \\ 0 & 0 & n_3 & 0 & n_1 & n_2 \end{bmatrix}^T & \text{in 3D,} \end{cases} \tag{22}$$

which accounts for the normal direction to the boundary.

For the sake of completeness, the matrix $\mathbf{T} \in \mathbb{R}^{\mathtt{n_{rr}} \times \mathtt{n_{sd}}}$ denoting the tangential direction $\boldsymbol{\tau}$ to a line/surface is introduced

$$\mathbf{T} := \begin{cases} \begin{bmatrix} -n_2, & n_1 \end{bmatrix} & \text{in 2D,} \\ \begin{bmatrix} 0 & -n_3 & n_2 \\ n_3 & 0 & -n_1 \\ -n_2 & n_1 & 0 \end{bmatrix} & \text{in 3D,} \end{cases} \tag{23}$$

and the projection of a vector $\boldsymbol{u}$ along such direction is given by $\boldsymbol{u} \cdot \boldsymbol{\tau} = \mathbf{T}\boldsymbol{u}$.

Voigt notation is thus exploited to rewrite the Oseen equations in (3). First, recall that the divergence operator applied to a symmetric tensor is equal to the transpose of the matrix $\boldsymbol{\nabla}_{\mathtt{S}}$ introduced in (18). Moreover, the vector $\mathbf{E}$ in (21) is utilized to compactly express the trace of a symmetric tensor required in the incompressibility equation, namely $\boldsymbol{\nabla} \cdot \boldsymbol{u} = \operatorname{tr}(\boldsymbol{\nabla}^{\mathtt{s}} \boldsymbol{u}) = \mathbf{E}^T \boldsymbol{\nabla}_{\mathtt{S}} \boldsymbol{u} = 0$. Finally, the formulation of the Oseen problem equivalent to (3) using Voigt notation is obtained by the combining the above matrix equations

$$\begin{cases} -\boldsymbol{\nabla}_{\mathtt{S}}^T (\mathbf{D} \boldsymbol{\nabla}_{\mathtt{S}} \boldsymbol{u} - \mathbf{E} p) + \boldsymbol{\nabla} \cdot (\boldsymbol{u} \otimes \boldsymbol{a}) = \boldsymbol{s} & \text{in } \Omega, \\ \mathbf{E}^T \boldsymbol{\nabla}_{\mathtt{S}} \boldsymbol{u} = 0 & \text{in } \Omega, \\ \boldsymbol{u} = \boldsymbol{u}_D & \text{on } \Gamma_D, \\ \mathbf{N}^T (\mathbf{D} \boldsymbol{\nabla}_{\mathtt{S}} \boldsymbol{u} - \mathbf{E}\, p) - (\boldsymbol{u} \otimes \boldsymbol{a}) \boldsymbol{n} = \boldsymbol{t} & \text{on } \Gamma_N. \end{cases} \tag{24}$$

Of course, as previously observed, replacing $\boldsymbol{a}$ by $\boldsymbol{u}$ or $\mathbf{0}$ leads to the Navier-Stokes, see (1), and the Stokes, see (4), problems, respectively.

### HDG-Voigt Strong Forms

The HDG local and global problems introduced in Section "Strong Forms of the Local and Global Problems" are now rewritten using Voigt notation. The local problem in (9) becomes

$$\begin{cases} \boldsymbol{L}_e + \mathbf{D}^{1/2} \boldsymbol{\nabla}_{\mathtt{S}} \boldsymbol{u}_e = \mathbf{0} & \text{in } \Omega_e, \\ \boldsymbol{\nabla}_{\mathtt{S}}^T \mathbf{D}^{1/2} \boldsymbol{L}_e + \boldsymbol{\nabla}_{\mathtt{S}}^T \mathbf{E}\, p_e + \boldsymbol{\nabla} \cdot (\boldsymbol{u} \otimes \boldsymbol{a}) = \boldsymbol{s} & \text{in } \Omega_e, \\ \mathbf{E}^T \boldsymbol{\nabla}_{\mathtt{S}} \boldsymbol{u}_e = 0 & \text{in } \Omega_e, \\ \boldsymbol{u}_e = \boldsymbol{u}_D & \text{on } \partial\Omega_e \cap \Gamma_D, \\ \boldsymbol{u}_e = \hat{\boldsymbol{u}} & \text{on } \partial\Omega_e \setminus \Gamma_D, \\ \langle p_e, 1 \rangle_{\partial\Omega_e} = \rho_e, \end{cases} \tag{25}$$

where the additional constraint described in (9b) to remove the indeterminacy of pressure is included. Note that $\boldsymbol{L}_e$ is the restriction to element $\Omega_e$ of the HDG mixed variable introduced to represent the strain-rate tensor in Voigt notation scaled via the matrix $\mathbf{D}^{1/2}$, featuring the square root of the eigenvalues of the diagonal matrix $\mathbf{D}$, see Eq. (21).

Similarly, the global problem using Voigt notation, see (10), is

$$\begin{cases} [\![\mathbf{N}^T(\mathbf{D}^{1/2}\boldsymbol{L} + \mathbf{E}\,p) + (\boldsymbol{u} \otimes \boldsymbol{a})\boldsymbol{n}]\!] = \mathbf{0} & \text{on } \Gamma, \\ \mathbf{N}^T(\mathbf{D}^{1/2}\boldsymbol{L} + \mathbf{E}\,p) + (\boldsymbol{u} \otimes \boldsymbol{a})\boldsymbol{n} = -\boldsymbol{t} & \text{on } \Gamma_N, \\ \langle \mathbf{E}^T\mathbf{N}_e\hat{\boldsymbol{u}}, 1\rangle_{\partial\Omega_e\setminus\Gamma_D} + \langle \mathbf{E}^T\mathbf{N}_e\boldsymbol{u}_D, 1\rangle_{\partial\Omega_e\cap\Gamma_D} = 0 & \text{for } e = 1, \dots, \mathtt{n_{el}}, \end{cases} \tag{26}$$

where the compatibility condition to close the global problem, see (10b), is also written in Voigt notation.

Moreover, by exploiting Voigt notation, the definition of the trace of the diffusive numerical flux, see (11a), becomes

$$\mathbf{N}_e^T\widehat{(\mathbf{D}^{1/2}\boldsymbol{L}_e + \mathbf{E}\,p_e)} := \begin{cases} \mathbf{N}_e^T(\mathbf{D}^{1/2}\boldsymbol{L}_e + \mathbf{E}\,p_e) + \tau^d(\boldsymbol{u}_e - \boldsymbol{u}_D) & \text{on } \partial\Omega_e \cap \Gamma_D, \\ \mathbf{N}_e^T(\mathbf{D}^{1/2}\boldsymbol{L}_e + \mathbf{E}\,p_e) + \tau^d(\boldsymbol{u}_e - \hat{\boldsymbol{u}}) & \text{elsewhere.} \end{cases} \tag{27}$$

Of course, the trace of the convective numerical flux, since it follows the standard tensorial notation, is the one defined in (11b), with $\tilde{\boldsymbol{u}} = \hat{\boldsymbol{u}}$.

### HDG-Voigt Discrete Weak Forms

Following the derivation in Section "Weak Forms of the Local and Global Problems", the HDG formulation of the Oseen equations with Voigt notation is obtained. The discrete weak formulation of the local problems proposed in (25) is as follows: for $e = 1, \dots, \mathtt{n_{el}}$, given $\boldsymbol{u}_D$ on $\Gamma_D$ and $\hat{\boldsymbol{u}}$ on $\Gamma \cup \Gamma_N$, find $(\boldsymbol{L}_e, \boldsymbol{u}_e, p_e) \in [\mathcal{V}^h(\Omega_e)]^{\mathtt{m_{sd}}} \times [\mathcal{V}^h(\Omega_e)]^{\mathtt{n_{sd}}} \times \mathcal{V}^h(\Omega_e)$ such that

$$-(\boldsymbol{v}, \boldsymbol{L}_e)_{\Omega_e} + (\boldsymbol{\nabla}_{\mathtt{S}}^T\mathbf{D}^{1/2}\boldsymbol{v}, \boldsymbol{u}_e)_{\Omega_e} = \langle \mathbf{N}_e^T\mathbf{D}^{1/2}\boldsymbol{v}, \boldsymbol{u}_D\rangle_{\partial\Omega_e\cap\Gamma_D} + \langle \mathbf{N}_e^T\mathbf{D}^{1/2}\boldsymbol{v}, \hat{\boldsymbol{u}}\rangle_{\partial\Omega_e\setminus\Gamma_D}, \tag{28a}$$

$$\begin{aligned} \big(\boldsymbol{w}, \boldsymbol{\nabla}_{\mathtt{S}}^T\mathbf{D}^{1/2}\boldsymbol{L}_e\big)_{\Omega_e} + \big(\boldsymbol{w}, \boldsymbol{\nabla}_{\mathtt{S}}^T\mathbf{E}\,p_e\big)_{\Omega_e} - (\boldsymbol{\nabla}\boldsymbol{w}, \boldsymbol{u}_e \otimes \boldsymbol{a})_{\Omega_e} + \langle \boldsymbol{w}, \tau\boldsymbol{u}_e\rangle_{\partial\Omega_e} & \\ = (\boldsymbol{w}, \boldsymbol{s})_{\Omega_e} + \langle \boldsymbol{w}, (\tau - \hat{\boldsymbol{a}}\cdot\boldsymbol{n}_e)\boldsymbol{u}_D\rangle_{\partial\Omega_e\cap\Gamma_D} & \\ + \langle \boldsymbol{w}, (\tau - \hat{\boldsymbol{a}}\cdot\boldsymbol{n}_e)\hat{\boldsymbol{u}}\rangle_{\partial\Omega_e\setminus\Gamma_D}, & \end{aligned} \tag{28b}$$

$$(\boldsymbol{\nabla}_{\mathtt{S}}^T\mathbf{E}\,q, \boldsymbol{u}_e)_{\Omega_e} = \langle \mathbf{N}_e^T\mathbf{E}\,q, \boldsymbol{u}_D\rangle_{\partial\Omega_e\cap\Gamma_D} + \langle \mathbf{N}_e^T\mathbf{E}\,q, \hat{\boldsymbol{u}}\rangle_{\partial\Omega_e\setminus\Gamma_D}, \tag{28c}$$

$$\langle p_e, 1\rangle_{\partial\Omega_e} = \rho_e, \tag{28d}$$

for all $(\boldsymbol{v}, \boldsymbol{w}, q) \in [\mathcal{V}^h(\Omega_e)]^{\mathtt{m_{sd}}} \times [\mathcal{V}^h(\Omega_e)]^{\mathtt{n_{sd}}} \times \mathcal{V}^h(\Omega_e)$, where, as done previously, the stabilization parameter is defined as $\tau = \tau^d + \tau^a$.

The global problem computes the hybrid variable $\hat{\boldsymbol{u}} \in [\hat{\mathcal{V}}^h(\Gamma \cup \Gamma_N)]^{\mathtt{n_{sd}}}$ and the scaled mean pressure on each element boundary $\boldsymbol{\rho} \in \mathbb{R}^{\mathtt{n_{el}}}$. The global problem is obtained starting from (26), recall also Remark 3.4, and the definitions of the numerical flux given in (27) and (11b): find $\hat{\boldsymbol{u}} \in [\hat{\mathcal{V}}^h(\Gamma \cup \Gamma_N)]^{\mathtt{n_{sd}}}$ and $\boldsymbol{\rho} \in \mathbb{R}^{\mathtt{n_{el}}}$, such that

$$\sum_{e=1}^{\mathtt{n_{el}}} \Big\{ \langle \widehat{\boldsymbol{w}}, \mathbf{N}_e^T (\mathbf{D}^{1/2} \boldsymbol{L}_e + \mathbf{E}\, p_e) \rangle_{\partial\Omega_e \setminus \Gamma_D} + \langle \widehat{\boldsymbol{w}}, \tau\, \boldsymbol{u}_e \rangle_{\partial\Omega_e \setminus \Gamma_D}$$

$$- \langle \widehat{\boldsymbol{w}}, \tau \hat{\boldsymbol{u}} \rangle_{\partial\Omega_e \cap \Gamma} - \langle \widehat{\boldsymbol{w}}, (\tau - \hat{\boldsymbol{a}} \cdot \boldsymbol{n}_e) \hat{\boldsymbol{u}} \rangle_{\partial\Omega_e \cap \Gamma_N} \Big\} = - \sum_{e=1}^{\mathtt{n_{el}}} \langle \widehat{\boldsymbol{w}}, \boldsymbol{t} \rangle_{\partial\Omega_e \cap \Gamma_N}, \tag{29a}$$

$$\langle \mathbf{E}^T \mathbf{N}_e \hat{\boldsymbol{u}}, 1 \rangle_{\partial\Omega_e \setminus \Gamma_D} = - \langle \mathbf{E}^T \mathbf{N}_e \boldsymbol{u}_D, 1 \rangle_{\partial\Omega_e \cap \Gamma_D} \quad \text{for } e = 1, \dots, \mathtt{n_{el}}, \tag{29b}$$

for all $\widehat{\boldsymbol{w}} \in [\hat{\mathcal{V}}^h(\Gamma \cup \Gamma_N)]^{\mathtt{n_{sd}}}$.

### HDG Linear System

The discretization of the weak form of the local problem given by (28) using an isoparametric formulation for the primal, mixed and hybrid variables leads to a linear system with the following structure

$$\begin{bmatrix} \mathbf{A}_{LL} & \mathbf{A}_{Lu} & \mathbf{0} & \mathbf{0} \\ \mathbf{A}_{Lu}^T & \mathbf{A}_{uu} & \mathbf{A}_{pu}^T & \mathbf{0} \\ \mathbf{0} & \mathbf{A}_{pu} & \mathbf{0} & \mathbf{a}_{\rho p}^T \\ \mathbf{0} & \mathbf{0} & \mathbf{a}_{\rho p} & 0 \end{bmatrix}_e \begin{Bmatrix} \mathbf{L}_e \\ \mathbf{u}_e \\ \mathbf{p}_e \\ \zeta \end{Bmatrix} = \begin{Bmatrix} \mathbf{f}_L \\ \mathbf{f}_u \\ \mathbf{f}_p \\ 0 \end{Bmatrix}_e + \begin{bmatrix} \mathbf{A}_{L\hat{u}} \\ \mathbf{A}_{u\hat{u}} \\ \mathbf{A}_{p\hat{u}} \\ \mathbf{0} \end{bmatrix}_e \hat{\mathbf{u}}_e + \begin{Bmatrix} 0 \\ 0 \\ 0 \\ 1 \end{Bmatrix}_e \rho_e, \tag{30}$$

for $e = 1, \dots, \mathtt{n_{el}}$. It is worth noticing that the last equation in (28) is the restriction (9b) imposed via the Lagrange multiplier $\zeta$ in the system of equations above. Recalling the dimensions of the unknown variables, the solution of this system implies inverting a matrix of dimension $\big((\mathtt{m_{sd}} + \mathtt{n_{sd}} + 1)\mathtt{n_{en}} + 1\big)$ for each element in the mesh, being $\mathtt{n_{en}}$ the number of element nodes of $\Omega_e$. Note that $\mathtt{n_{en}}$ is directly related

**Table 1** Dimension of the local problem

| Order of interpolation | 1 | 2 | 3 | 4 | 5 | 7 | 9 | 11 |
|---|---|---|---|---|---|---|---|---|
| Simplexes | | | | | | | | |
| 2D | 19 | 37 | 61 | 91 | 127 | 217 | 331 | 469 |
| 3D | 41 | 101 | 201 | 351 | 561 | 1201 | 2201 | 3641 |
| Parallelepipeds | | | | | | | | |
| 2D | 25 | 55 | 97 | 151 | 217 | 385 | 601 | 865 |
| 3D | 81 | 271 | 641 | 1251 | 2161 | 5121 | 10,001 | 17,281 |

to the degree $k$ of polynomial approximations. Table 1 shows the dimension of the local problem stated in (30) for Lagrange elements on simplexes and parallelepipeds for 2D and 3D and different orders of approximation.

Similarly, the following system of equations is obtained for the global problem

$$\begin{aligned}\sum_{e=1}^{\mathtt{n_{el}}} \Bigg\{ \big[\mathbf{A}_{L\hat{u}}^T \; \mathbf{A}_{\hat{u}u} \; \mathbf{A}_{p\hat{u}}^T\big]_e \begin{Bmatrix} \mathbf{L}_e \\ \mathbf{u}_e \\ \mathbf{p}_e \end{Bmatrix} + [\mathbf{A}_{\hat{u}\hat{u}}]_e \, \hat{\mathbf{u}}_e \Bigg\} &= \sum_{i=e}^{\mathtt{n_{el}}} [\mathbf{f}_{\hat{u}}]_e, \\ \mathbf{1}^T \, [\mathbf{A}_{p\hat{u}}]_e \hat{\mathbf{u}}_e &= -\mathbf{1}^T \, [\mathbf{f}_p]_e\end{aligned} \tag{31}$$

The expressions of the matrices and vectors appearing in (30) and (31) are detailed in Appendix "Appendix: Implementation Details".

After replacing the solution of the local problem (30) in (31), the global problem becomes

$$\begin{bmatrix} \widehat{\mathbf{K}} & \mathbf{G} \\ \mathbf{G}^T & \mathbf{0} \end{bmatrix} \begin{Bmatrix} \hat{\mathbf{u}} \\ \rho \end{Bmatrix} = \begin{Bmatrix} \hat{\mathbf{f}}_{\hat{u}} \\ \hat{\mathbf{f}}_{\rho} \end{Bmatrix}, \tag{32}$$

where

$$\widehat{\mathbf{K}} = \mathbf{A}_{e=1}^{\mathtt{n_{el}}} \big[\mathbf{A}_{L\hat{u}}^T \; \mathbf{A}_{\hat{u}u} \; \mathbf{A}_{p\hat{u}}^T \; \mathbf{0}\big]_e \begin{bmatrix} \mathbf{A}_{LL} & \mathbf{A}_{Lu} & \mathbf{0} & \mathbf{0} \\ \mathbf{A}_{Lu}^T & \mathbf{A}_{uu} & \mathbf{A}_{pu}^T & \mathbf{0} \\ \mathbf{0} & \mathbf{A}_{pu} & \mathbf{0} & \mathbf{a}_{\rho p}^T \\ \mathbf{0} & \mathbf{0} & \mathbf{a}_{\rho p} & 0 \end{bmatrix}_e^{-1} \begin{bmatrix} \mathbf{A}_{L\hat{u}} \\ \mathbf{A}_{u\hat{u}} \\ \mathbf{A}_{p\hat{u}} \\ \mathbf{0} \end{bmatrix}_e + [\mathbf{A}_{\hat{u}\hat{u}}]_e, \tag{33a}$$

$$\mathbf{G}^T = \begin{bmatrix} \mathbf{1}^T \, [\mathbf{A}_{p\hat{u}}]_1 \\ \mathbf{1}^T \, [\mathbf{A}_{p\hat{u}}]_2 \\ \cdots \\ \mathbf{1}^T \, [\mathbf{A}_{p\hat{u}}]_{\mathtt{n_{el}}} \end{bmatrix}, \tag{33b}$$

$$\hat{\mathbf{f}}_{\hat{u}} = \mathbf{A}_{e=1}^{\mathtt{n_{el}}} [\mathbf{f}_{\hat{u}}]_e - \big[\mathbf{A}_{L\hat{u}}^T \; \mathbf{A}_{\hat{u}u} \; \mathbf{A}_{p\hat{u}}^T \; \mathbf{0}\big]_e \begin{bmatrix} \mathbf{A}_{LL} & \mathbf{A}_{Lu} & \mathbf{0} & \mathbf{0} \\ \mathbf{A}_{Lu}^T & \mathbf{A}_{uu} & \mathbf{A}_{pu}^T & \mathbf{0} \\ \mathbf{0} & \mathbf{A}_{pu} & \mathbf{0} & \mathbf{a}_{\rho p}^T \\ \mathbf{0} & \mathbf{0} & \mathbf{a}_{\rho p} & 0 \end{bmatrix}_e^{-1} \begin{Bmatrix} \mathbf{f}_L \\ \mathbf{f}_u \\ \mathbf{f}_p \\ 0 \end{Bmatrix}_e, \tag{33c}$$

$$\hat{\mathbf{f}}_{\rho} = - \begin{bmatrix} \mathbf{1}^T \, [\mathbf{f}_p]_1 \\ \mathbf{1}^T \, [\mathbf{f}_p]_2 \\ \cdots \\ \mathbf{1}^T \, [\mathbf{f}_p]_{\mathtt{n_{el}}} \end{bmatrix}. \tag{33d}$$

It is worth noticing that the resulting system in (32) has a saddle-point structure, as it is classical in the discretization of incompressible flow problems (see Donea and Huerta 2003, Sect. 6.5). The proof of the symmetry of the global system in (32) is presented in Appendix "Appendix: Saddle-Point Structure of the Global Problem" for the stokes problem.

## Local Postprocess of the Primal Variable

HDG methods feature the possibility of constructing a superconvergent approximation of the primal variable via a local postprocess performed element-by-element. Several strategies have been proposed in the literature, either to recover a globally $H(\text{div})$-conforming and pointwise solenoidal velocity field (see Cockburn et al. 2011, 2010) or simply to improve the approximation of the velocity field (see Nguyen et al. 2010b or Sevilla and Huerta 2016) and then, derive an error indicator for degree adaptive procedures (see Giorgiani et al. 2014 and Sevilla and Huerta 2018). The former approach stem from the Brezzi-Douglas-Marini (BDM) projection operator, see Brezzi and Fortin (1991), whereas the latter is inspired by the work of Stenberg (1990).

The superconvergent property of the HDG postprocessed solution depends, on the one hand, on the optimal convergence of order $k+1$ of the mixed variable and, on the other hand, on a procedure to resolve the indeterminacy due to rigid body motions. This is especially critical when the Cauchy stress formulation of incompressible flow problems is considered since a loss of optimal convergence and superconvergence is experienced for low-order polynomial approximations, as noted by Cockburn et al. (2010). Cockburn and coworkers (see Cockburn et al. 2017; Cockburn and Fu 2017, b) proposed the $\boldsymbol{M}$-decomposition to enrich the discrete space of the mixed variable, whereas Qiu and Shi (2016) suggested an HDG formulation with different polynomial degrees of approximation for primal, mixed and hybrid variables. Using a pointwise symmetric mixed variable and equal-order of approximation for all the variables, HDG-Voigt formulation has proved its capability to retrieve optimal convergence of the mixed variable and superconvergence of the postprocessed primal one, as shown by Sevilla et al. (2018) and Giacomini et al. (2018).

To determine the postprocessed velocity field $\boldsymbol{u}_e^\star$ in each element $\Omega_e$, a local problem to be solved element-by-element is devised. Applying the divergence operator to the first equation in (9a), for each element $\Omega_e,\ e=1,\dots,\mathtt{n_{el}}$, it follows

$$\boldsymbol{\nabla}\cdot\left(\sqrt{2\nu}\boldsymbol{\nabla}^{\mathtt{S}}\boldsymbol{u}_e^\star\right)=-\boldsymbol{\nabla}\cdot\boldsymbol{L}_e, \tag{34a}$$

whereas boundary conditions enforcing equilibrated fluxes on $\partial\Omega_e$ are imposed, namely

$$\left(\sqrt{2\nu}\boldsymbol{\nabla}^{\mathtt{S}}\boldsymbol{u}_e^\star\right)\boldsymbol{n}=-\boldsymbol{L}_e\boldsymbol{n}. \tag{34b}$$

The elliptic problem (34) admits a solution up to rigid motions, that is $\mathtt{n_{sd}}$ translations (i.e. two in 2D and three in 3D) and $\mathtt{n_{rr}}$ rotations (i.e. one in 2D and three in 3D). First, the indeterminacy associated with the $\mathtt{n_{sd}}$ rigid translational modes is handled by means of a constraint on the mean value of the velocity, namely

$$(\boldsymbol{u}_e^\star,1)_{\Omega_e}=(\boldsymbol{u}_e,1)_{\Omega_e}. \tag{35}$$

Second, a condition on the curl of the velocity (i.e. the vorticity) is introduced to resolve the $\mathtt{n_{rr}}$ rigid rotational modes

$$(\nabla \times \boldsymbol{u}_e^\star, 1)_{\Omega_e} = \langle \boldsymbol{u}_D \cdot \boldsymbol{\tau}, 1 \rangle_{\partial\Omega_e \cap \Gamma_D} + \langle \hat{\boldsymbol{u}} \cdot \boldsymbol{\tau}, 1 \rangle_{\partial\Omega_e \setminus \Gamma_D}, \tag{36}$$

where $\boldsymbol{\tau}$ is the tangential direction to the boundary $\partial\Omega_e$ and the right-hand side follows from Stokes' theorem and the Dirichlet conditions imposed in the local problem (9a).

As previously done for the local and global problems, the discrete form of the postprocessing procedure is obtained by exploiting Voigt notation to represent the involved symmetric second-order tensors, that is

$$\begin{cases} \boldsymbol{\nabla}_{\mathtt{S}}^T \mathbf{D}^{1/2} \boldsymbol{\nabla}_{\mathtt{S}} \boldsymbol{u}_e^\star = -\boldsymbol{\nabla}_{\mathtt{S}}^T \boldsymbol{L}_e & \text{in } \Omega_e, \\ \mathbf{N}_e^T \mathbf{D}^{1/2} \boldsymbol{\nabla}_{\mathtt{S}} \boldsymbol{u}_e^\star = -\mathbf{N}_e^T \boldsymbol{L}_e & \text{on } \partial\Omega_e. \end{cases} \tag{37}$$

Similarly, condition (36) is equivalent to

$$(\boldsymbol{\nabla}_{\mathtt{W}} \boldsymbol{u}_e^\star, 1)_{\Omega_e} = \langle \mathbf{T} \boldsymbol{u}_D, 1 \rangle_{\partial\Omega_e \cap \Gamma_D} + \langle \mathbf{T} \hat{\boldsymbol{u}}, 1 \rangle_{\partial\Omega_e \setminus \Gamma_D}, \tag{38}$$

where $\boldsymbol{\nabla}_{\mathtt{W}}$ and $\mathbf{T}$ are defined in (20) and (23), respectively.

Thus, following Giacomini et al. (2018), the local postprocess needed to compute a superconvergent velocity in the HDG-Voigt formulation is as follows. First, consider a space for the velocity components with one degree more than the previous one defined in (7a), namely

$$\mathcal{V}_\star^h(\Omega) := \left\{ v \in \mathcal{L}_2(\Omega) : v|_{\Omega_e} \in \mathcal{P}^{k+1}(\Omega_e) \; \forall \Omega_e \,,\; e = 1, \ldots, \mathtt{n_{el}} \right\},$$

where $\mathcal{P}^{k+1}(\Omega_e)$ denotes the space of polynomial functions of complete degree at most $k+1$ in $\Omega_e$. Then, for each element $\Omega_e$, $e = 1, \ldots, \mathtt{n_{el}}$, the discrete local postprocess is: given $\boldsymbol{L}_e$, $\boldsymbol{u}_e$ and $\hat{\boldsymbol{u}}$ solutions of (28) and (29) with optimal rate of converge $k+1$, find the velocity $\boldsymbol{u}_e^\star \in \left[\mathcal{V}_\star^h(\Omega_e)\right]^{\mathtt{n_{sd}}}$ such that

$$\begin{cases} -(\boldsymbol{\nabla}_{\mathtt{S}} \boldsymbol{v}^\star, \mathbf{D}^{1/2} \boldsymbol{\nabla}_{\mathtt{S}} \boldsymbol{u}_e^\star)_{\Omega_e} = (\boldsymbol{\nabla}_{\mathtt{S}} \boldsymbol{v}^\star, \boldsymbol{L}_e)_{\Omega_e}, \\ (\boldsymbol{u}_e^\star, 1)_{\Omega_e} = (\boldsymbol{u}_e, 1)_{\Omega_e}, \\ (\boldsymbol{\nabla}_{\mathtt{W}} \boldsymbol{u}_e^\star, 1)_{\Omega_e} = \langle \mathbf{T} \boldsymbol{u}_D, 1 \rangle_{\partial\Omega_e \cap \Gamma_D} + \langle \mathbf{T} \hat{\boldsymbol{u}}, 1 \rangle_{\partial\Omega_e \setminus \Gamma_D}, \end{cases} \tag{39}$$

for all $\boldsymbol{v}^\star \in \left[\mathcal{V}_\star^h(\Omega_e)\right]^{\mathtt{n_{sd}}}$. It is worth emphasizing that in the first equation in (39), the boundary terms appearing after integration by parts are naturally equilibrated by the boundary condition on $\partial\Omega_e$, see (37). Moreover, it is also important to note that the right-hand-sides of the last two equations in (39) converge with orders larger than $k+1$. Finally, the rate of convergence of $\boldsymbol{u}_e^\star$ in diffusion dominated areas is $k+2$.

## Numerical Examples

### *Stokes Flow*

The first numerical example involves the solution of the so-called Wang flow (see Wang 1991) in $\Omega = [0, 1]^2$. This problem has a known analytical velocity field given by

$$\boldsymbol{u}(\boldsymbol{x}) = \begin{Bmatrix} 2ax_2 - b\lambda\cos(\lambda x_1)\exp\{-\lambda x_2\} \\ b\lambda\sin(\lambda x_1)\exp\{-\lambda x_2\} \end{Bmatrix}. \tag{40}$$

The value of the parameters is selected as $a = b = 1$ and $\lambda = 10$ and the kinematic viscosity $\nu$ is taken equal to 1. The pressure is selected to be uniformly zero in the domain.

Neumann boundary conditions are imposed on the bottom part of the domain, $\Gamma_N = \{(x_1, x_2) \in \Omega \mid x_2 = 0\}$, whereas Dirichlet boundary conditions are imposed on $\Gamma_D = \partial\Omega \backslash \Gamma_N$.

A sequence of uniform triangular meshes is generated. Figure 1 shows the magnitude of the velocity computed in the first two uniform triangular meshes with a degree of approximation $k = 3$. The results clearly illustrate the gain in accuracy provided by a single level of mesh refinement.

To validate the implementation of the HDG-Voigt formulation, a mesh convergence study is performed. Figure 2 shows the relative error in the $\mathcal{L}_2(\Omega)$ norm as a function of the characteristic element size $h$ for the velocity ($\boldsymbol{u}$), postprocessed velocity ($\boldsymbol{u}^\star$), scaled strain-rate tensor ($\boldsymbol{L}$) and pressure ($p$). Optimal rate of convergence, equal to $k + 1$, is observed for the velocity, scaled strain-rate tensor and pressure and optimal rate, equal to $k + 2$, is observed for the postprocessed velocity. In all the examples, the stabilization parameter is taken as $\tau^d = \kappa\nu/\ell$, with $\kappa = 3$.

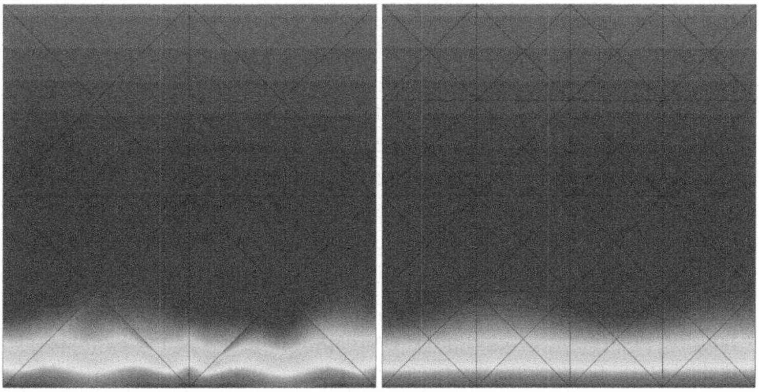

**Fig. 1** Stokes equations—Wang flow: magnitude of the velocity field computed with the HDG-Voigt method using a degree of approximation $k = 3$ on two triangular meshes

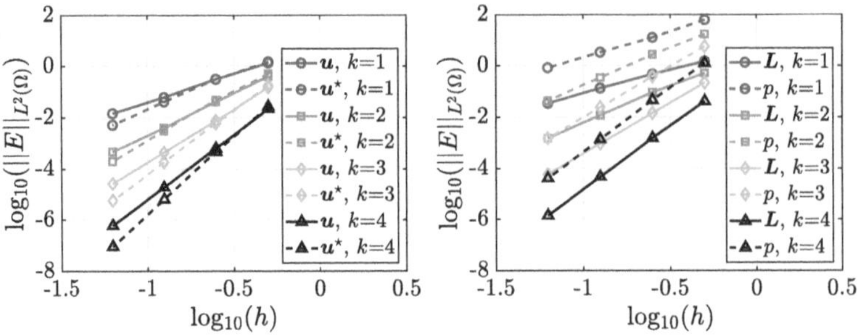


**Fig. 2** Stokes equations—Wang flow: error of the velocity ($\boldsymbol{u}$), postprocessed velocity ($\boldsymbol{u}^\star$), scaled strain-rate tensor ($\boldsymbol{L}$) and pressure ($p$) in the $\mathcal{L}_2(\Omega)$ norm as a function of the characteristic element size $h$

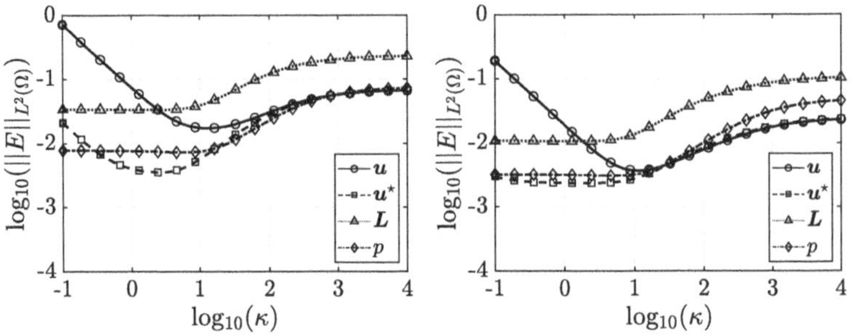


**Fig. 3** Stokes equations—Wang flow: error of the velocity ($\boldsymbol{u}$), postprocessed velocity ($\boldsymbol{u}^\star$), scaled strain-rate tensor ($\boldsymbol{L}$) and pressure ($p$) in the $\mathcal{L}_2(\Omega)$ norm as a function of the parameter $\kappa$ used to define $\tau^d$. Third mesh with $k = 1$ (left) and second mesh with $k = 2$ (right)

It is worth emphasizing that the optimal rate of convergence is observed even for linear and quadratic approximations. This is in contrast with the suboptimal convergence of the mixed variable, and a loss of superconvergence of the postprocessed velocity, using low-order approximations with the classical HDG equal-order approximation for the Cauchy formulation as shown by Cockburn et al. (2010).

The influence of the stabilization parameter $\kappa$ in the accuracy of the HDG-Voigt formulation is studied numerically. Figure 3 shows the relative error in the $\mathcal{L}_2(\Omega)$ norm, as a function of the parameter $\kappa$ used to define $\tau^d = \kappa\nu/\ell$. The error for the velocity, postprocessed velocity, scaled strain-rate tensor and pressure is considered. The results are displayed for two different levels of mesh refinement and two different degrees of approximation. In both cases, it can be observed that there is an optimal value of the stabilization parameter that provides the maximum accuracy for the velocity and the postprocessed velocity, corresponding to $\kappa \in [1, 10]$. The error for the scaled strain-rate tensor and the pressure is less sensitive to the value of $\kappa$. It can be observed that low values of the stabilization parameter substantially decrease the accuracy of the velocity field whereas large values produce less accurate results for all variables.

For more numerical examples in two and three dimensions and using different element types, the reader is referred to Giacomini et al. (2018), where the HDG-Voigt formulation with a strong imposition of the symmetry of the stress tensor via Voigt notation was originally introduced.

## *Oseen Flow*

The numerical study of the Oseen problem involves the solution of the so-called Kovasznay flow (see Kovasznay 1948) in $\Omega = [0, 1]^2$. This problem has a known analytical solution given by

$$\boldsymbol{u}(\boldsymbol{x}) = \begin{Bmatrix} 1 - \exp(2\lambda x_1)\cos\big((4x_2 - 1)\pi\big) \\ (\lambda/2\pi)\exp(2\lambda x_1)\sin\big((4x_2 - 1)\pi\big) \end{Bmatrix}, \quad p(\boldsymbol{x}) = -\frac{1}{2}\exp(4\lambda x_1) + C, \tag{41}$$

where $\lambda := Re/2 - \sqrt{Re^2/4 + 4\pi^2}$, $C = \big[1 + \exp(4\lambda) - (1/2\lambda)\big(1 - \exp(4\lambda)\big)\big]/8$. The Reynolds number is taken as $Re = 100$ and the convection velocity field $\boldsymbol{a}$ is taken as the exact velocity.

As done in the previous example, Neumann boundary conditions are imposed on the bottom part of the domain, $\Gamma_N = \{(x_1, x_2) \in \Omega \mid x_2 = 0\}$, whereas Dirichlet boundary conditions are imposed on $\Gamma_D = \partial\Omega \backslash \Gamma_N$.

Figure 4 shows the magnitude of the velocity computed in the first triangular mesh with a degree of approximation $k = 4$ and the postprocessed velocity computed by using the strategy described in Section "Local Postprocess of the Primal Variable". The results clearly illustrate the increased accuracy provided by the element-by-element postprocess of the velocity field.

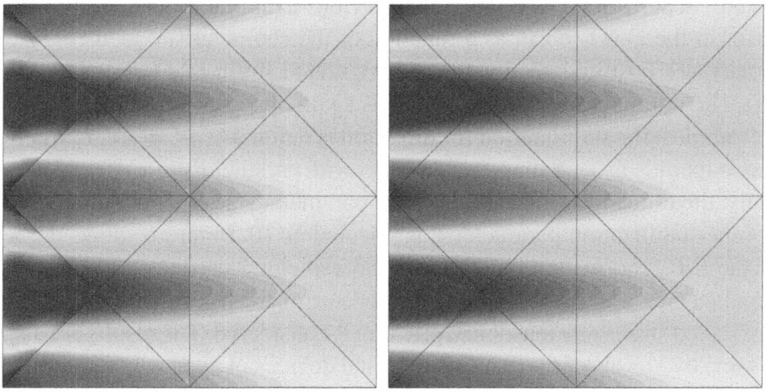

**Fig. 4** Oseen equations—Kovasznay flow: magnitude of the velocity field computed with the HDG-Voigt method using a degree of approximation $k = 4$ (left) and postprocessed velocity (right)

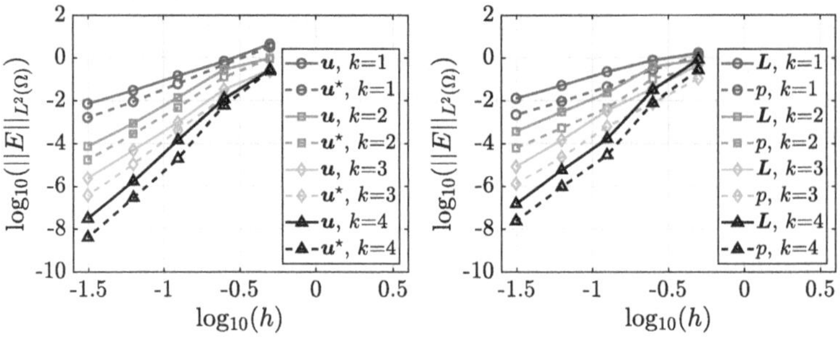


**Fig. 5** Oseen equations—Kovasznay flow: error of the velocity ($\boldsymbol{u}$), postprocessed velocity ($\boldsymbol{u}^\star$), scaled strain-rate tensor ($\boldsymbol{L}$) and pressure ($p$) in the $\mathcal{L}_2(\Omega)$ norm as a function of the characteristic element size $h$

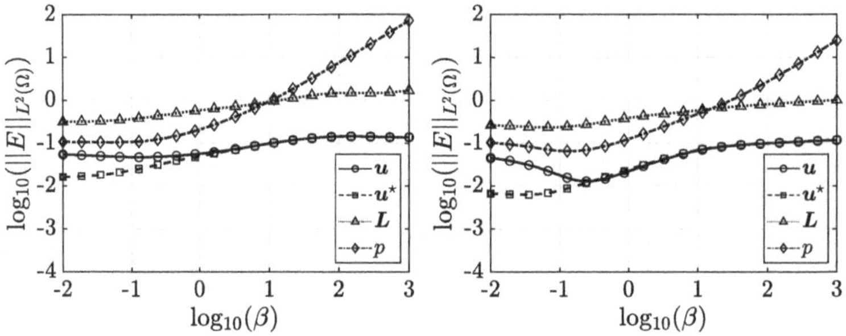


**Fig. 6** Oseen equations—Kovasznay flow: error of the velocity ($\boldsymbol{u}$), postprocessed velocity ($\boldsymbol{u}^\star$), scaled strain-rate tensor ($\boldsymbol{L}$) and pressure ($p$) in the $\mathcal{L}_2(\Omega)$ norm as a function of the parameter $\beta$ used to define $\tau^a = \beta\|\boldsymbol{a}\|_\infty$. Third mesh with $k = 1$ (left) and second mesh with $k = 2$ (right)

To validate the implementation of the HDG-Voigt formulation, a mesh convergence study is performed as done in the previous example. Figure 5 shows the relative error in the $\mathcal{L}_2(\Omega)$ norm as a function of the characteristic element size $h$ for the velocity ($\boldsymbol{u}$), postprocessed velocity ($\boldsymbol{u}^\star$), scaled strain-rate tensor ($\boldsymbol{L}$) and pressure ($p$). The optimal rate of convergence is again observed for all the variables. In all the examples, the stabilization for diffusion is defined as $\tau^d = \kappa\nu/\ell$, whereas for convection it holds $\tau^a = \beta \max_{\boldsymbol{x}\in\mathcal{N}_\Omega} \|\boldsymbol{a}(\boldsymbol{x})\|_2$ with parameters $\kappa = 10$ and $\beta = 0.02$.

Finally, the influence of the stabilization parameter $\beta$ is studied numerically. The value of the stabilization parameter $\kappa$ is selected as 10. Figure 6 shows the relative error in the $\mathcal{L}_2(\Omega)$ norm, as a function of the parameter $\beta$ used to define the stabilization parameter $\tau^a = \beta \max_{\boldsymbol{x}\in\mathcal{N}_\Omega} \|\boldsymbol{a}(\boldsymbol{x})\|_2$. The error for the velocity, postprocessed velocity, scaled strain-rate tensor and pressure is considered. The results are displayed for the same two levels of mesh refinement and degrees of approximation utilized for the sensitivity analysis in the Stokes flow problem. It can be observed that there is an optimal value of the stabilization parameter that provides the maximum accuracy for the velocity and the postprocessed velocity, corresponding to $\beta \in [0.01, 0.1]$. The pressure is the variable that shows a higher dependence in terms of the selected value of the stabilization parameter $\beta$. In particular, large values of $\beta$ lead to sizeable errors in the pressure.

## *Navier-Stokes Flow*

The solution of the nonlinear incompressible Navier-Stokes equations is considered. The HDG-Voigt formulation follows the rationale presented in Section "HDG Method for Oseen Flows" with $\boldsymbol{a}$ being $\boldsymbol{u}$ and $\hat{\boldsymbol{a}}$ being $\hat{\boldsymbol{u}}$. The resulting nonlinear local and global problems are solved by using a standard Newton-Raphson linearization. It is worth noticing that by substituting $\boldsymbol{u}$ for $\boldsymbol{a}$ and $\hat{\boldsymbol{u}}$ for $\hat{\boldsymbol{a}}$ in (13), the resulting stabilization parameter $\tau^a$ is now a function of the unknown primal, $\tau^a = \tau^a(\boldsymbol{u})$ or hybrid, $\tau^a = \tau^a(\hat{\boldsymbol{u}})$, variable. In order to avoid introducing an additional nonlinearity in the problem, the stabilization parameter $\tau^a$ is thus evaluated in the previous iteration of the Newton-Raphson algorithm. For the following numerical experiments, at step $r+1$ of the Newton-Raphson method, the definition of $\tau^a = \beta \max_{\boldsymbol{x}\in\mathcal{N}_\Omega} \|\boldsymbol{u}^r(\boldsymbol{x})\|_2$ is considered, $\boldsymbol{u}^r$ being the velocity field computed at iteration $r$ of the nonlinear procedure.

### Navier-Stokes Kovasznay Flow

First, the Kovasznay flow described in Section "Oseen Flow" is considered with the same boundary conditions and using the same set of meshes used to solve the Ossen equations. Figure 7 shows the magnitude of the velocity computed in the first triangular mesh with a degree of approximation $k = 4$ and the postprocessed velocity computed by using the strategy described in Section "Local Postprocess of the Primal Variable". A visual comparison of the results between Navier-Stokes and Ossen solutions, computed using the same coarse mesh, illustrates the extra level of difficulty induced by the nonlinearity of the Navier-Stokes equations.

To validate the implementation of the HDG-Voigt formulation, a mesh convergence study is performed as done in the previous examples. Figure 8 shows the relative error in the $\mathcal{L}_2(\Omega)$ norm as a function of the characteristic element size $h$ for the velocity ($\boldsymbol{u}$), postprocessed velocity ($\boldsymbol{u}^\star$), scaled strain-rate tensor ($\boldsymbol{L}$) and pressure ($p$). The optimal rate of convergence is again observed for all the variables, with a slightly higher rate for the primal variables when $k = 4$. In all the examples, the stabilization parameters are taken as $\tau^d = \kappa\nu/\ell$ and $\tau^a = \beta \max_{\boldsymbol{x}\in\mathcal{N}_\Omega} \|\boldsymbol{u}^r(\boldsymbol{x})\|_2$, with $\kappa = 10$ and $\beta = 0.1$, $\boldsymbol{u}^r$ being the velocity field computed in the previous Newton-Raphson iteration.

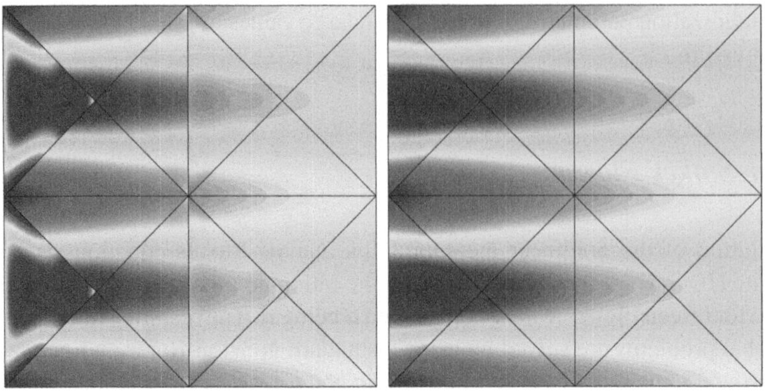

**Fig. 7** Navier-Stokes equations—Kovasznay flow: magnitude of the velocity field computed with the HDG-Voigt method using a degree of approximation $k = 4$ (left) and postprocessed velocity (right)

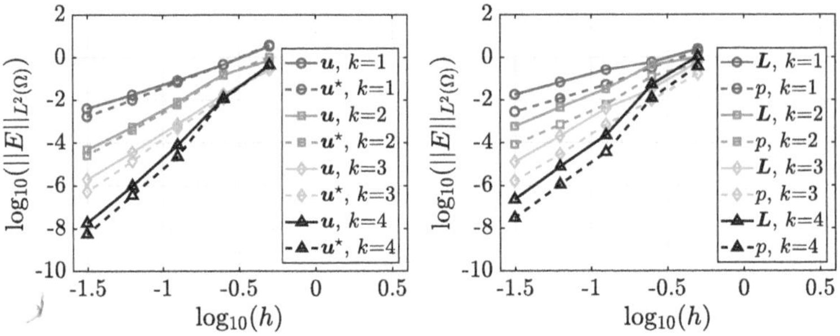


**Fig. 8** Navier-Stokes equations—Kovasznay flow: error of the velocity ($\boldsymbol{u}$), postprocessed velocity ($\boldsymbol{u}^\star$), scaled strain-rate tensor ($\boldsymbol{L}$) and pressure ($p$) in the $\mathcal{L}_2(\Omega)$ norm as a function of the characteristic element size $h$

## Navier-Stokes Poiseuille Flow

The following example considers another well-known problem with analytical solution to illustrate the effect of the boundary condition imposed on an outflow part of the boundary, as discussed in Section "Incompressible Flows: Problem Statement". The Poiseuille flow in a rectangular channel $\Omega = [0, 10] \times [0, 1]$ is considered, with Dirichlet boundary conditions on the left, top and bottom part of the boundary and different boundary conditions, described in Remark 2.3, on the right-end of the boundary. The exact solution is given by

$$\boldsymbol{u}(\boldsymbol{x}) = \begin{Bmatrix} 4Vx_2(1-x_2) \\ 0 \end{Bmatrix}, \quad p(\boldsymbol{x}) = -8\nu V x_1 + C, \tag{42}$$

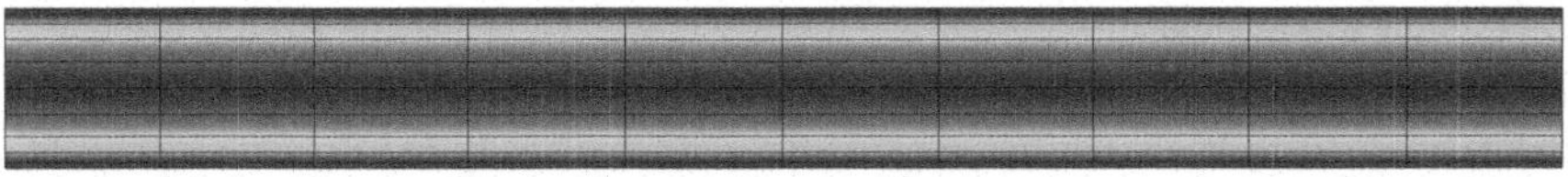

**Fig. 9** Navier-Stokes equations—Poiseuille flow: magnitude of the velocity field computed with the HDG-Voigt method using a degree of approximation $k = 2$ with the outflow boundary condition in (2c)

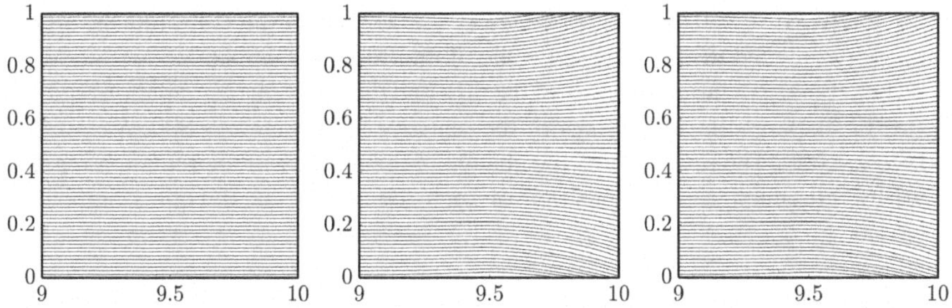


**Fig. 10** Navier-Stokes equations—Poiseuille flow: magnitude of the velocity field and streamlines $k = 2$ with the outflow boundary condition (left) homogeneous Neumann boundary condition in (2b) (middle) and traction-free boundary condition in (2a) (right)

where $V$ is the maximum value of the velocity profile, achieved at the centerline of the channel, that is for $x_2 = 1/2$, and $C = 80\nu V$.

The solution computed with a structured quadrilateral mesh with $10 \times 10$ biquadratic elements is shown in Figure 9, by imposing the outflow boundary condition given in (2c). As expected, the computed solution reproduces the exact solution (with machine accuracy) as the latter belongs to the polynomial space used to define the functional approximation. The error of the velocity measured in the $\mathcal{L}^\infty(\Omega)$ norm is $3.6 \times 10^{-13}$. In contrast, when the homogeneous Neumann boundary condition given in (2b) is utilized, the error of the velocity measured in the $\mathcal{L}^\infty(\Omega)$ norm is 1.11. Finally, when the traction-free boundary condition in (2a) is imposed on the right part of the boundary, the error of the velocity measured in the $\mathcal{L}^\infty(\Omega)$ norm is 1.29.

A detailed view of the solution near the outflow boundary for the three different boundary conditions considered is shown in Figure 10. The figure displays the streamlines. It can be clearly observed that, using the outflow boundary condition, the streamlines are parallel to the $x$-axis and the motion of a fluid between two parallel infinite plates is correctly reproduced. On the contrary, with the homogeneous Neumann boundary condition and the traction-free boundary condition the isolines present non-physical artifacts.

## Backward Facing Step

The last example considers another well-known test case for the incompressible Navier-Stokes equations, the so-called backward facing step. This problem is traditionally employed to test the ability of a numerical scheme to capture the recirculation zones and position of the reattachment point (see Armaly et al. 1983; Erturk 2008).

Figure 11 shows the magnitude of the velocity for three different Reynolds numbers, namely $Re = 800$, $Re = 1{,}900$ and $Re = 3{,}000$. The results illustrate the higher complexity induced by an increased value of the Reynolds number.

To better observe the complexity of the flow and the different recirculation regions, Figure 12 displays the streamlines of the velocity field for $Re = 800$, $Re = 1{,}900$ and $Re = 3{,}000$. The results show the change in the number of recirculation regions as well as the change in the position of such regions as the Reynolds number is increased.

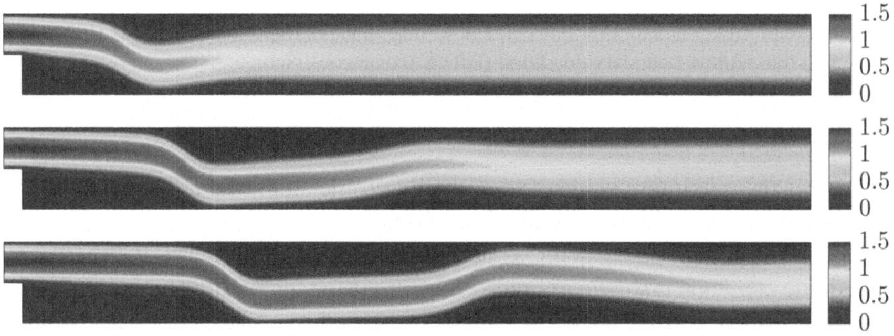


**Fig. 11** Navier-Stokes equations—Backward facing step: magnitude of the velocity field for $Re = 800$ (top), $Re = 1{,}900$ (middle) and $Re = 3{,}000$ (bottom)

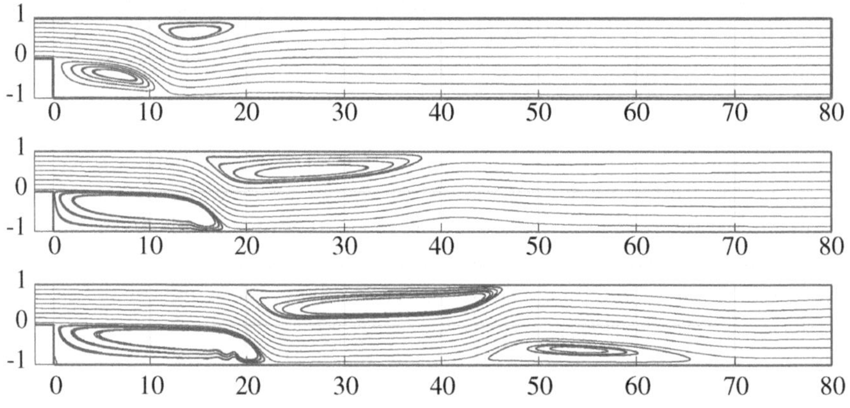


**Fig. 12** Navier-Stokes equations—Backward facing step: streamlines for $Re = 800$ (top), $Re = 1{,}900$ (middle) and $Re = 3{,}000$ (bottom)

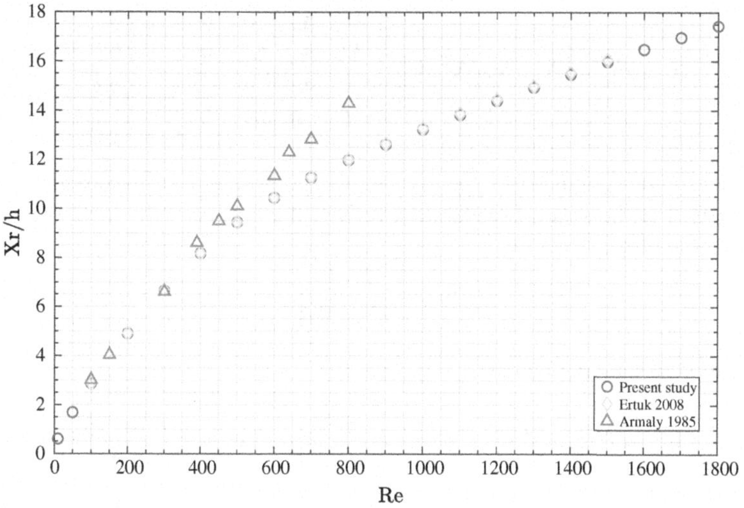


**Fig. 13** Navier-Stokes equations—Backward facing step: Position of the reattachment point as a function of the Reynolds number and comparison with published results

Finally, Figure 13 displays the position of the reattachment point as a function of the Reynolds number. The results obtained using the presented HDG-Voigt formulation are compared with the results in Armaly et al. (1983) and Erturk (2008), showing an excellent agreement with the more recent results of Erturk (2008).

**Acknowledgements** This work is partially supported by the European Union's Horizon 2020 research and innovation programme under the Marie Skłodowska-Curie actions (Grant No. 675919 and 764636) and the Spanish Ministry of Economy and Competitiveness (Grant No. DPI2017-85139-C2-2-R). The first and third author also gratefully acknowledge the financial support provided by Generalitat de Catalunya (Grant No. 2017-SGR-1278).

## Appendix: Saddle-Point Structure of the Global Problem

In this Appendix, the symmetry of the global system in (32) is demonstrated. First, rewrite (32) as

$$\begin{bmatrix} \widehat{\mathbf{K}} & \mathbf{H} \\ \mathbf{G}^T & \mathbf{0} \end{bmatrix} \begin{Bmatrix} \hat{\mathbf{u}} \\ \rho \end{Bmatrix} = \begin{Bmatrix} \hat{\mathbf{f}}_{\hat{u}} \\ \hat{\mathbf{f}}_{\rho} \end{Bmatrix}, \tag{43}$$

where the block $\mathbf{H}$ is obtained by the solution of the local problem in (30) and has the following form

$$\mathbf{H} = \mathsf{A}_{e=1}^{\mathtt{n_{el}}} \left[\mathbf{A}_{L\hat{u}}^T \, \mathbf{A}_{\hat{u}u} \, \mathbf{A}_{p\hat{u}}^T \, \mathbf{0}\right]_e \begin{bmatrix} \mathbf{A}_{LL} & \mathbf{A}_{Lu} & \mathbf{0} & \mathbf{0} \\ \mathbf{A}_{Lu}^T & \mathbf{A}_{uu} & \mathbf{A}_{pu}^T & \mathbf{0} \\ \mathbf{0} & \mathbf{A}_{pu} & \mathbf{0} & \mathbf{a}_{\rho p}^T \\ \mathbf{0} & \mathbf{0} & \mathbf{a}_{\rho p} & 0 \end{bmatrix}_e^{-1} \begin{Bmatrix} \mathbf{0} \\ \mathbf{0} \\ \mathbf{0} \\ 1 \end{Bmatrix}_e . \tag{44}$$

In order for the system in (43) to have a saddle-point structure, it needs to be proved that $\mathbf{H} = \mathbf{G}$. For the sake of readability, rewrite the matrix of the local problem in (30) using the block structure

$$\mathbf{K}_e := \begin{bmatrix} \mathbf{B}_e & \mathbf{C}_e \\ \mathbf{C}_e^T & \mathbf{D}_e \end{bmatrix} \tag{45}$$

where the blocks are defined as

$$\mathbf{B}_e := \begin{bmatrix} \mathbf{A}_{LL} & \mathbf{A}_{Lu} \\ \mathbf{A}_{Lu}^T & \mathbf{A}_{uu} \end{bmatrix}, \quad \mathbf{C}_e := \begin{bmatrix} \mathbf{0} & \mathbf{0} \\ \mathbf{A}_{pu}^T & \mathbf{0} \end{bmatrix}, \quad \mathbf{D}_e := \begin{bmatrix} \mathbf{0} & \mathbf{a}_{\rho p}^T \\ \mathbf{a}_{\rho p} & \mathbf{0} \end{bmatrix}.$$

**Proposition 5.1** *For each element $\Omega_e$, it holds*

$$\mathbf{K}_e^{-1} \begin{Bmatrix} \mathbf{0} \\ \mathbf{0} \\ \mathbf{0} \\ 1 \end{Bmatrix}_e = \begin{bmatrix} -\left[\mathbf{B}_e^{-1}\right]_{12} \mathbf{A}_{pu}^T \left[\mathbf{S}_e^{-1}\right]_{12} \\ -\left[\mathbf{B}_e^{-1}\right]_{22} \mathbf{A}_{pu}^T \left[\mathbf{S}_e^{-1}\right]_{12} \\ \left[\mathbf{S}_e^{-1}\right]_{12} \\ 0 \end{bmatrix}_e , \tag{46}$$

*where*

$$\begin{aligned} \left[\mathbf{B}_e^{-1}\right]_{12} &= -\mathbf{A}_{LL}^{-1}\mathbf{A}_{Lu}(\mathbf{A}_{uu} - \mathbf{A}_{Lu}^T\mathbf{A}_{LL}^{-1}\mathbf{A}_{Lu})^{-1}, \\ \left[\mathbf{B}_e^{-1}\right]_{22} &= (\mathbf{A}_{uu} - \mathbf{A}_{Lu}^T\mathbf{A}_{LL}^{-1}\mathbf{A}_{Lu})^{-1}, \\ \left[\mathbf{S}_e^{-1}\right]_{12} &= \Big(\mathbf{a}_{\rho p}\big(\mathbf{I} - (\mathbf{A}_{pu}(\mathbf{A}_{uu} - \mathbf{A}_{Lu}^T\mathbf{A}_{LL}^{-1}\mathbf{A}_{Lu})^{-1}\mathbf{A}_{pu}^T)^{\dagger} \\ &\qquad (\mathbf{A}_{pu}(\mathbf{A}_{uu} - \mathbf{A}_{Lu}^T\mathbf{A}_{LL}^{-1}\mathbf{A}_{Lu})^{-1}\mathbf{A}_{pu}^T)\big)\Big)^{\dagger}. \end{aligned}$$

***Proof*** The inverse of the block matrix in (45), written using Schur-Banachiewicz form (see Bernstein 2009, Sect. 2.17), is

$$\mathbf{K}_e^{-1} := \begin{bmatrix} \mathbf{B}_e^{-1}(\mathbf{I} + \mathbf{C}_e(\mathbf{D}_e - \mathbf{C}_e^T\mathbf{B}_e^{-1}\mathbf{C}_e)^{-1}\mathbf{C}_e^T\mathbf{B}_e^{-1}) & -\mathbf{B}_e^{-1}\mathbf{C}_e(\mathbf{D}_e - \mathbf{C}_e^T\mathbf{B}_e^{-1}\mathbf{C}_e)^{-1} \\ -(\mathbf{D}_e - \mathbf{C}_e^T\mathbf{B}_e^{-1}\mathbf{C}_e)^{-1}\mathbf{C}_e^T\mathbf{B}_e^{-1} & (\mathbf{D}_e - \mathbf{C}_e^T\mathbf{B}_e^{-1}\mathbf{C}_e)^{-1} \end{bmatrix},$$

where the block $(2, 2)$ is the inverse of the Schur complement

$$\mathbf{S}_e := \mathbf{D}_e - \mathbf{C}_e^T\mathbf{B}_e^{-1}\mathbf{C}_e = \begin{bmatrix} -\mathbf{A}_{pu}\left[\mathbf{B}_e^{-1}\right]_{22}\mathbf{A}_{pu}^T & \mathbf{a}_{\rho p}^T \\ \mathbf{a}_{\rho p} & 0 \end{bmatrix} \tag{47}$$

of block $\mathbf{B}_e$ of the matrix $\mathbf{K}_e$. Moreover, the block $(1, 2)$ of the inverse matrix $\mathbf{K}_e^{-1}$ has the form

$$
\begin{aligned}
\left[\mathbf{K}_e^{-1}\right]_{12} &= -\mathbf{B}_e^{-1}\mathbf{C}_e\mathbf{S}_e^{-1} \\
&= -\begin{bmatrix} \left[\mathbf{B}_e^{-1}\right]_{12}\mathbf{A}_{pu}^T\left[\mathbf{S}_e^{-1}\right]_{11} & \left[\mathbf{B}_e^{-1}\right]_{12}\mathbf{A}_{pu}^T\left[\mathbf{S}_e^{-1}\right]_{12} \\ \left[\mathbf{B}_e^{-1}\right]_{22}\mathbf{A}_{pu}^T\left[\mathbf{S}_e^{-1}\right]_{11} & \left[\mathbf{B}_e^{-1}\right]_{22}\mathbf{A}_{pu}^T\left[\mathbf{S}_e^{-1}\right]_{12} \end{bmatrix}.
\end{aligned} \tag{48}
$$

To fully determine the inverse matrix $\mathbf{K}_e^{-1}$, the blocks of $\mathbf{B}_e^{-1}$ and $\mathbf{S}_e^{-1}$ need to be computed. Following the Schur-Banachiewicz rationale utilized above, the blocks of $\mathbf{B}_e^{-1}$ have the form

$$
\begin{aligned}
\left[\mathbf{B}_e^{-1}\right]_{11} &:= \mathbf{A}_{LL}^{-1}(\mathbf{I} + \mathbf{A}_{Lu}(\mathbf{A}_{uu} - \mathbf{A}_{Lu}^T\mathbf{A}_{LL}^{-1}\mathbf{A}_{Lu})^{-1}\mathbf{A}_{Lu}^T\mathbf{A}_{LL}^{-1}), \\
\left[\mathbf{B}_e^{-1}\right]_{12} &:= -\mathbf{A}_{LL}^{-1}\mathbf{A}_{Lu}(\mathbf{A}_{uu} - \mathbf{A}_{Lu}^T\mathbf{A}_{LL}^{-1}\mathbf{A}_{Lu})^{-1}, \\
\left[\mathbf{B}_e^{-1}\right]_{22} &:= (\mathbf{A}_{uu} - \mathbf{A}_{Lu}^T\mathbf{A}_{LL}^{-1}\mathbf{A}_{Lu})^{-1},
\end{aligned} \tag{49}
$$

and, from the symmetry of $\mathbf{B}_e$, it follows that $\left[\mathbf{B}_e^{-1}\right]_{21} = \left[\mathbf{B}_e^{-1}\right]_{12}^T$.

Plugging the expression of $\left[\mathbf{B}_e^{-1}\right]_{22}$, see (49), into the definition of $\mathbf{S}_e$ in (47), it follows that the block (1, 1) of such matrix is

$$
\left[\mathbf{S}_e\right]_{11} := -\mathbf{A}_{pu}(\mathbf{A}_{uu} - \mathbf{A}_{Lu}^T\mathbf{A}_{LL}^{-1}\mathbf{A}_{Lu})^{-1}\mathbf{A}_{pu}^T. \tag{50}
$$

It is straightforward to observe that this matrix is the Schur complement of block

$$
\begin{bmatrix} \mathbf{A}_{LL} & \mathbf{A}_{Lu} \\ \mathbf{A}_{Lu}^T & \mathbf{A}_{uu} \end{bmatrix}
$$

of the matrix

$$
\begin{bmatrix} \mathbf{A}_{LL} & \mathbf{A}_{Lu} & \mathbf{0} \\ \mathbf{A}_{Lu}^T & \mathbf{A}_{uu} & \mathbf{A}_{pu}^T \\ \mathbf{0} & \mathbf{A}_{pu} & \mathbf{0} \end{bmatrix},
$$

which is singular, since it is obtained from the discretization of an incompressible flow problem with purely Dirichlet boundary conditions. Hence, to compute the blocks of $\mathbf{S}_e^{-1}$, the framework of the generalized inverse of a partitioned matrix is exploited (see Miao 1991) leading to

$$
\begin{aligned}
\left[\mathbf{S}_e^{-1}\right]_{11} &:= \Big(\mathbf{I} - (\mathbf{a}_{\rho p}(\mathbf{I} - \left[\mathbf{S}_e\right]_{11}^\dagger\left[\mathbf{S}_e\right]_{11}))^\dagger\mathbf{a}_{\rho p}\Big) \\
&\qquad\qquad \left[\mathbf{S}_e\right]_{11}^\dagger\Big(\mathbf{I} - \mathbf{a}_{\rho p}^T((\mathbf{I} - \left[\mathbf{S}_e\right]_{11}\left[\mathbf{S}_e\right]_{11}^\dagger)\mathbf{a}_{\rho p}^T)^\dagger\Big), \\
\left[\mathbf{S}_e^{-1}\right]_{12} &:= (\mathbf{a}_{\rho p}(\mathbf{I} - \left[\mathbf{S}_e\right]_{11}^\dagger\left[\mathbf{S}_e\right]_{11}))^\dagger, \\
\left[\mathbf{S}_e^{-1}\right]_{21} &:= ((\mathbf{I} - \left[\mathbf{S}_e\right]_{11}\left[\mathbf{S}_e\right]_{11}^\dagger)\mathbf{a}_{\rho p}^T)^\dagger, \\
\left[\mathbf{S}_e^{-1}\right]_{22} &:= 0
\end{aligned} \tag{51}
$$

where the Moore-Penrose pseudoinverse $\left[\mathbf{S}_e\right]_{11}^{\dagger}$ of the singular matrix $\left[\mathbf{S}_e\right]_{11}$ has the form

$$\left[\mathbf{S}_e\right]_{11}^{\dagger} := -\left(\mathbf{A}_{pu}(\mathbf{A}_{uu} - \mathbf{A}_{Lu}^T \mathbf{A}_{LL}^{-1} \mathbf{A}_{Lu})^{-1} \mathbf{A}_{pu}^T\right)^{\dagger}. \tag{52}$$

From (44), it is straightforward to observe that only the blocks in the last column of the inverse matrix are involved in the definition of the product in (46). The result (46) follows directly from (48), (51) and (52). □

**Proposition 5.2** *Given* $\mathbf{H}$ *and* $\mathbf{G}^T$ *from* (44) *and (33b) respectively, it holds that* $\mathbf{H} = \mathbf{G}$.

***Proof*** From (44) and (46), it follows that

$$\mathbf{H}_e = -(\mathbf{A}_{L\hat{u}}^T \left[\mathbf{B}_e^{-1}\right]_{12} + \mathbf{A}_{\hat{u}u} \left[\mathbf{B}_e^{-1}\right]_{22}) \mathbf{A}_{pu}^T \left[\mathbf{S}_e^{-1}\right]_{12} + \mathbf{A}_{p\hat{u}}^T \left[\mathbf{S}_e^{-1}\right]_{12}, \tag{53}$$

for each element $\Omega_e$.

First, recall that the matrix $\mathbf{I} - \left[\mathbf{S}_e\right]_{11}^{\dagger} \left[\mathbf{S}_e\right]_{11}$ defines an orthogonal projector onto the kernel of $\left[\mathbf{S}_e\right]_{11}$ (see Bernstein 2009, Sect. 6.1). As observed in the previous proposition, see (50), $\left[\mathbf{S}_e\right]_{11}$ is the Schur complement of the velocity block of the matrix obtained from the discretization of an incompressible flow problem with purely Dirichlet boundary conditions. Thus, the kernel of $\left[\mathbf{S}_e\right]_{11}$ contains all constant vectors representing the mean value of pressure. It follows that

$$\mathbf{a}_{pp}\left(\mathbf{I} - \left[\mathbf{S}_e\right]_{11}^{\dagger} \left[\mathbf{S}_e\right]_{11}\right) = \frac{1}{\mathtt{n_{en}}} \mathbf{1}^T$$

is the constant vector obtained as the average of 1 over the $\mathtt{n_{en}}$ element nodes of $\Omega_e$ and, consequently, $\left[\mathbf{S}_e^{-1}\right]_{12} = \mathbf{1}$. Moreover, since the kernel of the matrix $\mathbf{A}_{pu}^T$ also includes all constant vectors, $\mathbf{A}_{pu}^T \left[\mathbf{S}_e^{-1}\right]_{12} = \mathbf{0}$. Hence, from (53), it follows that

$$\mathbf{H} = \left[[\mathbf{H}_1] \; [\mathbf{H}_2] \ldots [\mathbf{H}_{\mathtt{n_{el}}}]\right] = \left[[\mathbf{A}_{p\hat{u}}^T]_1 \mathbf{1} \; [\mathbf{A}_{p\hat{u}}^T]_2 \mathbf{1} \ldots [\mathbf{A}_{p\hat{u}}^T]_{\mathtt{n_{el}}} \mathbf{1}\right]$$

which proves the statement. □

When convection phenomena are neglected (Stokes flow), $\mathbf{A}_{\hat{u}u} = \mathbf{A}_{u\hat{u}}^T$ and the symmetry of $\widehat{\mathbf{K}}$ and the global matrix in (32) follows straightforwardly. For general incompressible flow problems, the matrix $\widehat{\mathbf{K}}$ is not symmetric but the off-diagonal blocks $\mathbf{G}$ and $\mathbf{G}^T$ are one the transpose of the other and the resulting global matrix maintains the above displayed saddle-point structure (Benzi et al. 2005).

## Appendix: Implementation Details

In this Appendix, the matrices and vectors appearing in the discrete form of the HDG-Voigt approximation of the Oseen equations are detailed. The elemental variables

$\boldsymbol{u}$, $p$ and $\boldsymbol{L}$ are defined in a reference element $\tilde{\Omega}(\boldsymbol{\xi})$, $\boldsymbol{\xi} = (\xi_1, \ldots, \xi_{\mathtt{n_{sd}}})$ whereas the face variable $\hat{\boldsymbol{u}}$, is defined on a reference face $\tilde{\Gamma}(\boldsymbol{\eta})$, $\boldsymbol{\eta} = (\eta_1, \ldots, \eta_{\mathtt{n_{sd}}-1})$ as

$$\boldsymbol{u}(\boldsymbol{\xi}) \simeq \sum_{j=1}^{\mathtt{n_{en}}} \mathbf{u}_j N_j(\boldsymbol{\xi}), \qquad p(\boldsymbol{\xi}) \simeq \sum_{j=1}^{\mathtt{n_{en}}} \mathrm{p}_j N_j(\boldsymbol{\xi}),$$

$$\boldsymbol{L}(\boldsymbol{\xi}) \simeq \sum_{j=1}^{\mathtt{n_{en}}} \mathbf{L}_j N_j(\boldsymbol{\xi}), \qquad \hat{\boldsymbol{u}}(\boldsymbol{\eta}) \simeq \sum_{j=1}^{\mathtt{n_{fn}}} \hat{\mathbf{u}}_j \hat{N}_j(\boldsymbol{\eta}),$$

where $\mathbf{u}_j$, $\mathrm{p}_j$, $\mathbf{L}_j$ and $\hat{\mathbf{u}}_j$ are the nodal values of the approximation, $\mathtt{n_{en}}$ and $\mathtt{n_{fn}}$ the number of nodes in the element and face, respectively and $N_j$ and $\hat{N}_j$ the polynomial shape functions in the reference element and face, respectively.

An isoparametric formulation is considered and the following transformation is used to map reference and local coordinates

$$\boldsymbol{x}(\boldsymbol{\xi}) = \sum_{k=1}^{\mathtt{n_{en}}} \boldsymbol{x}_k N_k(\boldsymbol{\xi}),$$

where the vector $\{\boldsymbol{x}_k\}_{k=1,\ldots,\mathtt{n_{en}}}$ denotes the elemental nodal coordinates.

Following Sevilla et al. (2018), the matrices $\boldsymbol{\nabla}_{\mathrm{S}}$ and $\mathbf{N}$ in (18) and (22), respectively, are expressed in compact form as

$$\boldsymbol{\nabla}_{\mathrm{S}} = \sum_{k=1}^{\mathtt{n_{sd}}} \mathbf{F}_k \frac{\partial}{\partial x_k}, \qquad \mathbf{N} = \sum_{k=1}^{\mathtt{n_{sd}}} \mathbf{F}_k n_k,$$

where the matrices $\mathbf{F}_k$ are defined as

$$\mathbf{F}_1 = \begin{bmatrix} 1\,0\,0 \\ 0\,0\,1 \end{bmatrix}^T, \quad \mathbf{F}_2 = \begin{bmatrix} 0\,0\,1 \\ 0\,1\,0 \end{bmatrix}^T$$

in two dimensions and

$$\mathbf{F}_1 = \begin{bmatrix} 1\,0\,0\,0\,0\,0 \\ 0\,0\,0\,1\,0\,0 \\ 0\,0\,0\,0\,1\,0 \end{bmatrix}^T, \quad \mathbf{F}_2 = \begin{bmatrix} 0\,0\,0\,1\,0\,0 \\ 0\,1\,0\,0\,0\,0 \\ 0\,0\,0\,0\,0\,1 \end{bmatrix}^T, \quad \mathbf{F}_3 = \begin{bmatrix} 0\,0\,0\,0\,1\,0 \\ 0\,0\,0\,0\,0\,1 \\ 0\,0\,1\,0\,0\,0 \end{bmatrix}^T$$

in three dimensions. Moreover, from the definition of $\mathbf{E}$ in (21), it holds $\mathbf{N}^T\mathbf{E} = \boldsymbol{n}$ and $\boldsymbol{\nabla}_{\mathrm{S}}^T\mathbf{E} = \boldsymbol{\nabla}$ for the gradient operator applied to a scalar function. The following compact forms of the shape functions and their derivatives are introduced

$$
\begin{aligned}
\boldsymbol{\mathcal{N}} &:= \left[N_1 \mathbf{I}_{\mathtt{n_{sd}}} \; N_2 \mathbf{I}_{\mathtt{n_{sd}}} \dots N_{\mathtt{n_{en}}} \mathbf{I}_{\mathtt{n_{sd}}}\right]^T, \\
\widehat{\boldsymbol{\mathcal{N}}} &:= \left[\hat{N}_1 \mathbf{I}_{\mathtt{n_{sd}}} \; \hat{N}_2 \mathbf{I}_{\mathtt{n_{sd}}} \dots \hat{N}_{\mathtt{n_{fn}}} \mathbf{I}_{\mathtt{n_{sd}}}\right]^T, \\
\boldsymbol{\mathcal{N}}^\tau &:= \left[N_1 \tau \mathbf{I}_{\mathtt{n_{sd}}} \; N_2 \tau \mathbf{I}_{\mathtt{n_{sd}}} \dots N_{\mathtt{n_{en}}} \tau \mathbf{I}_{\mathtt{n_{sd}}}\right]^T, \\
\widehat{\boldsymbol{\mathcal{N}}}^\tau &:= \left[\hat{N}_1 \tau \mathbf{I}_{\mathtt{n_{sd}}} \; \hat{N}_2 \tau \mathbf{I}_{\mathtt{n_{sd}}} \dots \hat{N}_{\mathtt{n_{fn}}} \tau \mathbf{I}_{\mathtt{n_{sd}}}\right]^T, \\
\boldsymbol{\mathcal{N}}^n &:= \left[N_1 \boldsymbol{n} \; N_2 \boldsymbol{n} \dots N_{\mathtt{n_{en}}} \boldsymbol{n}\right]^T, \\
\boldsymbol{\mathcal{M}} &:= \left[N_1 \mathbf{I}_{\mathtt{m_{sd}}} \; N_2 \mathbf{I}_{\mathtt{m_{sd}}} \dots N_{\mathtt{n_{en}}} \mathbf{I}_{\mathtt{m_{sd}}}\right]^T, \\
\boldsymbol{\mathcal{N}}^a &:= \left[N_1 (\tau - \hat{\boldsymbol{a}} \cdot \boldsymbol{n}) \mathbf{I}_{\mathtt{n_{sd}}} \; N_2 (\tau - \hat{\boldsymbol{a}} \cdot \boldsymbol{n}) \mathbf{I}_{\mathtt{n_{sd}}} \dots N_{\mathtt{n_{en}}} (\tau - \hat{\boldsymbol{a}} \cdot \boldsymbol{n}) \mathbf{I}_{\mathtt{n_{sd}}}\right]^T, \\
\widehat{\boldsymbol{\mathcal{N}}}^a &:= \left[\hat{N}_1 (\tau - \hat{\boldsymbol{a}} \cdot \boldsymbol{n}) \mathbf{I}_{\mathtt{n_{sd}}} \; \hat{N}_2 (\tau - \hat{\boldsymbol{a}} \cdot \boldsymbol{n}) \mathbf{I}_{\mathtt{n_{sd}}} \dots \hat{N}_{\mathtt{n_{fn}}} (\tau - \hat{\boldsymbol{a}} \cdot \boldsymbol{n}) \mathbf{I}_{\mathtt{n_{sd}}}\right]^T, \\
\boldsymbol{\mathcal{Q}} &:= \left[(\mathbf{J}^{-1} \nabla N_1)^T \; (\mathbf{J}^{-1} \nabla N_2)^T \dots (\mathbf{J}^{-1} \nabla N_{\mathtt{n_{en}}})^T\right]^T, \\
\boldsymbol{\mathcal{Q}}^a &:= \left[\boldsymbol{a} \cdot (\mathbf{J}^{-1} \nabla N_1) \; \boldsymbol{a} \cdot (\mathbf{J}^{-1} \nabla N_2) \dots \boldsymbol{a} \cdot (\mathbf{J}^{-1} \nabla N_{\mathtt{n_{en}}})\right]^T,
\end{aligned}
$$

where $\boldsymbol{n}$ is the outward unit normal vector to a face, $\boldsymbol{a}$ is the convection field evaluated in the reference element, and $\hat{\boldsymbol{a}}$ is the convection field evaluated on the reference face. Moreover, for each spatial dimension, that is, for $k = 1, \dots, \mathtt{n_{sd}}$, define

$$
\begin{aligned}
\boldsymbol{\mathcal{N}}_k^{\mathrm{D}} &:= \left[N_1 n_k \mathbf{F}_k^T \mathbf{D}^{1/2} \; N_2 n_k \mathbf{F}_k^T \mathbf{D}^{1/2} \dots N_{\mathtt{n_{fn}}} n_k \mathbf{F}_k^T \mathbf{D}^{1/2}\right]^T, \\
\widehat{\boldsymbol{\mathcal{N}}}_k^{\mathrm{D}} &:= \left[\hat{N}_1 n_k \mathbf{F}_k^T \mathbf{D}^{1/2} \; \hat{N}_2 n_k \mathbf{F}_k^T \mathbf{D}^{1/2} \dots \hat{N}_{\mathtt{n_{fn}}} n_k \mathbf{F}_k^T \mathbf{D}^{1/2}\right]^T, \\
\boldsymbol{\mathcal{Q}}_k^{\mathrm{D}} &:= \left[[\mathbf{J}^{-1} \nabla N_1]_k \mathbf{F}_k^T \mathbf{D}^{1/2} \; [\mathbf{J}^{-1} \nabla N_2]_k \mathbf{F}_k^T \mathbf{D}^{1/2} \dots [\mathbf{J}^{-1} \nabla N_{\mathtt{n_{en}}}]_k \mathbf{F}_k^T \mathbf{D}^{1/2}\right]^T,
\end{aligned}
$$

where $n_k$ is the $k$-th components of the outward unit normal vector $\boldsymbol{n}$ to a face and $\mathbf{J}$ is the Jacobian of the isoparametric transformation.

The discretization of (28a) leads to the following matrices and vector

$$
\begin{aligned}
[\mathbf{A}_{LL}]_e &= -\sum_{\mathtt{g}=1}^{\mathtt{n_{ip}^e}} \boldsymbol{\mathcal{M}}(\boldsymbol{\xi}_{\mathtt{g}}^{\mathtt{e}}) \boldsymbol{\mathcal{M}}^T(\boldsymbol{\xi}_{\mathtt{g}}^{\mathtt{e}}) |\mathbf{J}(\boldsymbol{\xi}_{\mathtt{g}}^{\mathtt{e}})| w_{\mathtt{g}}^{\mathtt{e}}, \\
[\mathbf{A}_{Lu}]_e &= \sum_{k=1}^{\mathtt{n_{sd}}} \sum_{\mathtt{g}=1}^{\mathtt{n_{ip}^e}} \boldsymbol{\mathcal{Q}}_k^{\mathrm{D}}(\boldsymbol{\xi}_{\mathtt{g}}^{\mathtt{e}}) \boldsymbol{\mathcal{N}}^T(\boldsymbol{\xi}_{\mathtt{g}}^{\mathtt{e}}) |\mathbf{J}(\boldsymbol{\xi}_{\mathtt{g}}^{\mathtt{e}})| w_{\mathtt{g}}^{\mathtt{e}}, \\
[\mathbf{A}_{L\hat{u}}]_e &= \sum_{f=1}^{\mathtt{n_{fa}^e}} \left( \sum_{k=1}^{\mathtt{n_{sd}}} \sum_{\mathtt{g}=1}^{\mathtt{n_{ip}^f}} \boldsymbol{\mathcal{N}}_k^{\mathrm{D}}(\boldsymbol{\xi}_{\mathtt{g}}^{\mathtt{f}}) \widehat{\boldsymbol{\mathcal{N}}}^T(\boldsymbol{\xi}_{\mathtt{g}}^{\mathtt{f}}) |\mathbf{J}(\boldsymbol{\xi}_{\mathtt{g}}^{\mathtt{f}})| w_{\mathtt{g}}^{\mathtt{f}} \right) \left(1 - \chi_{\Gamma_D}(f)\right), \\
[\mathbf{f}_L]_e &= \sum_{f=1}^{\mathtt{n_{fa}^e}} \left( \sum_{k=1}^{\mathtt{n_{sd}}} \sum_{\mathtt{g}=1}^{\mathtt{n_{ip}^f}} \boldsymbol{\mathcal{N}}_k^{\mathrm{D}}(\boldsymbol{\xi}_{\mathtt{g}}^{\mathtt{f}}) \boldsymbol{u}_D\big(\boldsymbol{x}(\boldsymbol{\xi}_{\mathtt{g}}^{\mathtt{f}})\big) |\mathbf{J}(\boldsymbol{\xi}_{\mathtt{g}}^{\mathtt{f}})| w_{\mathtt{g}}^{\mathtt{f}} \right) \chi_{\Gamma_D}(f),
\end{aligned}
$$

where $\mathtt{n_{fa}^e}$ is the number of faces $\Gamma_{e,f}$, $f = 1, \dots, \mathtt{n_{fa}^e}$, of the element $\Omega_e$, $\boldsymbol{\xi}_{\mathtt{g}}^{\mathtt{e}}$ and $w_{\mathtt{g}}^{\mathtt{e}}$ (resp., $\boldsymbol{\xi}_{\mathtt{g}}^{\mathtt{f}}$ and $w_{\mathtt{g}}^{\mathtt{f}}$) are the $\mathtt{n_{ip}^e}$ (resp., $\mathtt{n_{ip}^f}$) integration points and weights defined on the reference element (resp., face) and $\chi_{\Gamma_D}$ is the indicator function of the boundary $\Gamma_D$, namely

$$\chi_{\Gamma_D}(f) = \begin{cases} 1 & \text{if } \Gamma_{e,f} \cap \Gamma_D \neq \emptyset \\ 0 & \text{otherwise} \end{cases}$$

Similarly, from the discretization of (28b) the following matrices and vectors are obtained

$$\begin{aligned}
[\mathbf{A}_{uu}]_e &= -\sum_{\mathtt{g}=1}^{\mathtt{n}_{\mathtt{ip}}^{\mathtt{e}}} \boldsymbol{\mathcal{Q}}^a(\boldsymbol{\xi}_{\mathtt{g}}^{\mathtt{e}}) \boldsymbol{\mathcal{N}}^T(\boldsymbol{\xi}_{\mathtt{g}}^{\mathtt{e}}) |\mathbf{J}(\boldsymbol{\xi}_{\mathtt{g}}^{\mathtt{e}})| w_{\mathtt{g}}^{\mathtt{e}} \\
&\quad + \sum_{f=1}^{\mathtt{n}_{\mathtt{fa}}^{e}} \sum_{\mathtt{g}=1}^{\mathtt{n}_{\mathtt{ip}}^{\mathtt{f}}} \boldsymbol{\mathcal{N}}^\tau(\boldsymbol{\xi}_{\mathtt{g}}^{\mathtt{f}}) \boldsymbol{\mathcal{N}}^T(\boldsymbol{\xi}_{\mathtt{g}}^{\mathtt{f}}) |\mathbf{J}(\boldsymbol{\xi}_{\mathtt{g}}^{\mathtt{f}})| w_{\mathtt{g}}^{\mathtt{f}}, \\
[\mathbf{A}_{u\hat{u}}]_e &= \sum_{f=1}^{\mathtt{n}_{\mathtt{fa}}^{e}} \left( \sum_{\mathtt{g}=1}^{\mathtt{n}_{\mathtt{ip}}^{\mathtt{f}}} \boldsymbol{\mathcal{N}}^a(\boldsymbol{\xi}_{\mathtt{g}}^{\mathtt{f}}) \widehat{\boldsymbol{\mathcal{N}}}^T(\boldsymbol{\xi}_{\mathtt{g}}^{\mathtt{f}}) |\mathbf{J}(\boldsymbol{\xi}_{\mathtt{g}}^{\mathtt{f}})| w_{\mathtt{g}}^{\mathtt{f}} \right) (1 - \chi_{\Gamma_D}(f)), \\
[\mathbf{f}_u]_e &= \sum_{\mathtt{g}=1}^{\mathtt{n}_{\mathtt{ip}}^{\mathtt{e}}} \boldsymbol{\mathcal{N}}(\boldsymbol{\xi}_{\mathtt{g}}^{\mathtt{e}}) \boldsymbol{s}(\boldsymbol{x}(\boldsymbol{\xi}_{\mathtt{g}}^{\mathtt{e}})) |\mathbf{J}(\boldsymbol{\xi}_{\mathtt{g}}^{\mathtt{e}})| w_{\mathtt{g}}^{\mathtt{e}} \\
&\quad + \sum_{f=1}^{\mathtt{n}_{\mathtt{fa}}^{e}} \left( \sum_{\mathtt{g}=1}^{\mathtt{n}_{\mathtt{ip}}^{\mathtt{f}}} \boldsymbol{\mathcal{N}}^a(\boldsymbol{\xi}_{\mathtt{g}}^{\mathtt{f}}) \boldsymbol{u}_D(\boldsymbol{x}(\boldsymbol{\xi}_{\mathtt{g}}^{\mathtt{f}})) |\mathbf{J}(\boldsymbol{\xi}_{\mathtt{g}}^{\mathtt{f}})| w_{\mathtt{g}}^{\mathtt{f}} \right) \chi_{\Gamma_D}(f).
\end{aligned}$$

The discrete forms of the incompressibility constraint in (28c) and the restriction in (28d) feature the following matrices and vector

$$\begin{aligned}
[\mathbf{A}_{pu}]_e &= \sum_{\mathtt{g}=1}^{\mathtt{n}_{\mathtt{ip}}^{\mathtt{e}}} \boldsymbol{\mathcal{N}}(\boldsymbol{\xi}_{\mathtt{g}}^{\mathtt{e}}) \boldsymbol{\mathcal{Q}}^T(\boldsymbol{\xi}_{\mathtt{g}}^{\mathtt{e}}) |\mathbf{J}(\boldsymbol{\xi}_{\mathtt{g}}^{\mathtt{e}})| w_{\mathtt{g}}^{\mathtt{e}}, \\
[\mathbf{A}_{p\hat{u}}]_e &= \sum_{f=1}^{\mathtt{n}_{\mathtt{fa}}^{e}} \left( \sum_{\mathtt{g}=1}^{\mathtt{n}_{\mathtt{ip}}^{\mathtt{f}}} \boldsymbol{\mathcal{N}}^n(\boldsymbol{\xi}_{\mathtt{g}}^{\mathtt{f}}) \widehat{\boldsymbol{\mathcal{N}}}^T(\boldsymbol{\xi}_{\mathtt{g}}^{\mathtt{f}}) |\mathbf{J}(\boldsymbol{\xi}_{\mathtt{g}}^{\mathtt{f}})| w_{\mathtt{g}}^{\mathtt{f}} \right) (1 - \chi_{\Gamma_D}(f)), \\
[\mathbf{f}_p]_e &= \sum_{f=1}^{\mathtt{n}_{\mathtt{fa}}^{e}} \left( \sum_{\mathtt{g}=1}^{\mathtt{n}_{\mathtt{ip}}^{\mathtt{f}}} \boldsymbol{\mathcal{N}}^n(\boldsymbol{\xi}_{\mathtt{g}}^{\mathtt{f}}) \boldsymbol{u}_D(\boldsymbol{x}(\boldsymbol{\xi}_{\mathtt{g}}^{\mathtt{f}})) |\mathbf{J}(\boldsymbol{\xi}_{\mathtt{g}}^{\mathtt{f}})| w_{\mathtt{g}}^{\mathtt{f}} \right) \chi_{\Gamma_D}(f), \\
[\mathbf{a}_{\rho p}]_e &= \sum_{f=1}^{\mathtt{n}_{\mathtt{fa}}^{e}} \sum_{\mathtt{g}=1}^{\mathtt{n}_{\mathtt{ip}}^{\mathtt{f}}} \boldsymbol{\mathcal{N}}(\boldsymbol{\xi}_{\mathtt{g}}^{\mathtt{f}}) \mathbf{1} |\mathbf{J}(\boldsymbol{\xi}_{\mathtt{g}}^{\mathtt{f}})| w_{\mathtt{g}}^{\mathtt{f}},
\end{aligned}$$

Finally, the matrices and vectors resulting from the discretization of the global problem in (29a) are

$$[\mathbf{A}_{\hat{u}\hat{u}}]_e = -\sum_{f=1}^{\mathtt{n}_{\mathtt{fa}}^e}\left(\sum_{\mathtt{g}=1}^{\mathtt{n}_{\mathtt{ip}}^{\mathtt{f}}} \widehat{\boldsymbol{\mathcal{N}}}^{\tau}(\boldsymbol{\xi}_{\mathtt{g}}^{\mathtt{f}})\widehat{\boldsymbol{\mathcal{N}}}^{T}(\boldsymbol{\xi}_{\mathtt{g}}^{\mathtt{f}})|\mathbf{J}(\boldsymbol{\xi}_{\mathtt{g}}^{\mathtt{f}})|w_{\mathtt{g}}^{\mathtt{f}}\right)\chi_{\Gamma}(f)$$

$$-\sum_{f=1}^{\mathtt{n}_{\mathtt{fa}}^e}\left(\sum_{\mathtt{g}=1}^{\mathtt{n}_{\mathtt{ip}}^{\mathtt{f}}} \widehat{\boldsymbol{\mathcal{N}}}^{a}(\boldsymbol{\xi}_{\mathtt{g}}^{\mathtt{f}})\widehat{\boldsymbol{\mathcal{N}}}^{T}(\boldsymbol{\xi}_{\mathtt{g}}^{\mathtt{f}})|\mathbf{J}(\boldsymbol{\xi}_{\mathtt{g}}^{\mathtt{f}})|w_{\mathtt{g}}^{\mathtt{f}}\right)\chi_{\Gamma_N}(f),$$

$$[\mathbf{A}_{\hat{u}u}]_e = \sum_{f=1}^{\mathtt{n}_{\mathtt{fa}}^e}\left(\sum_{\mathtt{g}=1}^{\mathtt{n}_{\mathtt{ip}}^{\mathtt{f}}} \widehat{\boldsymbol{\mathcal{N}}}^{\tau}(\boldsymbol{\xi}_{\mathtt{g}}^{\mathtt{f}})\boldsymbol{\mathcal{N}}^{T}(\boldsymbol{\xi}_{\mathtt{g}}^{\mathtt{f}})|\mathbf{J}(\boldsymbol{\xi}_{\mathtt{g}}^{\mathtt{f}})|w_{\mathtt{g}}^{\mathtt{f}}\right)\left(1-\chi_{\Gamma_D}(f)\right),$$

$$[\mathbf{f}_{\hat{u}}]_e = -\sum_{f=1}^{\mathtt{n}_{\mathtt{fa}}^e}\left(\sum_{\mathtt{g}=1}^{\mathtt{n}_{\mathtt{ip}}^{\mathtt{f}}} \widehat{\boldsymbol{\mathcal{N}}}(\boldsymbol{\xi}_{\mathtt{g}}^{\mathtt{f}})\boldsymbol{t}(\boldsymbol{x}(\boldsymbol{\xi}_{\mathtt{g}}^{\mathtt{f}}))|\mathbf{J}(\boldsymbol{\xi}_{\mathtt{g}}^{\mathtt{f}})|w_{\mathtt{g}}^{\mathtt{f}}\right)\chi_{\Gamma_N}(f),$$

where $\chi_\Gamma$ and $\chi_{\Gamma_N}$ are the indicator functions of the internal skeleton $\Gamma$ and the Neumann boundary $\Gamma_N$, respectively.

## References

Armaly, B. F., Durst, F., Pereira, J., & Schönung, B. (1983). Experimental and theoretical investigation of backward-facing step flow. *Journal of Fluid Mechanics*, *127*, 473–496.

Benzi, M., Golub, G. H., & Liesen, J. (2005). Numerical solution of saddle point problems. *Acta Numerica*, *14*, 1–137.

Bernstein, D. S. (2009). *Scalar, vector, and matrix mathematics*. Princeton, NJ: Princeton University Press.

Brezzi, F., & Fortin, M. (1991). *Mixed and hybrid finite elements methods*. Springer series in computational mathematics. Springer.

Cesmelioglu, A., Cockburn, B., Nguyen, N. C., & Peraire, J. (2013). Analysis of HDG methods for Oseen equations. *Journal of Scientific Computing*, *55*(2), 392–431.

Cesmelioglu, A., Cockburn, B., & Qiu, W. (2017). Analysis of a hybridizable discontinuous Galerkin method for the steady-state incompressible Navier-Stokes equations. *Mathematics of Computation*, *86*(306), 1643–1670.

Ciarlet, P. G. (2002). *The finite element method for elliptic problems*, volume 40 of *Classics in Applied Mathematics*. Society for Industrial and Applied Mathematics (SIAM), Philadelphia, PA. (Reprint of the 1978 original [North-Holland, Amsterdam])

Cockburn, B. (2017). Discontinuous Galerkin methods for computational fluid dynamics. In E. Stein, R. de Borst, & T. J. R. Hughes (Eds.), *Encyclopedia of computational mechanics second edition, volume Part 1 fluids, Chap. 5*. Chichester: Wiley Ltd.

Cockburn, B., & Fu, G. (2017a). Superconvergence by $M$-decompositions. Part III: Construction of three-dimensional finite elements. *ESAIM: Mathematical Modelling and Numerical Analysis*, *51*(1), 365–398.

Cockburn, B., & Fu, G. (2017a). Superconvergence by $M$-decompositions. Part II: Construction of two-dimensional finite elements. *ESAIM: Mathematical Modelling and Numerical Analysis*, *51*(1), 165–186.

Cockburn, B., & Gopalakrishnan, J. (2004). A characterization of hybridized mixed methods for second order elliptic problems. *The SIAM Journal on Numerical Analysis*, *42*(1), 283–301.

Cockburn, B., & Gopalakrishnan, J. (2005a). Incompressible finite elements via hybridization. I. The Stokes system in two space dimensions. *The SIAM Journal on Numerical Analysis*, *43*(4), 1627–1650.

Cockburn, B., & Gopalakrishnan, J. (2005b). New hybridization techniques. *GAMM-Mitt*, *28*(2), 154–182.

Cockburn, B., & Gopalakrishnan, J. (2009). The derivation of hybridizable discontinuous Galerkin methods for Stokes flow. *The SIAM Journal on Numerical Analysis*, *47*(2), 1092–1125.

Cockburn, B., & Shi, K. (2014). Devising HDG methods for Stokes flow: An overview. *Computers & Fluids*, *98*, 221–229.

Cockburn, B., Dong, B., & Guzmán, J. (2008). A superconvergent LDG-hybridizable Galerkin method for second-order elliptic problems. *Mathematics of Computation*, *77*(264), 1887–1916.

Cockburn, B., Dong, B., Guzmán, J., Restelli, M., & Sacco, R. (2009a). A hybridizable discontinuous Galerkin method for steady-state convection-diffusion-reaction problems. *SIAM Journal on Scientific Computing*, *31*(5), 3827–3846.

Cockburn, B., Gopalakrishnan, J., & Lazarov, R. (2009b). Unified hybridization of discontinuous Galerkin, mixed, and continuous Galerkin methods for second order elliptic problems. *The SIAM Journal on Numerical Analysis*, *47*(2), 1319–1365.

Cockburn, B., Nguyen, N. C., & Peraire, J. (2010). A comparison of HDG methods for Stokes flow. *Journal of Scientific Computing*, *45*(1–3), 215–237.

Cockburn, B., Gopalakrishnan, J., Nguyen, N. C., Peraire, J., & Sayas, F.-J. (2011). Analysis of HDG methods for Stokes flow. *Mathematics of Computation*, *80*(274), 723–760.

Cockburn, B., Fu, G., & Sayas, F.-J. (2017). Superconvergence by $M$-decompositions. Part I: General theory for HDG methods for diffusion. *Mathematics of Computation*, *86*(306), 1609–1641.

Di Pietro, D., & Ern, A. (2015). A hybrid high-order locking-free method for linear elasticity on general meshes. *Computer Methods in Applied Mechanics and Engineering*, *283*, 1–21.

Donea, J., & Huerta, A. (2003). *Finite element methods for flow problems*. Chichester: Wiley.

Egger, H., & Waluga, C. (2012). A hybrid mortar method for incompressible flow. *International Journal of Numerical Analysis and Modeling*, *9*(4), 793–812.

Erturk, E. (2008). Numerical solutions of 2-D steady incompressible flow over a backward-facing step, Part I: High Reynolds number solutions. *Computers & Fluids*, *37*(6), 633–655.

Fish, J., & Belytschko, T. (2007). *A first course in finite elements* (p. 0470035803). ISBN: Wiley.

Fraeijs de Veubeke, B. (1965). Displacement and equilibrium models in the finite element method. In O. C. Zienkiewicz & G. S. Holister (Eds.), *Stress analysis* (pp. 145–197). Wiley.

Giacomini, M., & Sevilla, R. (2019). Discontinuous Galerkin approximations in computational mechanics: Hybridization, exact geometry and degree adaptivity. *SN Applied Sciences*, *1*, 1047.

Giacomini, M., Karkoulias, A., Sevilla, R., & Huerta, A. (2018). A superconvergent HDG method for Stokes flow with strongly enforced symmetry of the stress tensor. *Journal of Scientific Computing*, *77*(3), 1679–1702.

Giorgiani, G., Fernández-Méndez, S., & Huerta, A. (2013a). Hybridizable Discontinuous Galerkin p-adaptivity for wave propagation problems. *International Journal for Numerical Methods in Fluids*, *72*(12), 1244–1262.

Giorgiani, G., Modesto, D., Fernández-Méndez, S., & Huerta, A. (2013b). High-order continuous and discontinuous Galerkin methods for wave problems. *International Journal for Numerical Methods in Fluids*, *73*(10), 883–903.

Giorgiani, G., Fernández-Méndez, S., & Huerta, A. (2014). Hybridizable Discontinuous Galerkin with degree adaptivity for the incompressible Navier-Stokes equations. *Computers & Fluids*, *98*, 196–208.

Guyan, R. (1965). Reduction of stiffness and mass matrices. *AIAA Journal*, *3*(2), 380–380.

Hesthaven, J. S. (2019). *Numerical methods for conservation laws: From analysis to algorithms*, volume 18 of *Computational Science and engineering series*. Society for Industrial and Applied Mathematics, Philadelphia. ISBN 978-1-611975-09-3.

Huerta, A., Angeloski, A., Roca, X., & Peraire, J. (2013). Efficiency of high-order elements for continuous and discontinuous Galerkin methods. *International Journal for Numerical Methods in Engineering*, *96*(9), 529–560.

Kabaria, H., Lew, A. J., & Cockburn, B. (2015). A hybridizable discontinuous Galerkin formulation for non-linear elasticity. *Computer Methods in Applied Mechanics and Engineering*, *283*, 303–329.

Kirby, R., Sherwin, S. J., & Cockburn, B. (2011). To CG or to HDG: A comparative study. *Journal of Scientific Computing*, *51*(1), 183–212.

Kovasznay, L. (1948). Laminar flow behind a two-dimensional grid. *Mathematical Proceedings of the Cambridge Philosophical Society*, *44*(1), 58–62.

Miao, J.-M. (1991). General expressions for the Moore-Penrose inverse of a $2 \times 2$ block matrix. *Linear Algebra and Its Applications*, *151*, 1–15.

Montlaur, A., Fernández-Méndez, S., & Huerta, A. (2008). Discontinuous Galerkin methods for the Stokes equations using divergence-free approximations. *International Journal for Numerical Methods in Fluids*, *57*(9), 1071–1092.

Nguyen, N. C., Peraire, J., & Cockburn, B. (2009a). An implicit high-order hybridizable discontinuous Galerkin method for linear convection-diffusion equations. *Journal of Computational Physics*, *228*(9), 3232–3254.

Nguyen, N. C., Peraire, J., & Cockburn, B. (2009b). An implicit high-order hybridizable discontinuous Galerkin method for nonlinear convection-diffusion equations. *Journal of Computational Physics*, *228*(23), 8841–8855.

Nguyen, N. C., Peraire, J., & Cockburn, B. (2010a). A hybridizable discontinuous Galerkin method for the incompressible Navier-Stokes equations. In *48th AIAA Aerospace Sciences meeting including the new horizons forum and aerospace exposition*, Orlando, FL. AIAA 2010-362.

Nguyen, N. C., Peraire, J., & Cockburn, B. (2010b). A hybridizable discontinuous Galerkin method for Stokes flow. *Computer Methods in Applied Mechanics and Engineering*, *199*(9–12), 582–597.

Nguyen, N. C., Peraire, J., & Cockburn, B. (2011a). Hybridizable discontinuous Galerkin methods for the time-harmonic Maxwell's equations. *Journal of Computational Physics*, *230*(19), 7151–7175.

Nguyen, N. C., Peraire, J., & Cockburn, B. (2011b). An implicit high-order hybridizable discontinuous Galerkin method for the incompressible Navier-Stokes equations. *Journal of Computational Physics*, *230*(4), 1147–1170.

Nguyen, N. C., Peraire, J., & Cockburn, B. (2011c). High-order implicit hybridizable discontinuous Galerkin methods for acoustics and elastodynamics. *Journal of Computational Physics*, *230*(10), 3695–3718.

Oikawa, I. (2015). A hybridized discontinuous Galerkin method with reduced stabilization. *Journal of Scientific Computing*, *65*(1), 327–340.

Peraire, J., Nguyen, N. C., & Cockburn, B. (2010). A hybridizable discontinuous Galerkin method for the compressible Euler and Navier-Stokes equations. In *48th AIAA Aerospace Sciences meeting including the New Horizons forum and aerospace exposition*, Orlando, FL. AIAA 2010-363.

Qiu, W., & Shi, K. (2016). A superconvergent HDG method for the incompressible Navier-Stokes equations on general polyhedral meshes. *IMA Journal of Numerical Analysis*, *36*(4), 1943–1967. ISSN 0272-4979.

Raviart, P.-A., & Thomas, J. M. (1977). A mixed finite element method for 2nd order elliptic problems. In *Mathematical aspects of finite element methods (Proc. Conf., Consiglio Naz. delle Ricerche (C.N.R.), Rome, 1975)*. Lecture notes in Mathematics, (pp. 292–315, Vol. 606). Springer, Berlin.

Sevilla, R. (2019). HDG-NEFEM for two dimensional linear elasticity. *Computers & Structures*, *220*, 69–80.

Sevilla, R., & Huerta, A. (2016). Tutorial on hybridizable discontinuous Galerkin (HDG) for second-order elliptic problems. In J. Schröder & P. Wriggers (Eds.), *Advanced finite element technologies*, volume 566 of *CISM International Centre for Mechanical Sciences* (pp. 105–129). Springer International Publishing.

Sevilla, R., & Huerta, A. (2018). HDG-NEFEM with degree adaptivity for Stokes flows. *Journal of Scientific Computing*, *77*(3), 1953–1980.

Sevilla, R., Giacomini, M., Karkoulias, A., & Huerta, A. (2018). A superconvergent hybridisable discontinuous Galerkin method for linear elasticity. *International Journal for Numerical Methods in Engineering*, *116*(2), 91–116.

Soon, S.-C., Cockburn, B., & Stolarski, H. K. (2009). A hybridizable discontinuous Galerkin method for linear elasticity. *International Journal for Numerical Methods in Engineering*, *80*(8), 1058–1092.

Stenberg, R. (1990). Some new families of finite elements for the Stokes equations. *Numerische Mathematik*, *56*(8), 827–838.

van de Vosse, F., de Hart, J., van Oijen, C., Bessems, D., Gunther, T., Segal, A., et al. (2003). Finite-element-based computational methods for cardiovascular fluid-structure interaction. *Journal of Engineering Mathematics*, *47*(3), 335–368.

Volker, J. (2002). Slip with friction and penetration with resistance boundary conditions for the Navier-Stokes equations—Numerical tests and aspects of the implementation. *Journal of Computational and Applied Mathematics*, *147*(2), 287–300.

Wang, C. Y. (1991). Exact solutions of the steady-state Navier-Stokes equations. *Annual Review of Fluid Mechanics*, *23*(1), 159–177.

# Non Intrusive Global/Local Coupling Techniques in Solid Mechanics: An Introduction to Different Coupling Strategies and Acceleration Techniques

**Olivier Allix and Pierre Gosselet**

## Introduction

In the last decade, many innovative modeling or solution techniques have been introduced in the field of computational mechanics. These techniques, such as enriched finite elements or multiscale modeling, enable to perform complex simulations that are out of the reach of conventional finite element analysis (FEA) tools, in terms of computational or human costs. Although these techniques have proved their performance by extensive testing on academic applications, they are scarcely applied on actual industrial problems because they cannot be conveniently implemented into commercial FEA software packages. Therefore, a scientific and practical challenge is to allow realistic simulation of complex industrial problems including all their physical and technological complexity. A view on this issue can be found in a prospective document of the NSF blue-ribbon panel (Oden et al. 2006):

> If an industry is to replace testing with simulation, the simulation tools must undergo robust verification and validation procedures for effectiveness. Overall, simulation in industry has yet to meet its full potential. The following list is a summary of its current limitations:
>
> 1. The development of models is very time-consuming, particularly for geometries of complex engineering systems [...]
> 2. Methods are needed for linking models at various scales and simulating multi-physics phenomena.

We are interested in the case where the complex phenomenon to be analyzed concerns one or several reduced parts (called local models) of the whole body (called

O. Allix (✉)
LMT, ENS Paris-Saclay/CNRS, Cachan, France
e-mail: olivier.allix@ens-paris-saclay.fr

P. Gosselet
LaMcube, Univ. Lille/CNRS/Centrale Lille, Lille, France

 
L. De Lorenzis and A. Düster (eds.), *Modeling in Engineering Using Innovative Numerical Methods for Solids and Fluids*, CISM International Centre for Mechanical Sciences 599,
https://doi.org/10.1007/978-3-030-37518-8_6

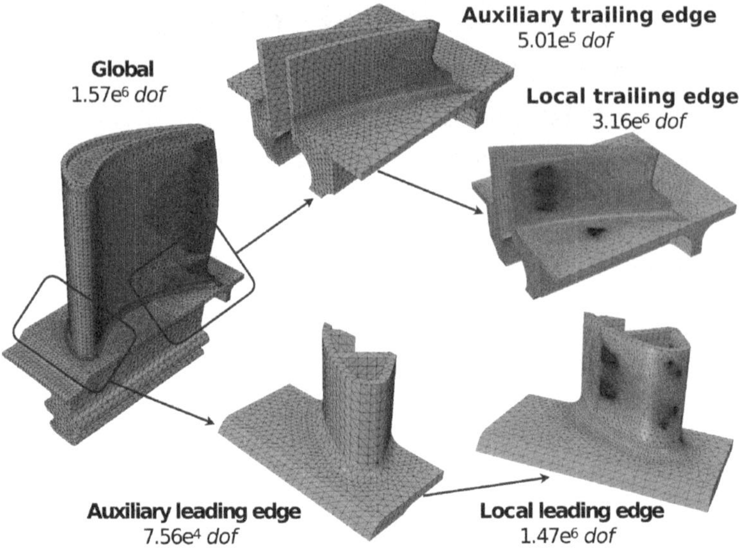


**Fig. 1** Illustration of global and locals models

global model). Those parts often correspond to very fine structural details that can not be taken into account in the global mesh of the structure, details which in turns can induce nonlinearity such as plasticity, visco-plasticity or damage. A typical example is presented in Fig. 1 where one can see that a proper description of the local parts may require meshes larger than the mesh used for the global model (Blanchard et al. 2019).

A method often used by engineers to tackle such problems is the submodeling approach (Cormier et al. 1999): after the *global* computation, structural zooms are applied on the *local* critical zones, with details represented exactly. An advantage of this approach is that it allows to easily connect research software and commercial code, as was done for example to deal with the prediction of delamination under low velocity impacts (Allix 2001). Unfortunately, it implies neglecting the influence of the local zones on the whole structure. This may lead in turn to quite important local errors. The problem becomes more crucial when nonlinearity initiated locally spreads over the whole structure.

To correct the drawback of submodeling while keeping its simplicity and flexibility, a non-invasive method was proposed in order to allow exact local/global analysis, embedding the same basic tools as those used in the submodeling inside an iterative procedure. The prerequisite of the proposed framework is to keep unchanged the global numerical model as well as the solver used for its treatment. Therefore, two or several models are used concurrently, the untouched global model and the local ones which are iteratively substituted where needed. The exchanges between the models are such that the data should be "natural" to the software, such as prescribed nodal reactions or displacements.

The proposed method aims therefore at converging by iterations toward the reference problem by means of submodeling-like steps. The formulation of the method and its numerical optimization have first been derived in the case of global linear models and local plasticity (Gendre et al. 2009, 2011). A number of other applications and extensions have been proposed: use of XFEM at the local scale (Passieux et al. 2013), treatment of non-matching interfaces (Liu et al. 2014), coupling between a global plate model and 3D parts for bolted assemblies (Guguin et al. 2014), geometrically non conforming coupling (Guinard et al. 2018), multiscale time and space computation in explicit dynamic (Bettinotti et al. 2014) with implementation in Abaqus Explicit for the analysis of delamination under impact (Bettinotti et al. 2017), non-invasive domain decomposition approach (Duval et al. 2016). Mesh refinement based on error estimation may also be cast in the proposed non-intrusive framework (Duval et al. 2018).

Alternative proposals exist, based on volume coupling as for example the Arlequin method (Dhia 1998). The implementation of such methods in a legacy code is not straightforward, mainly because the creation of the coupling operators between the two models in the transition zone requires complex integration operations. The volume coupling may also be performed by means of a non-invasive version of the Partition of Unity method (Plews et al. 2012; Fillmore and Duarte 2018), by using projection techniques between the local and global models (Temizer and Wriggers 2011; Holl et al. 2013) or by means of homogenization-like techniques (Hühne et al. 2016).

The aim of this paper is to provide, for a reader not familiar with the non-intrusive coupling method, the simplest possible example on which most of the different iterative coupling strategies used in the previously cited papers can be solved by hands. Among them, the basic algorithm, Aitken's method, mixed interface conditions …A drawback of this example is that, for some acceleration techniques, the convergence is achieved in one iteration after the initialization. Nevertheless, it allows to easily become acquainted with the different techniques. For this, we consider the case of a bar in tension as described on Fig. 2.

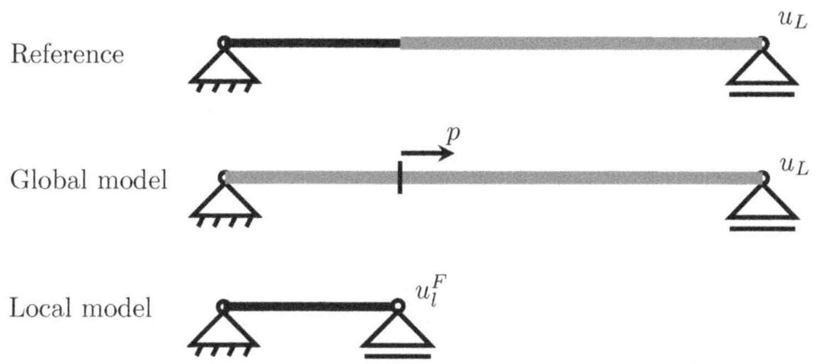


**Fig. 2** Reference, global and local models

## Reference Model

Let us consider a unit-section beam, clamped on its left side ($x = 0$) and with imposed traction displacement $u_L$ on its right side ($x = L$). The beam is made out of two components: the Young modulus of the left part $[0, l]$ is $E_F$, while it is $E_G$ on the right side $[l, L]$. We use the $'$ notation for the derivative with respect to the axial coordinate x. The subscript $R$ is used for the reference solution. In what follows we separate the left ($-$) and right ($+$) sides of the interface point $x = l$. The system of equations satisfied by the Reference displacement $u^R$ and the Reference tension $\sigma^R$, can be written as:

$$\begin{aligned} \sigma^R &= E^F (u^R)' \text{ in } ]0, l[ & \sigma^R &= E^G (u^R)' \text{ in } ]l, L[ \\ \sigma^{R'} &= 0 \text{ in } ]0, l[ \text{ and in } ]l, L[ & & \\ u^R(0) &= 0 & u^R(L) &= u_L \\ u^R_-(l) &= u^R_+(l) & \sigma^R_-(l) &= \sigma^R_+(l) \end{aligned} \tag{1}$$

The solution is uniform in tension, and continuous piecewise-linear in displacement:

$$\begin{aligned} u^R(x) &= u_l^R \frac{x}{l} \text{ in } [0, l] \\ u^R(x) &= u_l^R + (u_L - u_l^R) \frac{x - l}{L - l} \text{ in } [l, L] \\ u_l^R &= u_L \frac{1}{1 + \frac{E^F}{E^G} \frac{L-l}{l}} \\ \sigma^R &= \frac{E^F u_l^R}{l} = \frac{E^F E^G}{E^G l + E^F (L - l)} u_L \end{aligned} \tag{2}$$

## Iterative Techniques Using the Global and the Local Models Separately

As previously described, the principle of the method is to use the two models described on Fig. 2. The local model, to which a displacement coming from a previous global solution is imposed at the interface, and the global one where an extra load is prescribed at the interface. The different iterative techniques aim basically at determining the load $p^G$ to be prescribed to the global model which would lead to the exact solution, in the unchanged part of the global model (referred to as complement zone) and in the local model. To avoid possible misunderstanding we first define the two models. Let us also note that the convergence properties presented on this simple example can be generalized to structural problems introducing the Schur complement of the different domains (Gosselet et al. 2018a).

### *Local Model*

The Local model is the extraction of the left part of the Reference model. In order to avoid confusion between the point $L$ and the Local model, we use the F superscript for the Local model, meaning Fine model. A Dirichlet condition $u_l^F$ is imposed on the right side $x = l$ of the Fine model:

$$\begin{aligned} \sigma^F &= E^F u' \text{ in } ]0, l[ \quad && (\sigma^F)' = 0 \text{ in } ]0, l[ \\ u^F(0) &= 0 && u^F(l) = u_l^F \end{aligned} \tag{3}$$

and the solution is:

$$\begin{aligned} u^F(x) &= u_l^F \frac{x}{l} \text{ in } [0, l] \\ \sigma^F &= \frac{E^F u_l^F}{l} \text{ in } [0, l] \end{aligned} \tag{4}$$

### *Global Model*

The Global model is a simplification of the Reference model with a coarse representation of the zone of interest $[0, l]$. In our case, we chose a homogeneous beam with Young modulus $E^G$. An extra effort $p^G$ is imposed at the interface $x = l$:

$$\begin{aligned} &\sigma^G = E^G (u^G)' \text{ in } ]0, L[ \\ &(\sigma^G)' = 0 \text{ in } ]0, l[ \text{ and in } ]l, L[ \\ &u^G(0) = 0 && u^G(L) = u_L \\ &u_-^G(l) = u_+^G(l) && \sigma_-^G(l) = \sigma_+^G(l) + p^G \end{aligned} \tag{5}$$

The solution can be written as:

$$\begin{aligned} u^G(x) &= u_l^G \frac{x}{l} \text{ in } [0, l] \\ u^G(x) &= u_l^G + (u_L - u_l^G) \frac{x - l}{L - l} \text{ in } [l, L] \\ u_l^G &= \left( \frac{p^G (L - l)}{E^G} + u_L \right) \frac{l}{L} \\ \sigma_+^G &= \frac{E^G u_L}{L} - p^G \frac{l}{L} \end{aligned} \tag{6}$$

The coarse representation of the zone of interest in the Global model $[0, l]$ was often called Auxiliary model in previous papers, here it is noted with subscript $-$.

The zone $[l, L]$ where the Global and Reference models match is the Complement zone, here written with subscript $+$.

*Remark 3.1* (*Solution in term of* $p^G$) By comparing (6) and (2), we see that $\sigma_+^G = \sigma^R$ can be achieved for a specific value of $p^G$, named $p^R$:

$$\begin{aligned} \sigma^R &= \frac{E^F E^G}{E^G l + E^F(L-l)} u_L = \frac{E^G u_L}{L} - p^R \frac{l}{L} = \sigma_+^G \\ p^R &= \frac{(E^G - E^F)}{E^G l + E^F(L-l)} E^G u_L \end{aligned} \tag{7}$$

where we see that, of course, $p^R$ is proportional to the dissemblance between the Fine and the Coarse model in the zone of interest.

## Basic Fixed Point Iterative Technique

The basic iteration consists in a Global computation for a given $p_i^G$, from which we deduce the displacement to be imposed on the Local model. We then evaluate the lack of balance between the Local model and the Complement zone of the Global model, this residual, written $r$, is to be added to $p_i^G$ to define the next load $p_{i+1}^G$ of the Global model.

**Algorithm 1:** Non-invasive stationary iterations

```
Arbitrary initialization p_0^G
for j ∈ [0, ··· , m] do
    Solve Global model with extra Load p_j^G, extract displacement u_{l,j}^G and Traction σ_{+,j}^G
    Solve Fine model with imposed displacement u_{l,j}^G, extract Reaction σ_j^F
    Compute Residual r_j = (σ_{+,j}^G − σ_j^F)
    Update Global load p_{j+1}^G = p_j^G + r_j
end
```

Using previous formula, we have:

$$\begin{aligned} p_{j+1}^G &= p_j^G + r_j = p_j^G + (\sigma_{+,j}^G - \sigma_j^F) \\ &= p_j^G - \frac{E^F u_{l,j}^G}{l} - p_j^G \frac{l}{L} + \frac{E^G u_L}{L} \\ &= p_j^G - \frac{E^F \left( \frac{p_j^G (L-l)}{E^G} + u_L \right) \frac{l}{L}}{l} - p_j^G \frac{l}{L} + \frac{E^G u_L}{L} \end{aligned}$$

$$= p_j^G\left(1-\frac{E^F}{E^G}\right)\frac{L-l}{L}+u_L\left(\frac{E^G-E^F}{L}\right) \tag{8}$$

We recognise a fixed point iteration. We remind below the classical conditions of convergence of a fixed point algorithm; we will then make used of similar results for the other algorithms presented in this paper.

**Proposition 3.2** (Condition of convergence) *The iteration is a contraction and it converges if*

$$\rho = \left|\left(1-\frac{E^F}{E^G}\right)\frac{L-l}{L}\right| < 1 \tag{9}$$

In fact:

$$\begin{aligned} u_L\left(\frac{E^G-E^F}{L}\right) &= p_{j+1}^G - p_j^G\left(1-\frac{E^F}{E^G}\right)\frac{L-l}{L} \\ u_L\left(\frac{E^G-E^F}{L}\right) &= p_j^G - p_{j-1}^G\left(1-\frac{E^F}{E^G}\right)\frac{L-l}{L} \end{aligned} \tag{10}$$

Thus by subtracting the previous two relations we obtain, whatever $j \geq 1$:

$$p_{j+1}^G - p_j^G = \left(1-\frac{E^F}{E^G}\right)\frac{L-l}{L}(p_j^G - p_{j-1}^G) \tag{11}$$

Thus:

$$p_{j+1}^G - p_j^G = \left(\left(1-\frac{E^F}{E^G}\right)\frac{L-l}{L}\right)^j (p_1^G - p_0^G) \tag{12}$$

Thus, if $\rho < 1$, then $(p_j^G)$ is a Cauchy sequence. As we work in a complete space, $(p_j^G)$ tends to the limit $p_\infty^G$. The convergence is linear with rate $\rho$.

**Proposition 3.3** (Limit) *If the iteration is a contraction, we recover* (7) *when searching* $p_\infty^G = p^R$.

In fact, from the previous relation $p_\infty^G$ satisfies:

$$\begin{aligned} p_\infty^G &= p_\infty^G\left(1-\frac{E^F}{E^G}\right)\frac{L-l}{L}+u_L\left(\frac{E^G-E^F}{L}\right) \\ p_\infty^G &= \frac{(E^G-E^F)}{E^G l + E^F(L-l)}E^G u_L = p^R \end{aligned} \tag{13}$$

*Remark 3.4* The case where the basic fixed point iteration diverges corresponds to two cases:

$$\begin{aligned} \left(1-\frac{E^F}{E^G}\right)\frac{L-l}{L} > 1 &\Leftrightarrow -\frac{l}{L-l} > \frac{E^F}{E^G} \\ \left(1-\frac{E^F}{E^G}\right)\frac{L-l}{L} < -1 &\Leftrightarrow 1+\frac{L}{L-l} < \frac{E^F}{E^G} \end{aligned} \tag{14}$$

In those cases, it is required to make use of more refined algorithms. The first case corresponds practically to the possibility of softening. This is why relaxation was used in Gerasimov et al. (2018) where a local model prone to cracking was modeled by means of a phase field approach of fracture. The second case corresponds to a local material much stiffer than the global one.

## *Basic Fixed Point with Relaxation*

In order to improve the convergence rate, one can simply use relaxation. For a given $\omega \in \mathbb{R}^+$, the update formula is modified:

$$p_{j+1}^G = p_j^G + \omega r_j \tag{15}$$

and the rate of convergence can be computed, and minimized in order to obtain the optimal relaxation:

$$\begin{aligned} \rho &= 1 - \omega \frac{E^G l + E^F (L-l)}{E^G L} \\ \omega_{opt} &= \frac{E^G L}{E^F (L-l) + E^G l} \end{aligned} \tag{16}$$

### Aitken's Acceleration

Aitken's acceleration can be viewed as an automatic tuning of the relaxation parameter using the formula:

$$\omega_{j+1}^{aitken} = -\omega_j^{aitken} \frac{r_{j-1}(r_j - r_{j-1})}{\|r_j - r_{j-1}\|} \tag{17}$$

where an initial relaxation must be provided (in general equal to 1).

In this very simple case (scalar unknown and linear problem), it can be checked, using Eqs. 6 and 8, that Aitken's formula reaches the optimal relaxation after the first iteration.

### Robin Condition on the Fine Model

A possibility, in order to obtain fast convergence, is to improve the boundary condition applied to the Fine model, see Fig. 3. Indeed, a Dirichlet condition has the advantage of being easy to implement, and always available in legacy codes, but it provides an extremely simplified vision of the Complement domain.

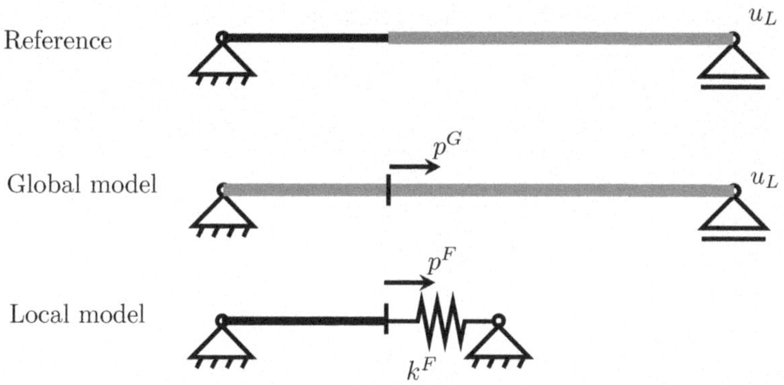


**Fig. 3** Reference, global and local models

In this approach, not only reactions are not balanced $(\sigma_+^G - \sigma^F) \neq 0$, but the local displacement $u^F(l)$ does not match the global displacement $u^G(l)$, unless convergence is reached. We gather the two conditions in an equivalent form:

$$\begin{aligned} (\sigma_+^G - \sigma^F) - k^F(u^F(l) - u^G(l)) = 0 \\ (\sigma_+^G - \sigma^F) - k^S(u^G(l) - u^F(l)) = 0 \end{aligned} \tag{18}$$

where $k^F$ and $k^S$ are parameters homogeneous to a stiffness. The first expression is used to define the boundary condition on the local model whereas the second is used to evaluate the residual which should take into account not only the lack of balance of forces but also the jump of displacements.

We consider the modified Fine model:

$$\begin{aligned} &\sigma^F = E^F(u^F)' \text{ in } ]0, l[ && (\sigma^F)' = 0 \text{ in } ]0, l[ \\ &u^F(0) = 0 && \sigma^F(l) + k^F u^F(l) = k^F u^G(l) + \sigma_+^G \end{aligned} \tag{19}$$

If we note $p^F = k^F u^G(l) + \sigma_+^G$, we recover the configuration of Fig. 3. The solution is:

$$\begin{aligned} u^F(x) &= \frac{p^F}{k^F l + E^F} x \text{ in } [0, l] \\ \sigma^F &= \frac{p^F E^F}{k^F l + E^F} \text{ in } [0, l] \end{aligned} \tag{20}$$

As said earlier, the second line of (18) is used to define the a residual which measures both the lack of balance and of continuity between the Local and Global models at the interface:

$$r_j = (\sigma_{+,j}^G - \sigma_j^F) + k^S(u_{l,j}^F - u_{l,j}^G) \tag{21}$$

This leads to Algorithm 2. If we analyze one iteration, we get:

$$
\begin{aligned}
p_{j+1}^G &= p_j^G + r_j \\
&= p_j^G \frac{l(E^G - E^F)((L-l)k^F - lk^S) - E^F L(k^S l - E^G)}{(k^F l + E^F)E^G L} \\
&\dots + \frac{(E^G - E^F)(k^F + k^S) l u_L}{(k^F l + E^F)L}
\end{aligned}
\tag{22}
$$

---

**Algorithm 2:** Non-invasive stationary iterations with Robin condition

---

Arbitrary initialization $p_0^G$
**for** $j \in [0, \cdots, m]$ **do**
  Solve Global model with extra Load $p_j^G$, extract displacement $u_{l,j}^G$ and Traction $\sigma_{+,j}^G$
  Solve Fine model with Robin condition $p_j^F = k^F u_{l,j}^G - \sigma_{+,j}^G$, extract Displacement $u_{l,j}^F$
  Reaction $\sigma_j^F$
  Compute Residual $r_j = (\sigma_{+,j}^G - \sigma_j^F) + k^S(u_{l,j}^F - u_{l,j}^G)$
  Update Global load $p_{j+1}^G = p_j^G + r_j$
**end**

---

The optimal parameters are $k^F = \frac{E^G}{L-l}$, that is to say the equivalent stiffness of the complement zone, and $k_S = \frac{E^G}{l}$ the stiffness of the Global representation of the zone of interest. With these parameters, convergence is obtained in one iteration. Indeed, choosing these values one obtains:

$$
p_{j+1}^G = 0 * p_j^G + p^R \tag{23}
$$

In fact there exists a whole range of admissible mixed parameters ensuring the convergence, which can be fully characterized using Proposition 3.2.

*Remark 3.5* It can be shown that all the properties of different versions of the coupling only depends on the equivalent stiffness of the domains: Fine model $S^F = E^F/l$, Complement domain $S^C = E^G/(L-l)$, Auxiliary model (coarse representation of the zone of interest in the Global model) $S^A = E^A/l$. More, the equivalent stiffness of a beam is generalized by the concept of Schur complement, and with minor caution, all the results above can be generalized to 2D ou 3D elasticity. Note that in practice using the optimal $k_S$ is not a problem whereas estimating the optimal $k^F$ is much more involved. The question of finding a good Robin condition mimicking the stiffness of a given domain has been addressed in many papers. In the frame of non-intrusive coupling, a two-scale approximation was proposed and tested in Gendre et al. (2011).

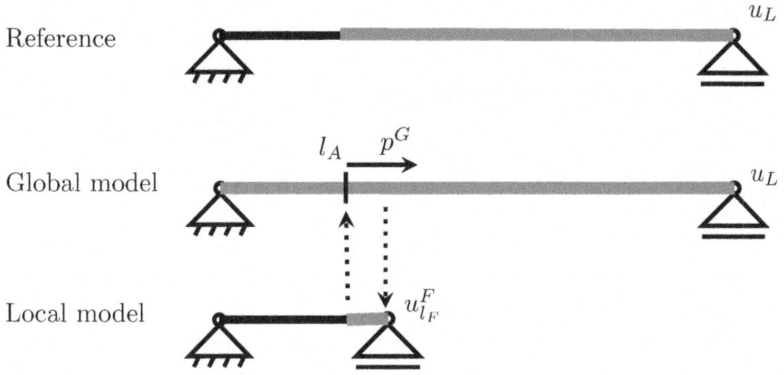


**Fig. 4** Reference, global and local models with overlap

## *Use of Overlap*

This version of the method was proposed to handle certain incompatibilities of the models: use of non-matching meshes (Gosselet et al. 2018b) or models with different dimensionality (Global plate vs Fine 3D) (Guguin et al. 2014).

We distinguish two inner boundaries: let $l_F$ be the limit of the Fine model and $l_A < l_F$ be such that the Global model in $[l_A, L]$ matches the Reference model. The extra traction on the Global model is imposed at position $l_A$ whereas the displacement to be imposed on the Fine model is obtained at position $l_F$. See Fig. 4.

Note that if the models do not match exactly in the overlap, then the limit of the iterations might differ from the Reference model. In the case where the Reference is well characterized this might be a problem. In many circumstances (non matching meshes, models of different dimensionality) there is no real reference and the limit of the iterations gives a mechanically sound coupled model.

**Algorithm 3:** Non-invasive stationary iterations with overlap

Arbitrary initialization $p^G_0$
**for** $j \in [0, \cdots, m]$ **do**
  Solve Global model with extra Load $p^G_j$ at position $l_A$, extract displacement $u^G_{l_F,j}$ and Traction $\sigma^G_{l_A+,j}$
  Solve Fine model with imposed displacement $u^G_{l_F,j}$, extract Reaction $\sigma^F_{l_A,j}$
  Compute Residual $r_j = (\sigma^G_{l_A+,j} - \sigma^F_{l_A,j})$ at position $l_A$
  Update Global load $p^G_{j+1} = p^G_j + r_j$
**end**

## Illustrations

### *Application of the Previous Results*

We illustrate previous study for the following values: $L = 1$, $u_L = 1/10$, $l = L/4$, $E^G = 1$, $E^F = .75$.

In Fig. 5, we compare the relaxation techniques. Note that in this simple case with 1D interface, optimal relaxation leads to convergence in on iteration, Aitken's formula finds the optimal relaxation as soon as it possibly can (iteration 2).

In Fig. 6, we compare the convergence for different values of the Robin parameters. The basic (primal) iteration is printed, it corresponds to $k^S = 0$ and $k^F = \infty$. The optimal setting leads to convergence at the first iteration. It appears that the convergence rate is more sensitive to variations in $k^S$ than in $k^F$, which is lucky since the computation of the optimal $k^S$ is actually feasible. As soon as $k^S$ is well-chosen, a wide range of values of $k^F$ leads to better convergence than basic iteration.

In Fig. 7, we compare the convergence for different lengths of the overlap. Large overlap is needed to ensure significant speedup. Using overlap is thus not a competitive acceleration technique. Its interest mostly lies in its capability to handle non-conforming meshes or models.

### *Comments on Other Iterative Algorithms*

This subsection briefly presents other iterations that have been tested but can not be illustrated on the simple 1D linear example.

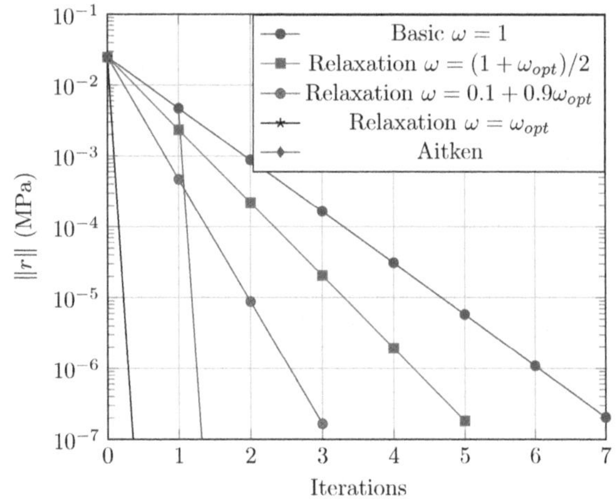


**Fig. 5** Comparison between relaxation techniques

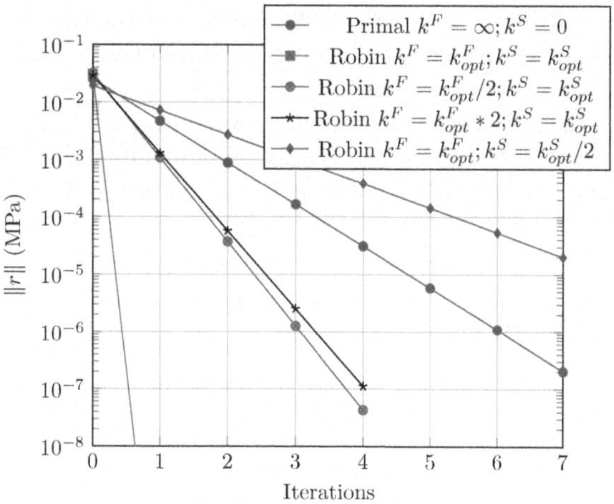


**Fig. 6** Comparison between various Robin techniques

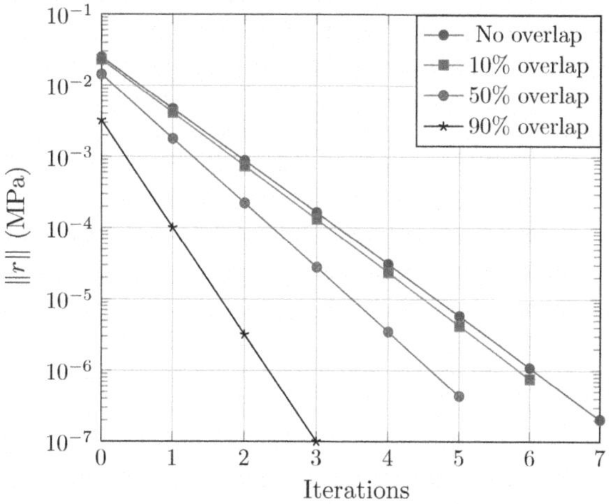


**Fig. 7** Comparison between different overlap sizes (basic iteration)

In the linear case, the fixed point can be accelerated by a Krylov solver. The particular structure of the fixed point operator, which can be written as the matrix of the reference problem preconditioned by the global problem, makes it possible to apply a conjugate gradient algorithm (Gosselet et al. 2018b). Since the Global problem often forms an excellent preconditioner for the reference problem, the convergence is extremely fast.

In the nonlinear case, if the Global model remains linear, quasi-Newton techniques like BFGS or SR1 can naturally be applied (Gendre et al. 2009). Again, because of the quality of the Global model, line-search appears not to be mandatory which enabled us to propose a fully nonlinear version of quasi-Newton (Gosselet et al. 2018b). Also, nonlinear conjugate gradient was tested with interesting performance if carefully configured (Gosselet et al. 2018b).

Anyhow, in almost all the cases we ever tested, it appeared that Aitken's $\Delta^2$ formula provided excellent performance in term of wall clock time, for an extremely simple implementation.

In the case of studies defined over large (pseudo)time intervals where the computation must be carried out on a succession of time steps. The use of partially decoupled time scales between the models was studied in Blanchard et al. (2019).

## 3D Example

In order to illustrate the method on a more significant test case, we propose to use the data from Gendre et al. (2011). The structure is the "sweded" turbine blade presented in Fig. 8, it was provided by Safran Aircraft Engines and it is representative of actual engineering work. In this particular case, only the mechanical behavior of the zone of interest is altered in the Global model: the Fine model is elastoplastic whereas the Global model is purely elastic. Meshes are unchanged in the different models. Again, the constitutive relations are representative of actual problems. In general industrialists address this kind of problems with submodeling or "structural zoom" techniques which can be interpreted as not iterating in the Global/Local coupling.

The Reference and Global meshes contain about 500000 degrees of freedom (dof), the zone of interest contains 80000 dof, and the interface is about 6400 dof large. Only one increment of load (pressure applied to one face) is considered.

The computation is managed by a python script which drives Abaqus software. The implementation is non-intrusive in the sense that the models are barely modified, only the extra interface load must be added to the global model. The two-scale Robin approach of Gendre et al. (2011) consists in extending the Fine model with 3 layers of (elastic) elements to approximate the surrounding stiffness; the long-scale effects are taken into account by a projector based on the response of the global structure to Saint-Venant loads on the interface. When needed, Sherman-Morrison is applied to take into account low-rank alteration to the stiffness matrix.

The problem was small enough in order to compute the Reference solution. Figure 9 presents the convergence of the method, measured by the error in terms of maximum accumulated plastic strain compared to the reference. The basic iteration takes 10 iterations to lower the error by 2 order of magnitudes. The accelerated version (SR1 quasi-Newton) needs 7 iterations to lower the error by 5 orders of magnitudes. Finally, the Robin version with acceleration only needs 4 iterations for a 6-order of magnitude decrease.

Note that a plateau can be observed around a relative error of $10^{-6}$. This is very common when using industrial software: some truncation was applied by Abaqus on strain and stress which makes it impossible to achieve better precision. Often, the estimation of the residual (nodal forces) does not suffer such limitation, so that the residual can be decreased up to machine zero whereas the actually attainable mechanical precision was reached much earlier.

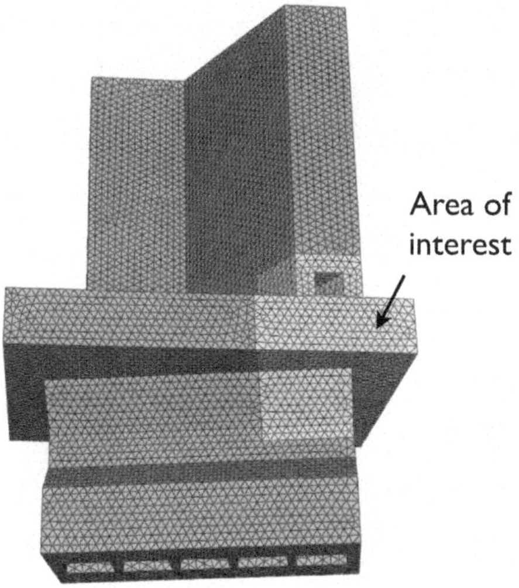


**Fig. 8** Sweded turbine blade

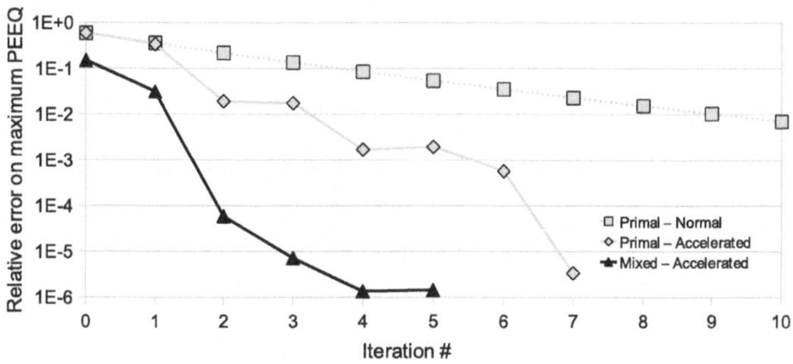


**Fig. 9** Convergence on the 3D test case—error in accumulated plastic strain

We end up the illustration by important mechanical considerations. Figures 10 and 11 present a comparison between the solution obtained by the classical submodeling approach and the Global/Local coupling. In the presented case, the classical submodeling, widely used by industrialists, provides a good estimation of the stress in the zone of interest. But it is really inaccurate in terms of plastic equivalent strain which, for the record, is one of the mechanical quantity used to estimate the lifespan of the structure under cyclic loading. Global/Local coupling is thus a powerful tool to achieve higher precision in the computation without impeding the industrialists' design chain, since usual tools can be employed without profound alteration.

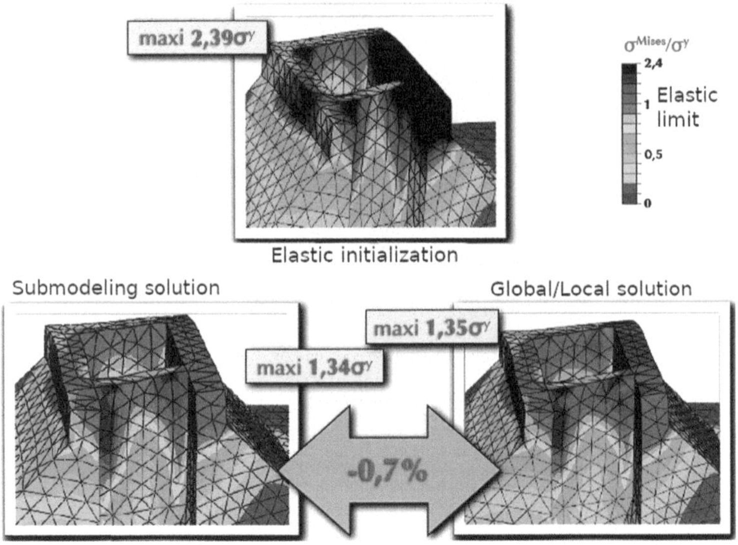


**Fig. 10** Comparison between submodeling of global/local coupling: von Mises stress

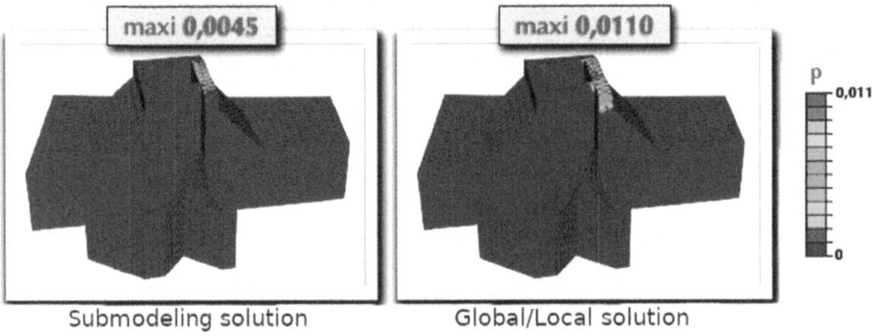


**Fig. 11** Comparison between submodeling of global/local coupling: accumulated plastic strain

Note that the Global/Local coupling can also provide a simple framework in order to introduce dedicated software for localized phenomena, see e.g. Guguin et al. (2016) for the use of a research code specialized in friction contact simulation to precisely model bolts in a composite plate assembly.

## Conclusion

So far, most of what has been done regarding the global/local non-invasive coupling technique concerns the development of the method within legacy codes. We hope that the very simple case which was analyzed in this paper will be useful for anyone, and possibly Ph.D. students, to get familiar with the method and to further develop it. We think in particular to its initial motivation: to make it possible realistic simulation of complex industrial problems including all their physical and technological complexity. The proposed method should allow, by an easy and fast coupling, to merge research software, with their enhanced physical capabilities, with industrial ones with their geometrical and technological capabilities.

## References

Allix, O. (2001). A composite damage meso-model for impact problems. *Composites Science and Technology*, *61*, 2193–2205.

Ben Dhia, H. (1998). Problèmes mécaniques multi-échelles: la méthode Arlequin. *Comptes Rendus de l'Académie des Sciences - Series IIB - Mechanics-Physics-Astronomy*, *326*(12), 899 – 904.

Bettinotti, O., Allix, O., & Malherbe, B. (2014). A coupling strategy for adaptive local refinement in space and time with a fixed global model in explicit dynamics. *Computational Mechanics*, *53*(4), 561–574.

Bettinotti, O., Allix, O., Perego, U., Oancea, V., & Malherbe, B. (2017). Simulation of delamination under impact using a global local method in explicit dynamics. *Finite Elements in Analysis and Design*, *125*(8), 1–13.

Blanchard, M., Allix, O., Gosselet, P., & Desmeure, G. (2019). Space/time global/local noninvasive coupling strategy: Application to viscoplastic structures. *Finite Elements in Analysis and Design*, *156*, 1–12. https://doi.org/10.1016/j.finel.2019.01.003.

Cormier, N. G., Smallwood, B. S., Sinclair, G. B., & Meda, G. (1999). Aggressive submodelling of stress concentrations. *International Journal for Numerical Methods in Engineering*, *46*(6), 889–909. ISSN 1097-0207.

Duval, M., Lozinski, A., Passieux, J.-C., & Salaün, M. (2018). Residual error based adaptive mesh refinement with the non-intrusive patch algorithm. *Computer Methods in Applied Mechanics and Engineering*, *329*, 118–143.

Duval, M., Passieux, J.-C., Salaün, M., & Guinard, S. (2016). Non-intrusive coupling: Recent advances and scalable nonlinear domain decomposition. *Archives of Computational Methods in Engineering*, *23*(1), 17–38.

Fillmore, T. B., & Armando Duarte, C. (2018). A hierarchical non-intrusive algorithm for the generalized finite element method. *Advanced Modeling and Simulation in Engineering Sciences*, *5*(2).

Gendre, L., Allix, O., & Gosselet, Pierre. (2011). A two-scale approximation of the Schur complement and its use for non-intrusive coupling. *International Journal for Numerical Methods in Engineering*, *87*(9), 889–905.

Gendre, L., Allix, O., Gosselet, P., & Comte, François. (2009). Non-intrusive and exact global/local techniques for structural problems with local plasticity. *Computational Mechanics*, *44*(2), 233–245.

Gerasimov, T., Noii, N., Allix O., et al. (2018). A non-intrusive global/local approach applied to phase-field modeling of brittle fracture. *Advanced modeling and simulation in engineering sciences*, *5*, 1.

Gosselet, P., Blanchard, M., & Allix, O. (2018a). Non-invasive global-local coupling as a schwarz domain decomposition method: Acceleration and generalization. *Advanced Modeling and Simulation in Engineering Sciences*, *5*(4). https://www.hal-01613966v1.

Gosselet, P., Blanchard, M., Allix, O., & Guguin, G. (2018b). Non-invasive global-local coupling as a Schwarz domain decomposition method: Acceleration and generalization. *Advanced Modeling and Simulation in Engineering Sciences*, *5*(4). https://doi.org/10.1186/s40323-018-0097-4.

Guguin, G., Allix, O., Gosselet, P., & Guinard, S. (2016). On the computation of plate assemblies using realistic 3d joint model: A non-intrusive approach. *Advanced Modeling and Simulation in Engineering Sciences, 3*.

Guguin, G., Allix, O., Gosselet, P., & Guinard, S. (2014). Nonintrusive coupling of 3d and 2d laminated composite models based on finite element 3d recovery. *International Journal for Numerical Methods in Engineering, 98*(5), 324–343.

Guinard, S., Bouclier, R., Toniolli, M., & Passieux, J.-C. (2018). Multiscale analysis of complex aeronautical structures using robust non-intrusive coupling. *Advanced Modeling and Simulation in Engineering Sciences, 5*(1).

Holl, M., Loehnert, S., & Wriggers, P. (2013). An adaptive multiscale method for crack propagation and crack coalescence. *International Journal for Numerical Methods in Engineering, 93*(1), 23–51.

Hühne, S., Reinoso, J., Jansen, E., et al. (2016). A two-way loose coupling procedure for investigating the buckling and damage behaviour of stiffened composite panels. *Composite Structures, 136*, 513–525.

Liu, Y. J., Sun, Q., & Fan, X. L. (2014). A non-intrusive global/local algorithm with non-matching interface: Derivation and numerical validation. *Computer Methods in Applied Mechanics and Engineering, 277*, 81–103.

Oden, J. T., Belytschko, T., Fish, J., Hughes, T. J. R., Johnson, C., Keyes, D., Laub, A., Petzold, L., Srolovitz, D., & Yip, S. (2006). Simulation-based engineering science: Revolutionizing engineering science through simulation. *NSF Blue Ribbon Panel on SBES*.

Passieux, J.-C., Réthoré, J., Gravouil, A., & Baietto, Marie-Christine. (2013). Local/global non-intrusive crack propagation simulation using a multigrid x-fem solver. *Computational Mechanics, 52*(6), 1381–1393.

Plews, J., Duarte, C. A., & Eason, T. (2012). An improved non-intrusive global-local approach for sharp thermal gradients in a standard fea platform. *International Journal for Numerical Methods in Engineering, 91*(4), 361–397.

Temizer, I., & Wriggers, P. (2011). An adaptive multiscale resolution strategy for the finite deformation analysis of microheterogeneous structures. *Computer Methods in Applied Mechanics and Engineering, 200*(37–40), 2639–2661. ISSN 0045-7825. Special Issue on Modeling Error Estimation and Adaptive Modeling.

The manufacturer's authorised representative in the EU is Springer
Nature Customer Service Centre GmbH, Europaplatz 3, 69115 Heidelberg,
Germany. If you have any concerns regarding our products, please
contact ProductSafety@springernature.com

Printed and bound by CPI Group (UK) Ltd, Croydon, CR0 4YY
05/10/2026
02227652-0001